T0245228

CAMBRIDGE LIBRARY COLLECTION
Books of enduring scholarly value

Earth Sciences

In the nineteenth century, geology emerged as a distinct academic discipline. It pointed the way towards the theory of evolution, as scientists including Gideon Mantell, Adam Sedgwick, Charles Lyell and Roderick Murchison began to use the evidence of minerals, rock formations and fossils to demonstrate that the earth was older by millions of years than the conventional, Bible-based wisdom had supposed. They argued convincingly that the climate, flora and fauna of the distant past could be deduced from geological evidence. Volcanic activity, the formation of mountains, and the action of glaciers and rivers, tides and ocean currents also became better understood. This series includes landmark publications by pioneers of the modern earth sciences, who advanced the scientific understanding of our planet and the processes by which it is constantly re-shaped.

Life, Letters and Journals of Sir Charles Lyell, Bart

Sir Charles Lyell (1797–1875) was one of the most renowned geologists of the nineteenth century. He was awarded the Copley Medal by the Royal Society in 1858 and the Wollaston Medal by the Geological Society of London in 1866 for his contributions to geology. Lyell's most important contribution to modern geology was his refining and popularising of the geological concept of uniformitarianism, the idea that the earth has been formed through slow-acting geological forces. This biography, first published in 1881 and edited by his sister-in-law K.M. Lyell, provides an intimate view of Lyell's personal and professional life through the inclusion of his correspondence with family, friends and academic peers. His changing ideas concerning the validity of the theory of natural selection and other geological ideas are also examined through the inclusion of extracts from his private journal. Volume 1 covers Lyell's early life and career until 1836.

Cambridge University Press has long been a pioneer in the reissuing of out-of-print titles from its own backlist, producing digital reprints of books that are still sought after by scholars and students but could not be reprinted economically using traditional technology. The Cambridge Library Collection extends this activity to a wider range of books which are still of importance to researchers and professionals, either for the source material they contain, or as landmarks in the history of their academic discipline.

Drawing from the world-renowned collections in the Cambridge University Library, and guided by the advice of experts in each subject area, Cambridge University Press is using state-of-the-art scanning machines in its own Printing House to capture the content of each book selected for inclusion. The files are processed to give a consistently clear, crisp image, and the books finished to the high quality standard for which the Press is recognised around the world. The latest print-on-demand technology ensures that the books will remain available indefinitely, and that orders for single or multiple copies can quickly be supplied.

The Cambridge Library Collection will bring back to life books of enduring scholarly value (including out-of-copyright works originally issued by other publishers) across a wide range of disciplines in the humanities and social sciences and in science and technology.

Life, Letters and Journals of Sir Charles Lyell, Bart

VOLUME 1

EDITED BY K.M. LYELL

CAMBRIDGE
UNIVERSITY PRESS

CAMBRIDGE UNIVERSITY PRESS

Cambridge, New York, Melbourne, Madrid, Cape Town, Singapore,
São Paolo, Delhi, Dubai, Tokyo, Mexico City

Published in the United States of America by Cambridge University Press, New York

www.cambridge.org
Information on this title: www.cambridge.org/9781108017848

© in this compilation Cambridge University Press 2010

This edition first published 1881
This digitally printed version 2010

ISBN 978-1-108-01784-8 Paperback

SIR CHARLES LYELL, BART.

VOL I.

Cha Lyell

Engraved by Chas Holl from a drawing by George Richmond. R.A.

London Published by John Murray, 50 Albemarle Street, 1881.

LIFE

LETTERS AND JOURNALS

OF

SIR CHARLES LYELL, BART.

AUTHOR OF 'PRINCIPLES OF GEOLOGY' &C.

EDITED BY HIS SISTER-IN-LAW, MRS. LYELL

IN TWO VOLUMES—VOL. I.

With Portraits

LONDON

JOHN MURRAY, ALBEMARLE STREET

1881

PREFACE.

THE Geological Works of SIR CHARLES LYELL are the best monument which he has left to the world as a record of his labours in Science, but something more may be desired to be known of the life of one who loved Nature's works with an intensity which was only equalled by his love of Truth.

These volumes contain a sketch of the early days of Geological Science, and of the Geological Society of London, with glimpses of some of the bright characters who adorned it.

An autobiographical account of his boyhood is given, and large extracts from the private journals and letters to his wife, his family and friends, as these record better than any panegyric the untiring energy and enthusiasm which never flagged during a long life, rendering it a useful and a very happy one.

His cultivated mind and classical taste, his keen interest in the world of politics and in the social progress and education of his country, and the many opportunities he enjoyed of friendly intercourse with the most leading characters of his age, make the letters abound in lively anecdotes and pictures of society, constantly interspersed with his enthusiastic devotion to Natural History.

The geological information is given, as he gathered it in, from his personal observations at the time, though he afterwards may have arrived at different conclusions as he advanced in the study of the subject.

Many thanks are due to all who have kindly lent me their letters, and have thus contributed to the completion of these memoirs.

CONTENTS

OF

THE FIRST VOLUME.

———•◦•———

CHAPTER I.

NOVEMBER 1797—1812-14.

CHAPTER II.

FEBRUARY 1816—NOVEMBER 1817.

CHAPTER III.

JUNE, JULY 1818.

LIST OF PLATES.

KINNORDY, FORFARSHIRE,

THE BIRTHPLACE OF SIR CHARLES LYELL.

LETTERS AND JOURNALS

OF

SIR CHARLES LYELL, BART.

———❖———

CHAPTER I.

NOVEMBER 1797—1812–14.

BIRTH AT KINNORDY—BARTLEY LODGE—SCHOOL AT RINGWOOD—ADVEN-
TURES — SCHOOL-LIFE AT SALISBURY — OLD SARUM — MORE SCHOOL
ADVENTURES—DR. FOWLER—ILL, AND SENT HOME—LOVE OF INSECTS
—DR. BAYLEY'S SCHOOL AT MIDHURST—REFLECTIONS ON SCHOOL LIFE
—CLASSICUS PAPER — TASTE FOR POETRY — LATIN VERSES — PETTY
GAMBLING—CHESS—MUSIC—THE ENRAGED MUSICIAN—COWDRY PARK
AND BIRD-NESTING.

[Charles Lyell was born November 14, 1797, at Kinnordy, the
family estate in Forfarshire. He was the eldest of ten children,
having two brothers and seven sisters, who all grew up. His father
was a man of cultivation and refinement, and had both literary and
scientific tastes. In early life he devoted himself to Botany, especially to the cryptogams, and latterly took up the study of Dante,
and published several works on, and translations from, the great poet.
His mother, the daughter of Thomas Smith, of Maker Hall, Swaledale,
Yorkshire, was gifted with strong sense, and a tender anxiety for the
welfare of her children.

In 1832 he married the eldest daughter of Mr. Leonard Horner,
F.R.S., and in his letters to her, when she was living abroad at Bonn
on the Rhine, he wrote at her request the following autobiography,
which only extends to near the end of his school-days.]

AUTOBIOGRAPHY.

I was born November 14, 1797, at Kinnordy.[1] 'The front of heaven was *not* full of fiery shapes, at my nativity,' but it was a remarkable winter and spring, so warm that my mother slept all night with her bedroom windows open—which no doubt portended something remarkable in the bairn; and sure enough he was pronounced to be the loudest and most indefatigable squaller of all the brats in Angus, and while he kept others awake all night by his noise, thrived the while most vigorously. Besides this, it was more than twelve months before I cut a single tooth, and some old woman in Southampton, finding that my gums were very hard, and that I could eat well, very considerately tried to persuade my mother that her first-born would *never have any teeth !*

But I have travelled to Hants too fast. At the age of three months, I was taken on a tour to Inverary, afterwards to Ilfracombe in Devon, then to Weymouth, and finally to Southampton, where Tom[2] was born. My father then took a fourteen years' lease of Bartley Lodge in the New Forest, Hants, and stayed there twenty-eight years. The first thing of which I have a very distinct remembrance is learning the alphabet. I was about three, I believe, at that time, and was four and a half when an event happened which was not likely to be forgotten. My father and mother were on their way to Kinnordy to spend a summer there, and were within a stage and a half of Edinburgh, with Tom and me in their carriage, and behind came a post-chaise with two nursemaids and the cook, with Fanny and Marianne[3] (then the baby). On a narrow road, with a steep brae above, and an equally precipitous one below, and no parapet by the roadside, a flock of sheep jumped down into the road, and frightened the horses. Away they ran, and with the chaise, man, horses and all, disappeared clean out of sight, over the brae in an instant. No one was hurt but one of the maids, whose arm was cut by the glass being broken. My father broke the upper pane of glass, and pulled Fanny and Marianne through the opening. All were afterwards got out, and while they sat by

[1] See vignette, on title-page.
[2] His next brother.
[3] His sisters.

the roadside, one maid fainting, and the other with my mother washing the blood from her arm, a chaise with two ladies and a gentleman came past, who stared on them, and went on without offering any assistance. A shepherd was more obliging. Meanwhile Tom and I were left in the carriage. We thought it fine pastime, and I am accused of having prompted Tom to assist in plundering the pockets of the carriage of all the buns and other eatables, which we demolished with great speed for fear of interruption. The next event I remember was a daft woman, who came and haunted Kirriemuir,[4] and often and often visited Kinnordy, swearing that she would not leave the country till she had one of ' the laird's bairns.' She attacked the nursery-maids one day most furiously, when they were out with us, who, frightened at her frantic questions, ran away. After several of these visits, which made a great impression on my imagination, they caught the poor creature, and locked her up in the stable, after which she never returned. I really cannot recall anything worth mentioning for nearly three years after. We lived at Bartley, and walked out among that beautiful wild forest scenery with the nurses, &c. My Grandmamma Lyell came from Scotland with her two daughters, and settled in Southampton, and their frequent visits to Bartley were agreeable incidents to the children, always with presents of toys, sweetmeats, &c. I was kept back nearly a year from going to school, in order to wait till Tom could go with me. Then, when I was seven and three-quarter years old, I was sent to Ringwood, to a Rev. R. S. Davies, about fifty boys there I suppose. What an event is this in a boy's life! It is completely a new world, and a rough one enough, if they have been humoured and petted.

It is a great amusement to me to recall scenes that seem so distant, and yet which are so vividly impressed upon my mind. Tom and I, and about three others, were the youngest in the school. One of these was a little fellow of the name of Montague, who must have been a complete ' Billy Bottom,' ready to undertake everything, and who undertook to direct us, teach us games, &c., and to whom we appealed for explanations. I went in the middle of a half year, and

[4] The neighbouring town in Forfarshire.

stayed, therefore, about ten weeks before the holidays. A few weeks after my arrival, a quarrel happened between the boys of the town, and ours. The former were familiarly called ' the blackguards,' and they termed us ' the Latin cats.' A challenge was carried, I know not from which party, to fight, accepted, day fixed, weapons (sticks) agreed upon, their size, and everything else. It was decided upon by a council of war that we five little ones were too small for soldiers, so we were to be left, but sworn to secrecy. Each one was called up, and told that if he blabbed, every bone of him should be broken. When the evening came, every one of the whole school marched and met the foe. They must have had a regular ' shindy,' as the Irishmen call it, for although the news of it spread in time for Davies and some of the tradesmen to rush in and separate them, a great number of heads were broken. All holidays were taken away for ten days, and extra tasks imposed, and a sermon read to them on their atrocious guilt. At that time smuggling was very active on the coast, and ' smugglers ' was our favourite game. One party were the custom-house officers, the other were the contrabandists, with the kegs, &c. The first time I went out, not knowing what was to happen, I was so frightened by the surprise and hallooing, and row of the sham fight, when we were set upon from behind a hedge in a dark lane, that I verily trembled for my life.

At that time the expectation of a French invasion was general, and ' the volunteers ' were incorporated. My father accepted a captaincy, and his corps were ordered to Ringwood, a pleasant event for us. Before my first return for the holidays, the rejoicings for the victory of Trafalgar were celebrated. There were bonfires on the summit of every hill round Ringwood, which had a grand effect at night : an illumination in the town, where almost every candle was blackened on the outside, in mourning for Nelson. The band of the volunteers played ' Rule Britannia ' and ' Battle of the Nile,' while the people sang standing round a great bonfire in the market-place. The volunteers had a spite against the Mayor of Ringwood, because he would neither subscribe to their funds, nor let them exercise in a field of

his, so they threw squibs into his windows, and set fire to
the drawing-room curtains, which were extinguished with
difficulty. I remember participating perfectly in the mixed
feeling of sorrow for the death of Nelson and triumph for
the great victory. The Government, by way of making the
volunteers on the alert, gave a sham alarm at Portsmouth
and Southampton that the French were landed. I have
since heard that this was a party ruse, to stir up the anti-
Gallican feeling. The drums beat to arms. My father's
troop turned out. My mother packed up a travelling trunk,
to fly into the *interior*. It nearly cost my grandmother her
life, she was near dead of fright.

The return to the liberty and luxuries of home, after
roughing it at a boys' school, is of all things the most
delightful, yet I always got rather weary of the idleness of
home, and was almost glad to go back. Soon after my
return to school, I saw two strange exhibitions. One was
the chopping off of a duck's head, and the bird running for
a short distance afterwards, which we thought good fun.
The other was the killing of a calf, which I remember
shocked me exceedingly. I think I see the animal stunned
by the first blow now, the only execution I have ever seen of
the kind. Little Montague, who let us into these good things,
took us the same half year to see a corpse. A female relation (not an old person) of Mrs. Cleater, no less a personage
than the accredited vendor of gingerbread and pies, &c., to
the school, had died next door to the school. We saw the
corpse laid out, covered with flowers; I recollect thinking it
a pleasing sight. Montague knew that I and Tom were
great pets of Rachel Davies, a girl of about thirteen,
daughter of our reverend schoolmaster, so he bethought him
to send me to Miss Rachel to ask, as a great favour of her
father, to let us little ones, M. included, go to the funeral.
This favour she obtained, and took us. Her father officiated.
We thought ourselves in great luck, for a private soldier was
buried the same day, and they fired over the grave. For
weeks afterwards there was nothing but acting funerals,
which superseded every game. Small graves were dug near
the corner of the playground, or to whatever place the boys
walked out, and some blocks of wood covered with a piece of

crape carried as coffins, and some going with their pocket-handkerchiefs to their eyes as mourners. I seem to see Montague now, with something like a gown on, lifting up his hands, and repeating, in exact mimicry of our old pedagogue, ' Ashes to ashes, and dust to dust.' When we buried soldiers, we made a grand row, to represent the musketry, and threw handfuls of dust for smoke.

Of all the dreaded penances which we had to undergo we thought going to church, sitting whole hours doing nothing, incomparably the worst; far more intolerable than lessons, in which I always had some mixture of pleasure. We took marbles in our pockets, to play at ' odd or even,' to relieve the tedium, and many other devices, for which we often got punished. I one day, as we returned from church, asked Montague 'what could be the meaning of "to beat down Satan under our feet"?' He was always ready with an answer, by which means he kept up his authority amongst us, and we gave implicit faith to his decisions and explanations. But this was nearly a poser. He replied, however, without hesitation: ' I cannot tell you, but I will show you by-and-by how to do it.' So he took our little squad to a shallow pool of water, and taking off his shoes and stockings, jumped about in it, splashing us all over, and then said he had been beating down Satan. I did not presume to doubt his interpretation, but for years afterwards thought it so odd that he had not jumped on a piece of satin, instead of water, which was only soft like satin.

I have not reached the beginning of my tenth year, and have not left Ringwood. The last half-year I was there, a set of strolling players set up a stage in a barn, and persuaded our master to let the whole school attend. They acted ' Hamlet.' I had never read a play, nor had any notion of such a thing before, yet I entered into it so much that the impression remained very vivid, and when, years afterwards, I read the play, I thought the effect much heightened from my former recollection. In almost every other instance when I have read Shakspere first, I have found the acting fall short of my conceptions of the play from reading; besides the natural indignation one feels for the liberties they take in altering the text, which I never could endure even as a

schoolboy. I remember being puzzled at the play within
play in ' Hamlet,' but my delight in the Ghost scene was very
great. I should like much to know what was the degree of
mediocrity of the poor itinerants, for I question whether I
have ever thought any acting since comparable to them.

It was now decided that Tom and I should leave Mr.
Davies's, who was no great scholar, and whose set of boys
was of a very mixed kind, and not so select as my father
thought he might get at Salisbury—at Dr. Radcliffe's. This
Dr. Radcliffe was a good man, and kept a school of about
fifty boys, and all of the very best families in Wilts, Dorset,
and part of Hants. We had the Portmans, Windhams,
Beresfords (Irish), Beckfords, Preston, &c. The Doctor
was a good Oxonian classical scholar, and had two good
ushers. We had a long holiday at home between the
schools.

Bartley belonged to a Major Gilbert, and consisted of
eighty English acres, seventy of which was laid down in
grass, and produced hay, which was celebrated as the best in
the county for the hunters—for Lyndhurst was only two
miles off where the New Forest hounds were kept. The
hay harvest, therefore, was a great concern at Bartley, and
when we returned home for each summer holiday the mow-
ing was always just beginning. We (Tom and I) used to
attend the mowers all day, and learnt to mow a little, and
used to tumble about with Marianne and Fanny in the hay-
cocks, and ride in the loaded waggons to the haystack. It
lasted the greater part of our five or six weeks' holidays, as
part of the land was always three weeks later in being cut.
I have always since thought haymaking the most delightful
of sights. We always had our small rakes and forks, and
used to try how much we could glean after the waggon. As
the whole of this grass land was unenclosed, and surrounded
the house, it made a good-sized park, and was full of fine
old oaks, some of which the Major cut down whenever he
wanted a few hundred pounds, for which I always owed him
a grudge, for I knew every tree, great and small, and used to
miss them as you might a piece of old furniture in a room.
To every clump and single tree in the park I gave a name.
One was ' Ringwood,' another ' Salisbury,' or ' London,' or

'Paris,' &c. Single trees were named after flowers, an odd
fancy: thus one was called 'Geranium.' These names were
afterwards adopted by the rest of the young ones. Although
I never forgave the Major for cutting the large oaks, yet
there was nothing I was so fond of as seeing trees felled.

We were now allowed to roam about a good deal by ourselves, and I always contrived to learn where the trees were
to be cut within reach of a walk, which was very frequent,
for the demand for the navy was then most pressing; and
besides the thousands of splendid oaks then cut near Lyndhurst, a great number of beech trees were also consumed,
for every house has a right to a certain share of firewood,
by old custom, from the king's forests. I have been reading
lately that the Government surveyors found that forty acres
of the best soil for oaks were required to build one seventy-
four gun-ship, the oak being a century old, and a fair proportion of young trees being left. In other words, that forty
acres of good oak soil would produce a seventy-four per century. The 'Agricultural Journal' adds: 'In all Scotland
there is not now ash timber enough to build two large men
of war.' You may suppose that when such numbers of
frigates and so many men of war were built at the time I
allude to, at Portsmouth, which was so near, and when the
price of timber was at its height, how annually, at 'one fell
swoop,' large spaces were denuded of their noblest trees.
In order that a large tree might not injure younger wood in
its fall, much skill was required in so driving in wedges, after
the tree was sawed through near the root, as to direct its
weight to some clear space. The last wedge being driven in,
the tree was supported by its roots, and when all were severed
but one strong root, and this half cut through, the axe used
to be given to us boys, and we had the pride, at one blow, of
finishing the work. The tree at first moved slowly, then fell
with great rapidity, and with a magnificent crash as the
boughs reached the ground.

I had only reached my tenth year when I went to Salisbury, sixteen miles from home, to Dr. Radcliffe's—a school
in the middle of a large town, whereas Ringwood school was
in the suburbs of a small one. In this respect we thought
the change for the worse. Instead of a large meadow for a

playground, near a river, where we could bathe, and where
we were in the country the moment we walked out, which
we enjoyed at Davies's, we had now a small yard surrounded
by walls, and only walked out twice or three times a week,
when it did not rain, and were obliged to keep in ranks
along the endless streets and dusty roads of the suburbs of
a city. It seemed a kind of prison by comparison, especially
to me, accustomed to liberty in such a wild place as the New
Forest. Our favourite walk on holidays was to Old Sarum,
the celebrated rotten borough where one alehouse, with its
tea-gardens attached, sends two members to Parliament, and
which Lord Caledon bought no great time since, at its full
price, little thinking of the evil day which approached. Sir
Rufane Donkin remarked to me that the fact of Alexander
Baring having given 42,000*l*. for a borough within a year
of Schedule A, had destroyed his (Sir R.'s) confidence in
the funds, for in this case too, he says, shrewd merchants
may equally miscalculate the approaching era of the *sponge*.
But I must not run into such digressions. This Old Sarum
is a singular isolated chalk-hill, on the summit of which is a
splendid old camp, with a deep triple trench. I have seen
many, but not one where the moats are so deep, and the
slope of the high banks so rapid. We used to heap piles of
large chalk flints on the opposite ridges, and set them running to the bottom, where they broke on meeting and dashing against each other, as they descended from the opposite
sides. Then we examined to see which had crystals of calce-
dony in the middle, or of sparkling quartz. Near the centre
of the flat space enclosed within the mounds and trenches
was a deep long subterranean tunnel, said to have been used
by the garrison to get water from a river in the plain below.
But whatever its former use, it was to us a great source of
fun. According to established custom, I and some other new
comers were taken to the mouth of a subterranean passage,
in which the air was very cold. While we stood looking down
and wondering, all sorts of tales were told of its enormous
depth, and how it grew steeper far down, and ended in a
pool of water, &c. Then some of the initiated got behind
us, and knocked off all our hats, which began to roll down,
and were soon out of sight. After a great laugh at our

expense, and when we were in dismay at the thoughts of parading the streets of Sarum in hatless plight, a guide was appointed, whose duty it was to take all of us down in search of our lost property. We went down many hundred yards, standing upright, then were obliged to stoop. Here I found my hat, but the others had gone farther, and their owners were obliged to crawl on, upon all fours, with their guide. They soon came to a large chamber, very dark, where, after groping about in the dark, they found their hats, greatly improved, of course, as was mine, by knocking about upon dripping chalk and hard flints in their descent. The passage went on, but how far, I know not.

At Ringwood I think my studies were confined to English, to writing and reading, and getting things by heart. At Sarum I began the Latin grammar, which, disagreeable as it is, is considered by most schoolboys as a piece of promotion when they first begin it. There was nothing at Radcliffe's for exciting the emulation of the boys, and as I had an aversion to labour, which nothing but such a stimulus could overcome, I learnt but little during the two years I was at this school.

Bartley appears still to me, and all of us, more a home than even this (Kinnordy). It is the home of our recollections, at least, and it was the place where as schoolboys Tom and I spent our holidays. Had I spent them here, I am sure I should have had a delight in Kinnordy which I shall never have now,—much as I like it. All the trees in the park were oaks, and in the surrounding hedge and ditch were oaks, beech, and a few clumps of fir. Near part of a group which I named ' London' there was an old brick kiln, overgrown with briars, which I may mention again. A clump of trees on the slope of the hill was ' Romsey,' and a single tree in front of the right wing of our house was always styled ' The Umbrella,' from its shape. It was a fine oak, spreading equally on any side.

To return to Salisbury, we called our pedagogue Dr. Radcliffe ' Bluebeard,' because he had then his *third* wife. She died soon after I left, and he married a fourth, who I believe is now alive. He was a diligent teacher and fair scholar, and had a very rare merit, that of impartiality, which many of his ushers had not. One of these, Monsieur Borelle, the French master, was much disliked for favouritism. He was an emigrant, had been in the army, and being very poor, Dr. R. gave him leave to sleep in the house. His room was within one in which I and eight others slept. One night, when we were very angry with him for having *spatted* us all round with a ruler, for a noise in the schoolroom, which only *one* had made, and no one would confess, we determined to be revenged. We balanced a great weight of heavy volumes on the top of the door, so that no one could open it without their falling on his head. He was caught like a mouse in a trap, and threw a book in a rage at each boy's head, as they lay shamming sound sleep. Another stratagem of mine and young Prescott (son of Sir G. P.) was to tie a string across the room from the legs of two beds, so as to trip him up: from this string others branched off, the ends of which were fixed to the great toes of two sound sleepers, so that when Monsieur drew the lines, they woke, making a great outcry. At last we wearied him out, and he went and slept elsewhere. I conclude that there were far too many hours allotted to sleep at this school, for at all others we were glad to sleep after the labours of the day, and got punished for late rising in the morning, and being too late for roll-call. Here, on the contrary, a great many of our best sports were at night, particularly one, which, as very unique, and one which lasted all the time I was there, I must describe. It consisted of fighting, either in single combat, or whole rooms against others, with *bolsters*. These were shaken until all the contents were at one end, and then they were kept there by a girth of string or stockings. This made a formidable weapon, the empty end being the handle, and the ball at the other would hit a good blow, or coil round a fellow's leg, and by a jerk pull him up so that he fell backwards. The rules of our military code for the ' bolster-fight,' which were observed with the rigour which boys always introduce into their sports, were as follows. No one could put on any clothes, save chemise and nightcap; no one might return kicks or fistycuffs for blows with a bolster; no more than two rooms which were contiguous, or leading into the others, could join forces either

for attack or defence. The invading party were always to
station a watch at the head of the stairs, to give notice of
the approach of 'Bluebeard,' for he was particularly severe
against this warfare, though he never succeeded in putting
it down. He used to come up with a cane, which, as none
were clothed, took dire effect on those caught out of bed.
He had a fortunate twist in his left foot, which made his
step recognisable at a distance, and his shoe to creak loudly.
This offence was high treason, not only because it led to
broken heads and made a horrible row in the night, but because Mrs. Radcliffe found that it made her *bolsters* wear
out most rapidly.

I shall never forget the confusion when the scout gave
notice that the enemy was approaching. The invaders ran
most risk, as they had to make their way back to their
room, but in compensation they were able, when they surprised a chamber of sleepers, to give them all a good blow
round before they began to fight. If some could not make
good their retreat, I have known them get up the sooty
chimney, and when the Doctor was in the inner of the two
rooms, descend and get to their bed before he came back; or
if they were so unlucky as to be in the inner room of the
two invaded, they could only conceal themselves, and their
friends at home would sometimes get them off by putting a
bolster into their bed with a nightcap on, to be mistaken for
them.

Among other amusements, was keeping field mice in
boxes, concealed under our beds. These were dug up in the
fields, with their stock of nuts and grain, a favourite sport.
The young mice were carefully educated in the use of a small
musket, for they were very docile, and would hold the small
piece of wood between their fore feet, and shoulder, ground,
and present arms, at word of command, and go through a
whole drill. You may suppose that by this odd feat being
taught the little animals, that we were at that time in the
habit of seeing a drill. In fact a sergeant was employed
twice a week to drill us, and we had all guns with tin barrels
and locks, a strange fancy rather, but as there were volun-
teer corps everywhere, I suppose Dr. R. had his head full of

the chances of invasion. The exercise, however, of the drill, was very proper at a school.

I had the measles at this school, my first year I think, very severely. I was in a high fever, I thought myself for two days going round a number of wheels in all directions, one of which transferred me to another, until I was giddy; and I thought there were thousands of others going round in the same manner, which appeared whichever way I turned. I had long a fear of going to sleep afterwards, lest this, which, however, was a *waking* dream, should return.

At this time, in my eleventh year, 1808, my Aunt Ann married Captain Heathcote, R.N. This connection brought us into still greater intimacy with Mr. Heathcote, then M.P. for Hants, afterwards Sir Thos. H., and during our holidays Tom and I paid some pleasant visits to Embley. My Grand-mamma Lyell had till now resided at Southampton with her two daughters, the eldest of whom, Mary, still lived with her there. They came over once to the music meeting at Sarum, a triennial fête, and took Tom and me to part of it. Soon after this I was ill, and Dr. Fowler of Sarum rather alarmed my mother about the air of Sarum being bad for me, and that I seemed falling into a consumption, and that my lungs were affected. I thought it all stuff at the time, though certainly I had some complaint on my lungs, which I have not had since; but as I did not like Sarum, I did not try to make light of it, and was taken home for three months. As they were afraid to overtask me, I began to get annoyed with *ennui,* which did not improve my health, for I was always most exceedingly miserable if unemployed, though I had an excessive aversion to work unless forced to it. It happened that a little before this time my father had for a short time exchanged botany for entomology, a fit which only lasted just long enough to induce him to purchase some books on the latter subject, after which he threw it up, principally I believe from a dislike to kill the insects. I did not like this *department* of the subject either, but soon satisfied myself that their feelings must somewhat resemble that of plants, that one part can live long without the other, that they will eat after many of their limbs are cut off, fight when

pins are stuck through them, fly away when all their insides
are emptied out, and other signs which imply a kind of
vitality too remote from that of the higher warm-blooded
animals to awaken any feeling of disgust in one who studies
them much, at the idea of giving them pain, by the length
of time which it takes to slay them, at least, unless the
operation be unskilfully performed. Collecting insects was
just the sort of desultory occupation which suited me at that
time, as it gave sufficient employment to my mind and body,
was full of variety, and to see a store continually increasing,
gratified what in the cant phrase of the phrenologist is termed
the 'accumulative propensity.' I soon began to know what
was rare, and to appreciate specimens by this test. In the
evenings I used to look over ' Donovan's Insects,' a work in
which a great number of the British species are well given
in coloured plates, but which has no scientific merit. This
was a royal road at arriving at the names, and required no
study, but mere looking at pictures. At first I confined my
attention to the Lepidoptera (butterflies, moths, &c.), as the
most beautiful, but soon became fond of watching the
singular habits of the aquatic insects, and used to sit whole
mornings by a pond, feeding them with flies, and catching
them if I could.

I had no companion to share this hobby with me, no one to
encourage me in following it up, yet my love for it continued
always to increase, and it afforded a most varied source of
amusement. I was chiefly attracted by the beauty of the Le-
pidopterous tribe; in common parlance, by the butterflies,
moths, hawk-moths (sphinxes), besides the procuring the chry-
salis and seeing its transformation; and the feeding and breeding of caterpillars was another reason for preferring this numerous and showy class. I soon, however, learnt to prefer the
rare to the brilliant species, and was not long in discovering,
by the comparison of one season with another, that each
species had its peculiar time for appearing, some twice, some
once only in the year, some by day, some in the evening, and
others at distant hours of the night.

The only other insects that engaged in the least my attention at that time were the aquatic. I was greatly surprised
to find every pond tenanted by water-beetles of different sizes

and shapes, and to observe them row themselves along by
the broad row of bristles attached to their legs. I threw flies
and moths into the water and observed them rise, and learnt
their relative strength, seeing some species relinquish the
booty on the appearance of others. The long spider-like flies
which run on the surface, the glimmer chafers which thread
the surface, in what we called a figure-of-eight movement, the
beetles which swim on their backs, and many others, such
as the red tick, used to be caught and brought in a basin
into my bedroom, and there kept, to the annoyance of the
housemaids when the water was none of the sweetest; and
then the whole were fed with window-flies, until some died,
and others took wing in the night, and flew back to their
native waters. I passed nearly the whole day alone fishing
for this small fry in some deserted gravel pit, or often in a
great pond called the brick-kiln, just by the fir-trees called by
us 'London.' This sport could be pursued even in winter, and
often I have found the large water-beetles frozen in the ice
which covered our ponds, and have brought home blocks and
melted them out. By this same brick-kiln (out of which
all the brick clay of which the house was built had been
taken) there were a great quantity of brambles, the flowers
of which in summer were usually ornamented by numbers
of the most beautiful and rarest of the English butterflies.
Among these, three varieties of Frittilary and the black
Admiral (Camilla) were conspicuous. I was very ill
supplied with instruments, both for taking and preserving
my prey, and many insects of real value to the entomologist,
as I afterwards discovered, were mutilated by being knocked
down by my hats, and then pressed between the leaves of a
white paper album. As my hats shared often in the mutila-
tion, I received sundry admonitions on that score, till at last
Miss Newlands, the governess, knit a small string net, a very
poor substitute for the gauze apparatus which would have
equally secured the flies, and kept them there when once in.
The larger ones, however, could not escape from these toils,
and the number caught was soon greater than I had the
patience to preserve. After a time my Aunt Mary gave me a
piece of furniture, henceforth called 'Charles's Secretary,' an
heirloom, which was not wanted in their new lodging when

they quitted that which my Grandmamma Lyell and her
unmarried daughter had previously lived in in Southampton.
This valuable acquisition was immediately turned into a
cabinet for insects, which, for want of a lining of cork in the
drawers, were impaled on the softish wood in the bottom of
each, which was nevertheless hard enough to turn the pins
sometimes. I still possess this piece of mahogany, and
although only fitted for a bedroom, I still look upon it as a
favourite, and used to have it in my sitting-room in the
Temple. Transfixing these insects upon the soft wood of these
drawers was a great step beyond pressing them between
paper, and some of the varieties thus preserved were afterwards useful to Curtis the entomologist. My father's head
livery servant, John Devinish, who had been employed to
botanise, and showed some skill in discovering some of the
rare species of that minute tribe the Jungermanniæ (allied
to the mosses), was the only associate I ever remember to
have obtained in my rambles through the Forest. With him I
formed as strict an alliance as his scanty leisure permitted,
but any sympathy was valuable, any one sufficiently familiar
with the *commoner* species to be capable of appreciating a
great treasure when carried to him on some triumphant occa
sion. Instead of sympathy, I received from almost every one
else beyond my home, either ridicule, or hints that the pursuits of other boys were more manly. Whether did I fancy that
insects had no feeling? what could be the use of them? The
contemptuous appellation of 'butterfly-hunting' applied to my
favourite employment always nettled me, yet since one class
of insects was sought to the exclusion of almost every other,
a naturalist even might have found some point in such a
disparaging designation of the hobby. It must be confessed,
too, that the organ of acquisitiveness, of which craniologists
talk, was more bent on the miserly delight of amassing
treasure, than in using it, for it was less for information than
the love of possessing the specimens that they were accumulated. Of their history I knew but little, and still less of
their structure, yet I knew most accurately to distinguish
several hundred species, some very minute ; and still retained
a very perfect recollection of nearly all, and could select the
English butterflies and moths out of a foreign collection,

and without the aid of books, gave names to certain tribes,
such as the 'fold-up-moths,' 'the yellow-underwings,' &c. &c.,
which I afterwards found were natural genera or families,
and my rule of thumb classification had thrown them into
natural groups. In the course of each holiday, for three or
four years successively, I fell back into my old haunts in the
woods, and became so keen before my return to school, that
I could seldom resist the temptation of employing my leisure
hours there in the same way. I could not see a rare moth without catching it, especially if not exposed to be laughed at by
any witnesses of such a queer fancy. When taken, they were
put between the leaves of my dictionary, as the only book
not taken up to lessons. When I looked out a word, I often
found the two pages firmly glued together by some moth, the
contents of whose body had been squeezed out between the
leaves. The disrepute in which my hobby was held had a
considerable effect on my character, for I was very sensitive
of the good opinion of others, and therefore followed it up
almost by stealth; so that although I never confessed to
myself that I was wrong, but always reasoned myself into a
belief that the generality of people were too stupid to comprehend the interest of such pursuits, yet I got too much in
the habit of avoiding being seen, as if I was ashamed of what
I did.

Among the things which supported me in my secret
estimation of entomology, was the number of expensive books
on the subject which I found in my father's library. Many
of these were full of plates, but these did not weigh so much
with me as those written in Latin, and full of dry descrip-
tions and hard terms, for it was evidence of the learned and
studious sages who had devoted their time to it. But with the
exception of a few scraps of Linnæus, I never troubled any
one of these learned writers, but whenever I got a new thing,
turned over all the plates, till I saw whether I could find the
name by that royal road to knowledge. If found, the first
inquiry was, whether it was rare.

I have now reached the eventful period when I was about
to leave Salisbury, and go to a new and last school. There
was an interregnum of near half a year, in which Tom and I
were taught by my father at home. I believe we got on

pretty well. We read at that time Virgil's ' Eclogues,' &c.,
and a French master came out twice a week. At length it
was decided that we should go to Winchester, but it was
found that our names should have been down on the College
lists for two years, to give us a chance of entry. So another
school, as a sort of preparation I believe, was decided upon,
kept by Dr. Bayley (once a master at Winchester, and an
Oxonian), who had about seventy boys, in a school of ancient
foundation, at the old (rotten) borough of Midhurst in
Sussex, a beautiful spot just opposite the old park of the
Castle of Cowdray, which belonged to the Montague family,
the title of which is now extinct, and which belongs to
Poyntz, who married the daughter and heiress. Dr. Bayley
had married a Goodenough, one of the Bishop of Hereford's
first cousins. He had been long in good society at college,
and then at Winchester, and it was natural perhaps enough
that he should feel the change to a small provincial place,
where there was no society but a few tradespeople, &c. The
school into which I was now ushered was a very different
world from either of the former. From the age and number
of the boys, and the system adopted, it had all the character
of one of our great public schools. I was now past the age
of twelve, and no longer reckoned one of the little boys.
Whatever some may say or sing of the happy recollections
of their school days, I believe the generality, if they told the
truth, would not like to have them over again, or would
consider them as less happy than those which follow. I
felt for the first time that I had to fight my own way in
a rough world, and must depend entirely on my own
resources.

You have never heard, perhaps, how much the different
boys struggle for power, and how exactly each learns at last
to know who is his master, a supremacy which depends
partly on spirit, or as they term it *pluck*, and in part on
physical strength. There are always some pugnacious fel-
lows who rather delight in affrays, and in setting others on,
and are the dread of the weak quiet-spirited youngsters.
They are often bullies, who are afraid of many of their own
age and strength, and knock under to them, but who tor-
ment the quiet souls, and are hated as tyrants by the weaker

and younger in the school. There are some from kindly
feeling, but many more who from a love of affording protection, and feeling their own consequence by shielding others,
take a certain number under their care. You often hear
some luckless wight who has been struck, exclaim, 'I'll tell
C. of you if you hit me again.' As to telling *the master*,
that is an appeal against the law whatever cruelty may prevail, and finds sympathy with none of the boys. One of
the oldest boys in the school, Deacon Senior (for there were
three of them, Senior, Secundus, and Junior), voluntarily
took me under his charge for nearly all the first half year,
out of pure kindness, and then told me I was old enough,
and had been there long enough, to shift for myself. No
sooner was it known that I could not resort to a patron any
longer, than I began to be cuffed about and persecuted.
Almost every boy has to fight one battle before he gains a
position—before it is settled who is to be afraid of him, and
of whom he has to be in fear. It required a good deal to
work me up to the point to defend my independence, although
I could not brook being put upon by my juniors, and about
the middle of the second half year I was set down by many
as incapable of resistance. At last I saw that not to resent
and to fight, was a sure way of suffering more than any
battle could entail, yet this was the result of reasoning, and
I was never ready to retort immediately. One day a vulgar
low sort of a boy, of the name of Tilt, had got into a quarrel
with me, and a good many of the complimentary epithets so
common among schoolboys, such as ass, fool, raff, and
finally coward, were exchanged, upon which he hit me a blow
which knocked me to the ground, and gave me a severe
bruise on my head. As Tilt was a full year younger and
half a head shorter than me (though stronger built), I was
annoyed beyond measure by this ignominious occurrence,
and after some hours mustered up courage to declare to
some boys who came to ask me if it was true that T. had
knocked me down, that 'I should like to see him try it
again.' Immediately the report was spread that I was
ready to fight. The delight which the intelligence of an im-
pending battle occasions is always great, and numbers came
to pat me on the back, to offer to be my 'bottle-holder,' to

advise me to have 'two minute rounds,' &c., and every expedient for prolonging the diversion. One was appointed to hold the watch and proclaim when the rounds were to begin. It was decided that instead of two, only one minute should be allowed between each bout, so if one was knocked down, and was not up again and ready in one minute (the other calling on him to fight), the affair was over, and the one who declined coming to the scratch was to ' receive the cowardly blow,' a gentle tap, followed by a shout of the friends of the victor, and then sometimes, but rarely, by a shaking of the hands of the two combatants. This fight lasted two days, five or six hours each day, for there was no half holiday, and we were pretty equal, Tilt having learned to box, and so being nearly equal, though weaker than me. Both his eyes were black and his side much hurt, and his head swollen for a week after. I was, I believe, very much hurt, but my wounds were not in the face; and though I ought to have been put to bed, as he was, my friends advised me to put a good face on it, and pretend not to be hurt, ' as it would make others twice as much afraid of me.' I suffered severely from sustaining this character for eight days. I recollect the pain at this moment—every bone ached, and I was black and blue all over my body, and so stiff, that when I walked up 'to hills' on half holiday, I was obliged to get the arm of a friend. After an affair of this kind, there is a discussion amongst the different boys of the same age and standing whether such and such a one ought to be afraid of the conqueror. Every disparaging circumstance is brought forward, and as I was a year older than Tilt, it was said that I ought to be ashamed of having taken two days to do him up; ' besides, Lyell's more damaged than he pretends to be, his is a skin that don't show it so much as Tilt's,' &c. In fact I gained less than most would have done from so severe a combat, yet every boy who was afraid of Tilt had from that day of course great respect for me and my *fist*; and this saved me from much molestation, and I felt I had gained a great point.

A short time after this, a most desperate battle took place between Carter (one of the cousins of the present M.P.), a Portsmouth boy, and Norris, two of the biggest in the

school in the second class.   It lasted three days, and was a
most savage concern.   The pluck with which they fought on,
after both had grown pale with the blood lost from the nose,
and black with bruises, and when each could scarce totter
on his legs from weakness, was a sight that I recollect thinking barbarous and brutal; but the eager enjoyment and
admiration of the school at their *pluck* was surprising.
You cannot imagine the extent to which the stronger and
more spirited, domineer over the milder and more timid.
You will see on a cold day, when a crowd of little boys are
round the fire shivering, three larger or more powerful walk
up, and, without any order or hint, the crowd steals off, and
the three monopolise the warmth for hours, and will let none
but themselves toast or boil or bake on the fire.   There is often,
too, a sort of tribute paid by the weaker to propitiate these
majestic powers.   They receive all kinds of présents, in
order to make them less free with their cuffs and exactions.
For though fagging was not allowed, as in some public
schools, yet there is always much hard service exacted by
the older and stronger from their inferiors in age and
prowess.   Some boys quite sink under this roughing, but in
general they become more manly and hardier.   If they have
come from a comfortable home, have been petted and spoiled
to some degree, are sensitive of insult and hard usage, and
yet not possessing force of character enough and active
courage to fight their way, they must go through a most
painful ordeal before they feel themselves at all comfortable, especially when about the middle of the school.   The
recollection of it makes me bless my stars I have not to go
through it again.

There were about seven classes, two of which were inferior to the one I was placed in, and at the bottom of which
I found myself.   Every year some from the top of each class
were draughted into the class above—not those who happened to be at the top when the half year was over, but, by a
much more judicious arrangement, those who during the
five months had been oftenest at the top or near it.   I must
describe to you how this was ascertained.   The 'Classicus
Paper,' an invention adopted I believe from time immemorial
at Winchester, and some other of our great public schools,

was kept by each class, in which document each boy's place at the end of the day was noted down.   Thus, suppose there were eight boys in the class.   Their names would be arranged according to the number which they had got the week before, when Forbes being least was Classicus, and had the fag of keeping the Classicus Paper.   But as he obtained the mark of 13 during that week, and topped Lyell Senior, the last-mentioned dunce would take his place, and be Classicus for the ensuing week.   The great satisfaction with which the said paper is handed over to another to keep, may be con-

Feb. 2. A.D. 1811.	Blake	Thresher	Bailey Tertius	Deacon Junior	Dunbar	Ramsay	Lyell Sen.	Forbes
Monday . .	8	7	6	5	4	3	2	1
Tuesday . .	7	8	5	6	3	4	1	2
Wednesday .	6	5	8	7	4	1	2	3
Thursday . .	8	6	5	7	1	4	3	2
Friday. . .	5	7	6	3	8	2	1	4
Saturday . .	8	7	6	5	4	3	2	1
Total . .	42	40	36	33	24	17	11	13

ceived, for it is considered as a place of some reproach. Every boy is put on to translate a passage in his turn, and as often as he fails, the word is passed on to the next below : if he knows it he takes the other's place.   In this manner the attention of all is kept alive, for the boys at the top of the class are liable immediately to lose ground unless they are constantly attending to what those at the bottom are doing ; for when the last cannot answer to a question passed down, it goes over to the first boy at the top, and if he is absent, he loses a place immediately, however easy the query.

I believe this and the Classicus Paper to be an admirable contrivance for exciting a constant spirit of emulation, for
checking favouritism in the teacher, for ascertaining the relative strength of different boys, knowing how to promote
them into other classes, fixing their attention, &c. At the
end of the half year, all the totals being added together, the
boy who had the highest number received a prize, and three
or four of the first or more were raised into the class above.
The competition was greatly increased, as boys rose to the
higher classes, because those of the first had many valuable
privileges. They were called 'The Seniors,' and constituted
the magistracy, being empowered by the head master to keep
the rest in order, to report on all boys found 'out of bounds'
or guilty of any other delinquency. Of course the entrust-
ing such powers to young boys of sixteen to eighteen was
liable to some abuse occasionally, yet it works well, and the
effect upon the whole is, that much of the tyranny that
would otherwise prevail, of mere physical force and bold
daring dispositions over the meek and tenderer bodies, was
kept down by the supremacy of responsible magistrates,
raised to their station by talent, and usually younger than
some of the idle bullying dunces in the inferior classes. To
strike a 'Senior' is viewed in the same light by the head
master as for a junior officer in the army or navy to strike
or challenge a senior, and such an act of insubordination
was quite unknown. Besides, there were almost always one
or two in the first class stronger than any other in the
school, who made a point of enforcing the most complete
respect to the weakest in the 'magistracy' from the greatest
of the junior boys. We used to see with great delight some
youths who had been kept under by certain dreaded characters, suddenly put over their heads by being promoted to the
first class. The Senior might retaliate then upon his old
enemy in a legitimate way, by being very strict in looking
after him, and reporting any breach of the law, which might
have been winked at in a less obnoxious individual.

My ambition during the second half year was excited by
finding myself rising near the top of a class of fifteen boys,
in which I was; and when miserable, as I often was, with
the kicks and cuffs I received, I got into a useful habit of

thinking myself happy when I got a high number in the
class-paper. So great an effect had this on me, that if I was
lower than usual, and we failed to get on certain occasions a
half holiday on Friday (an extra treat given now and then
from fine weather and other occasions), I used immediately
to comfort myself with the idea that I should probably get a
higher mark. On the other hand, if unusually high up, my
anxiety was great to get the holiday, as then I was secure
of that number. By this feeling, much of my natural
antipathy to work, and extreme absence of mind, was conquered in a great measure, and I acquired habits of attention,
which, however, were very painful to me, and only sustained
when I had an object in view.

At the end of the first year arrived the anniversary, or
what was called 'the speaking,' when certain boys recited
verses written by themselves, those in the first two classes;
and the rest different Greek, Latin, and English passages.
The rehearsal first began, at which every boy had to exhibit,
and then ten were selected to perform before the public. I
obtained one of the places for reciting English, and was
accordingly gifted with a prize, Milton's 'Paradise Lost,' of
which I was very proud. Every year afterwards I received
invariably a prize for speaking, until high enough to carry
off the prizes for Latin and English original composition.
My inventive talents were not quick, but to have any, is so
rare a qualification, that it is sure to obtain a boy at our
great schools (and afterwards as an author) some distinction.
I had a livelier sense than most of the boys of the beauty of
English poetry, Milton, Thomson, and Gray being my fa-
vourites; and even Virgil and Ovid gave me some real pleasure,
and I knew the most poetic passages in them. I was much
taken with Scott's 'Lady of the Lake' on holidays, when I
had risen to the second class, and presumed, when the prize
was given on 'Local Attachment' in English verse (it being
an understood thing that the metre was to be the usual ten-
syllabic rhyme), to venture on writing it in the versification
of Scott's 'Lady of the Lake.' The verses were the only ones
out of the first class which had any originality in them, or
poetry, so the Doctor was puzzled what to do. The innovation
was a bold one : my excuse was that he had not given out a

precise metre ; on which he determined that this case was not
to serve as a precedent, that in future the classical English
metre was to be adopted, but mine was to have the prize,
being eight-syllabic and irregular, and not in couplets.

When in the second class, I wrote a Latin copy of verses
(a weekly exercise required of all) on the fight between the
land-rats and the water-rats, suggested by reading Homer's
battle of the frogs and mice—a mock-heroic. Dr. Bayley
had just drained a pond much infested by water-rats, which
was on one side of our playground, and they used to forage
on not only our cakes and bread and cheese in the night, but
literally on our clothes and books. I am sure that from the
date of this early achievement to the present hour I have
never thought of this copy of verses ; but I can recall with
pleasure the incident, and it convinces me that I must very
early have felt a pleasure not usual among boys of about sixteen in exerting my inventive powers voluntarily. The plot
was begun with a consultation of water-rats, to each of whom
altisonant Greek names were given, after the plan of Homer
—cake-stealer, gin-dreader, book-eater, ditch-lover, &c.
The king began by describing a dream in which the water-
prophet covered with slimy reeds appeared to him, foretelling
that the delicious expanse of sweet-scented mud would soon
dry up, and foreboding woes. Part of the warning was
copied or paraphrased from the Sybil's song to the Trojans
in the ' Æneid ' of what should happen when they reached
Italy. The dream and warning, taken, I suppose, from
Agamemnon's to the Grecian chiefs, being communicated,
the others entered into the debate what they should do, and
it was agreed that, as the fates had decreed the drying up of
the waters, they should migrate to a neighbouring sewer,
and should destroy the house-rats, who consumed so much
provender in the schoolroom, and who had usurped their
rights. One passage, in which a chief was described as a
great map-eater, and having at one meal consumed Africa,
Europe, Asia, America, and the Ocean, was admired as a
good specimen of pompous description of mighty deeds, on
the first entrance of a hero in an epic poem. The verses
ran on to thirty-eight, and when done, there was great
discussion whether I should dare show up such a thing. It

was thought, however, a wondrous feat, till the second master, Mr. Ayling, a youth of nineteen, who heard of it, said, ' I dare say it's all nonsense and bad Latin.' I was requested, in vindication, to let him see it before it went up to Dr. Bayley. To justify his own anticipation, he cut it up as much as he could, pointing out all the grammatical errors and one false quantity. Though he thus made many think light of it, and checked my growing vanity not a little, it of course had the effect of my correcting the lines, and rewriting a copy. Dr. Bayley, when he saw it, was much surprised at the correctness of the Latin, and struck, more than he chose to admit to us, with the invention displayed in the whole thing. He told the class that it was such good Latin that I deserved great credit, but he did not wish them or me to send him up more mock-heroics.

From this time I took it into my head that I should one day do great things in a literary way, but my ambition was quenched afterwards, by failing in carrying off any prize at Oxford.

But I must not digress, but return to my school-days again. Indeed, I have already been running on to near the highest class in the school, whereas I must return to the long toilsome ascent of the ladder, and to my second half year, when I was just fighting my way among the boys, and getting out of the fifth into the fourth class by high numbers in the Classicus Paper. It was the custom of Dr. Bayley to have the highest classes chiefly under his immediate care, but once in six weeks each of the others came before him in review. This much dreaded inspection was intended not only for the boys, but also for the under-masters, and when the great Doctor called out in a thundering voice, ' This class is disgracefully backward,' not only every boy in it quaked with fear, but the under-masters, if they had been idle, participated in the feeling of reproach. Two of these were university men—Ayling and Belin, the latter the son of a French *émigré*, and who had been brought up at Winchester, and I believe New College, Oxford. Although I abhorred these reviews as much as any one, yet when they were over I had usually the comfort of feeling I had gained in the Classicus Paper. The reason of this was that my father used during every holiday to require me to read with him, a certain part of every day, on the books we were going to work at the half year following. In this exercise he always went out of the beaten path, and instead of confining me to the mere translation of the Latin and Greek, used to teach me the geography and mythology, &c. Now Dr. Bayley usually finished each day by a few questions quite out of the ordinary line, and I often astonished the class by answering them. Well do I recollect a day when I was within two of the bottom of a class of about fifteen, the first boy being asked what the Ægean Sea was now called? It was passed all the way down, every one staring with astonishment, and when I answered, 'The Grecian Archipelago,' and was ordered to take all their places, I sat down, feeling quite dizzy at the sudden elevation. From this height I should have been at once precipitated to my former place if I had been 'put on' afterwards, but it was the end of the lesson, and we were dismissed, I scoring fifteen, a number nearly equalling an ordinary week's work.

About the end of the first year a spirit of gambling began in the school, which reached its height about the middle of the succeeding half year. First it was draughts, a penny a game; then some got cards, but these being against law, the Seniors suppressed them. The draughts, however, and betting on the game, and on other players, continued, and when all our money was gone, we gambled like savages for our food, and would have played for our clothes, if they had been part of the disposable goods belonging to us. It was common to hear this challenge, ' I'll bet you my Wednesday's top slice, against your Friday's bottom slice, that I beat you most games out of five.' Now this you must know was wagering half a breakfast against another moiety, but as the two slices were of unequal value, the top, as we called it, of the round of the loaf being less than the bottom half, the better player who wished to tempt his antagonist by offering him odds, either staked his bottom slice against the other's top slice, or to-morrow's against one not to be paid for several days hence. Schoolboys generally are as ravenous as wolves, and always almost fancy they have too little to eat. They take much exercise, are usually

in fine vigorous health, not over-fed, as at home, and ready
to devour the allotted portion. You may imagine, therefore,
the misery of the famished gambler, who was obliged, when
the sound of the breakfast bell rang, to go and seat himself
in the midst of the others, who were eating their bread and
butter, and drinking each a small basin of warm milk and
water, without having himself more than half, and often
none, of his ration. Payment was always rigorously exacted
and made, but some were able to borrow when in distress,
though usually on usurious interest. ' If you'll lend me half
a slice to-day, I'll pay you a whole one on Saturday.' Some
of those who had lost all their breakfasts for several mornings running, of whom I was occasionally one, got into a
habit of taking out one of the iron bars of the window, and
getting into the breakfast-room, or stealing in when the door
was left open, while John Budd, the purveyor, was gone out
to get the milk, having laid out the bread and butter.
When in, they clipped small bits off the slices of the larger
rounds, which would best bear it, and picked the butter out
of all the holes in the spread slices. They were caught more
than once at this petty larceny, and made severe examples
of by the other boys, without the interference of judge or
jury. I was cunning enough to escape one of these summary
punishments, but was often in a dreadful fright ; and once,
when I heard Budd approach, had nothing left for it but to
disguise my face and rush out, knocking him over, with half
his milk streaming in the gutter. Upon this he complained
to the Seniors of the depredation, and on their instituting an
inquiry, they soon got to the bottom of the whole evil, and
made a regulation that no boy should pay away more than
half his breakfast on one morning. They also kept guard on
the breakfast-room.

A rage for chess succeeded to draughts, and I gave myself over to this with great devotion, to the great detriment
of my advancement in the school. Instead of wholesome
exercise, football, cricket, fives, &c., I confined myself with
others during the hours of play to the schoolroom, over a
chess-board, when the mind was excited with the same kind
of feelings which are aroused by gambling, and the lassitude
which follows is perhaps greater. Instead of allotting an

extra half-hour to work, to get a higher place in the Classicus Paper, I was ever contriving means to steal a game of chess, out of sight, during school hours. At the end of the second year, a new hobby took possession of me—music. One of the boys had a flageolet, another a flute, which they played tolerably. Mr. Ayling also played the flute pretty well. I had an ear, and was very desirous of accompanying, so I set to work for it on a small octave flute, and when I returned home my father gave me one of his large flutes, which he had once played very well. As I always took up every hobby with energy, I soon learned to play a certain number of tunes from notes, and to accompany. Some of the boys who had finer ears were good at setting our flutes, so as to be in perfect unison; and as there are no discordant sounds on wind instruments, at least flutes, the music we produced was by no means despicable, and among the boys greatly admired. Some theatricals having been got up, we had a regular orchestra to play between the acts—three flutes, two octaves, a tambourine, and, by way of melodious accompaniment, a triangle. The second year of my hobby I had eight flutes in *my band,* as it was called, of which, however, I was an unworthy leader, two of the others being much better players, and having taken lessons. Dr. Bayley became at last annoyed at the incessant flute playing, which was heard even at his house, but knew not how to interfere. One day I happened to observe that one of the hall tables, when set in a particular way, gave a sound very like a kettle-drum, the middle boards, which were long, vibrating, and producing a fine mellow drum-like sound. I accordingly got my band into the dining-hall, to the great delight of the boys, but it attracted a crowd in the street every evening, and gave the Doctor an excuse for putting a stop to it. Our first French master, Monsieur Flambard, an *émigré,* had now left, and in his place a Monsieur Simon (?) came, a believer in animal magnetism, and who looked like a thoroughbred fiddler. He played well, and having no person to listen to him, he wished to join our musicians, and to instruct us. But his notions of application were far beyond those which idle schoolboys could tolerate. Most of the boys liked our flutes better than the thin meagre sounds of his cat-gut, though he really played excellently, and I suspect had been professionally in the orchestra in the London theatres. So after several unsuccessful attempts to drill us into performing an overture with him, he cut us, protesting there was not one note in tune, from beginning to end of our performance. I suppose this was nearly true, but I for one felt this aspersion on our musical talents a vile libel, and we gave out that he was jealous of the superior melody of wind instruments ; and his miserable violin, and the attitudes in which he placed himself, and his cries to us not to play so loud in the 'tenderer passages,' and other directions, became a standing joke and subject of ridicule. At last we pretended that we were overwhelmed with contrition, and coaxed him back to an evening rehearsal. Just as he was performing a solo infinitely to his own satisfaction, we all, at an agreed signal, burst in with flutes, three fifes, triangles, tambourines, and every noise we could make, to the infinite delight of the bystanders ; and I am sure the Enraged Musician in Hogarth would not have supplied a better subject for a picture.

About the second year of my Midhurst campaign, before Tom went to sea, Lady Ramsay (of Bamff, Perthshire) put her two eldest sons to Dr. Bayley's, induced by my father having us there (Sir James R. and his brother George, who afterwards went to Harrow). She used to have us out on Sundays once a fortnight, a great treat, principally because we got off one of the *churches*, which all boys abhor, and then we got liberty to walk where we liked. We employed the time chiefly in hunting for the eggs of partridges and pheasants. It was a great game country, and the preserves of the Poyntzes of Cowdray, Lord Robert Seymour, Mr. Blake, and others, afforded us great sport. I remember once finding a nest of pheasants, in which there were fifteen chicks and two eggs. We broke the latter, and out stepped the two chicks, and immediately thrust their heads under the egg-shells, all the rest of the body being exposed, apparently supposing that if they could not see us they were concealed. We often brought home enough good eggs to fill two coffee-pots, which were boiled next morning, and were esteemed a great treat; besides a vague notion that if detected we

might be transported to Botany Bay for this kind of poach-
ing, added much in our estimation to the superior flavour of
these eggs over those of a barn-door hen.

One of the favourite amusements of many boys in the
school, during our walks, was birds'-nesting, and I learnt the
egg of almost every bird in that country, where there were
a great variety. I was able to climb some trees which no
other boys could scale, of which I was very proud, and this
accomplishment made me greatly in request on certain occasions. I remember in particular an owl's nest, which could
only be taken by a long-legged and long-armed boy, who could
stretch up from one bough to the next. Once I took the
young owls, which were petted, and fed for some time at
home. The old ones flew out in my face. My way of descending these difficult trees (fine aged beeches in Mr. Poyntz's
park), for the descent was the most dangerous part of the
task, was one which I found out myself. Instead of coming
down by the main trunk, I got along one of the long boughs
till it bent down within reach of six or eight feet of the
ground, and then I let myself off. But I could never take
the jackdaws' nests, which were on trees without branches,
and could only be mounted by boys who had power to grasp
with their arms and knees. There were many beeches of
great height, which ran up like palm trees, and up which
some . . .

[Here the autobiography ends.]

CHAPTER II.

FEBRUARY 1816—NOVEMBER 1817.

LETTERS FROM OXFORD—FELLOW-STUDENTS—GIFFORD—KIRKE WHITE—
CHRISTABEL—EXAMINATION—LORD EXMOUTH—AMMONITES BUCKLANDI
—VISIT TO MR. DAWSON TURNER AT YARMOUTH—DR. ARNOLD—GEOLOGY
—JOURNEY TO SCOTLAND—TRIP TO STAFFA—IONA—RETURN TO OXFORD
—LINES ON STAFFA.

HIS EARLY STUDIES.

[At the age of seventeen, Charles Lyell was matriculated at
Exeter College, Oxford. He was neither a hard student, nor idle at
college, but he took respectable honours in classics, being in the
second class. He gave occasionally some hours to entomology, with
the late well-known naturalist, the Rev. Lansdown Guilding of St.
Vincent's. Bakewell's 'Geology,' which he found in his father's
library, was the first book which gave him an idea of the existence of
such a science as geology, and something said in it about the antiquity
of the earth excited his imagination so much that he was well prepared to take interest in the lectures of Dr. Buckland, Professor of
Geology at Oxford, who was then at the height of his popularity.
Lyell attended a course of these lectures and took notes of them.

He paid a visit in July 1817 to Mr. Dawson Turner and his
family at Yarmouth, and the following month went to Forfarshire
with his father. He then accompanied two college friends on a trip
to Staffa.]

To CHARLES LYELL, ESQ., BARTLEY LODGE, STONY CROSS,
HANTS.

Exeter College : February 29, 1816.

My dear Father,—There is full as much necessary business here as I had at Midhurst the last year and a half. But
at Corpus it is beyond everything. Although Norris has
been diligent lately, fagging in the evenings and refusing

invitations, yet yesterday he and Richards, during a holiday which they had, sat five whole hours, from breakfast till dinner, working. Considering the expedition with which Norris gets over things, I think they have too much. It has prevented his writing for the Latin prize, which he was extremely desirous of. In spite of all this he succeeds to the utmost of his wishes in keeping up among his acquaintance the highly honourable and fashionable title of a complete idle fellow. And it amuses me not a little to hear a certain friend of his, who is not over desirous that N. should outshine him, so far as to take a first class, deplore the manner in which ' so clever a man *throws away his abilities!* '

I have become acquainted with several gentlemanly men here, but as I at present know nothing about them, I will not fill the letter with empty names. Mr. Selwood has this moment been here to ask me to breakfast with him to-morrow. Unfortunately I am engaged. I have subscribed to the music room three guineas, being told by my *out-college* acquaintances that I might be sure all the men here did, and that I should be forced to go now and then, which would amount to more than the subscription for the year. There is to be no grand music-meeting here, although it is the third year. I believe the Imperial visit[1] is to be thanked for it, which disburdened the University chest of rather more than 6,000*l.* It will be long, they say, before it is encumbered with so much again. I have been into the theatre, to see the magnificent chairs which the great people sat in, and the *Rostrum,* which I should like to be reduced to the awkward situation of being exposed in. They have begun a reform at Trinity, at least as far as the enormous expense. Everything which goes on there is inspected, and one of their gentleman commoners told Norris yesterday that he should take his name off positively, for the Principal has declared that not an *ice* or a *pine-apple* shall enter the college! Dreadful extremity!

Believe me, my dear father, your affectionate son,

CHARLES LYELL.

---

[1] The visit of the Allies to England in June 1814.

To CHARLES LYELL, ESQ., *Bartley Lodge, Stony Cross, Hants.*

Oxford : March 20, 1816.

My dear Father,—I have got Gifford's ' Juvenal.' A short life of the author, written by himself in the most unaffected style possible, which precedes the translation, will prejudice you much in his favour. His mother, he tells us, was a carpenter's daughter, and his father scarcely so respectable. He was bound apprentice to a cobbler, and his first attempt at versification, indeed the first time he had heard of it, was a competition between his brother apprentices about writing some lines for the sign of an alehouse. So true it appears that great obstacles are rather assistants than otherwise to genius ; indeed, he confesses that nothing would have ever induced him to finish his translation after he had arrived at a state of affluence, unless he had received a subscription while at this college, which obliged him from motives of honour to fulfil his engagement.

With Kirke White I was acquainted a long time ago, having met with his ' Remains ' at Midhurst. There is a delicious poetical, placid melancholy, which pervades almost all his writings, either in prose or verse, which is very interesting ; and nobody, we may venture to assert, ever joined a more amiable disposition to a more elegant mind. A small ode of his to ' An early Primrose ' is the only thing I know in English in the light way, which supports itself thoroughly without rhyme, so as to be as beautiful as any of Cowper's. I can conceive nothing which would give a person, who knew nothing of Latin, a better idea of what the Odes of Horace may be, though so distinct from our poetry.

I have gone to the expense of an Oxford Calendar, for fear of being too strictly examined in Hampshire. I shall now shelter my ignorance with ' read, and you'll know.'

The decision of the Property Tax has decided not a few bets this morning, and caused great joy *here*, though by no means general exaltation through Oxford. This is a complete Whig college, which you need not go twice into the common room to discover. They pretend that it was by far the greatest support of Lord Grenville, which is sufficiently improbable ; because, though we are strong in the number of

resident members, the number of names on our books is
very few, consequently there are some colleges much smaller
who have double and treble the quantity of votes. Most of
the men here come from Devon and Cornwall, and who of
course, when they have taken their degree, never think any
more of this part of the world.

My love to all at home, and believe me, my dear father,
your affectionate son,

CHARLES LYELL.

To CHARLES LYELL, ESQ., *Bartley Lodge, Stony Cross, Hants.*

Exeter College, Oxford : May 14, 1816.

My dear Father,—I have just been down to Dr. Williams.
He is not at home to-day, but I called there yesterday with
Tragitt, and thanked him for the roses. There is a most
beautiful wild flower lately opened in Magdalen meadows,
which reddens the whole meadow. I saw a plant of it in the
Botanic Garden, and there was written on the stick, if I
remember right, *Fritillaria Meleagris.* But it does not
flourish in the garden, for I suppose it to be some water
plant.

I was very glad to hear you were satisfied with the
verses,[2] as I shall now care less about a disappointment, as
they certainly have not been written in vain. I am afraid you
have no idea how backward I find myself in classics. Though
I did not come here puffed up with Deacon's high ideas
that I should shortly become the first scholar in Oxford, and
though I am not such a Cæsar as him, who, when he found
he could not reach the height of his ambition, determined
in despair never to look in a book again, yet I am confi-
dently sure, that without working very hard indeed, I shall
never arrive at the ' Mediocriter,' or at least never pass it.

I cannot help being sorry when I reflect that since last
June, when I left Midhurst, I can only say I have read the
Odes of Horace ! For as to Herodotus, that part which I
thought I knew perfect, I found the other day, in the schools,
I knew nothing of, so puzzling are their questions in
history. I was hearing a Corpus man examined. And here

[2] *The Horses of Lysippus,* by C. Lyell.

D 2

is June coming round again ! The list of books which they
take up for an *under-the-line* is very great, and the examination strict. C., the man of C. C. C., showed us what was
to be done by attending to the Corpus lectures. He has
always been otherwise an idle, dissipated, *gaming* man,
notorious in name, even to my knowledge. Yet he certainly
would have got a first class if he had taken up a little more
Aristotle.

I heard a story out of college the other day, which,
though probably made in the first instance, shows in what
estimation the *Devonshire*-Exeter-men were held when the
college was entirely provincial. Now the greatest part of
those West-countrymen that remain have rubbed off their
dross in our public schools. This story is, that one of our
men being examined for his degree, was asked, for the first
question in divinity, ' who Moses was ? ' ' Moses ? ' he
answered, ' knows nothing about Moses, but ax me about
St. Paul, *and there I has ye.*' There was a concert here last
night, Matthews on his way from Bath, and performed gratis.

My love to all at home, and believe me, dear father, your
affectionate son,

CHARLES LYELL.

To CHARLES LYELL, ESQ., *Bartley Lodge, Stony Cross, Hants.*

Exeter College : May 30, 1816.

My dear Father,—There is an address gone from Oxford,
as perhaps you saw by the papers. It takes away the Vice-
Chancellor, Proctors, &c. &c., by which means the prizes will
not be decided this fortnight perhaps. There were sent in
sixty-four copies for the English, double the usual number.
Ellison, the tutor of C. C. C., who was elected Fellow of Baliol
last term, and who is one of the cleverest men there, offers
to take any bet in favour of a Baliol man, whose verses he
thinks fully equal to any which have ever gained the Newdi-
gate. I have seen a copy of this college, which I am afraid
would be preferred to mine, as they lay more stress on
describing the horses. The men here have taken a good
deal of interest about ' Christabel,' as the author is an uncle
of our Coleridge.[3] The criticisms in the ' Times ' are so

------

[3] Lord Chief Justice Coleridge.

good in general on dramatic pieces, that I was surprised to
see anything so bad as that on ' Christabel.' You would take
it for a long poem, from that flaming account, instead of a
thing about as long as that column and a half in the 'Times'
about it. You would read it through at breakfast easily.

Its appearance now seems very strangely timed, though
the ' Times ' finds, as a cause, the 'Indolence of Genius.'
Ushered in by Lord Byron's recommendation, and supported
by this puff, I can't help considering it as a vehicle to raise
money. There are certainly poetical beauties in it, but it is
curious that when Scott had shown us all the powers of this
new style, the original discoverer of which certainly deserves
infinite praise, and when he had himself exhausted it, and
the ' Lake Poets ' had increased our fastidium, and the
Imitatores servum genus, had worn it out—when it had
flourished in youth, and died a natural death in old age—
then is produced its birth! It is rather like a prophecy
which is related after the event. Though you don't wish to
dispute the fact, one cannot help thinking it a pity that it
was not told you before the circumstance fell out. But the
evil is, the public must feel it in some manner their duty now,
when all the pleasure of novelty is fled, to give Mr. Coleridge
a great part of that unbounded applause which Scott carried
off. But my opinion of Scott is not at all diminished. I lay
so much stress on the style, because it strikes me in that
consists all the originality. In the story there is not a clue,
nothing to interest you but *singularity*.

I daresay you will laugh at my presuming to pass my
opinions on this, with all the insolence of the 'Edinburgh
Review,' unsupported by their soundness of judgment, but I
have my excuse. It is the common vice here, *Defendit
numerus*,—a latitudinarian principle, I confess.

Give my love to all at Bartley, and believe me, my dear
father, your affectionate son,

CHARLES LYELL.

To CHARLES LYELL, ESQ., *Bartley Lodge, Stony Cross, Hants.*

Exeter College: Thursday, October 31, 1816.

My dear Father,—You will be surprised to hear that the
examination is over, and, much to my satisfaction, I have

had but few days here, but I do not know what I should have done without them. As soon as I arrived here, and it was known that I meant to go up, I was sent for before the three tutors. Jones evidently was offended, that after attending his logic lecture for a term, I had thought myself so ignorant in it, as to choose mathematics; but Dalby was still more so, because, after refusing to attend his mathematics, I had taken them up. 'And how did you get them up? And when? Who assisted you?'

He told me to come to his rooms, and unless I was advanced a certain way, it was ridiculous to think of passing. He found me perfect in the propositions taken singly, but he puzzled me dreadfully in general questions about the relation of one to another, and he dismissed me peevishly. 'Well, you can go up, but you will make but a very, very poor show!'

I worked most perseveringly all the week, and when he examined me afterwards, the day before I went up, he declared me marvellously improved. I was the last on the list, that is of the eight who were examined the same day, and they gave me, to employ the time, three propositions to prove on paper which were not in Euclid, which, as some of my companions who took up six books could neither make head or tail of, they gave me great κῦδος for doing correctly, and gave me another stiffener, the proof of which I got at, to their surprise, though they said it was neither by the shortest road nor by the way they intended. Horace and Sophocles went off without a blunder, except in the declension of one unfortunate Greek noun. Dalby gave me the *viva voce* examination in mathematics, but he did not give me one of the questions which he did in preparing me. He paid me some flattering compliments this morning after an Aristotle lecture, but I am well aware that it was not solely because I did well, but because he thought I should have done so much worse.

Speaking of Aristotle, I was surprised to find what a number of deep scholars there are, even in this college, which has but till very lately not pretended to bear any great literary character. It is no presumption in me to say that Dalby has chosen the picked men of the College, though I am included, since I found myself so infinitely inferior to

the greater part. I was thunder-struck to find that Enys,
one of the idlest of our gentlemen commoners, a Winchester
man, who seems to attend the lecture as a mere lounge, and
who never intends to take a class, acquits himself better
than I do, when I get it up carefully. They all, however,
almost, are above my standing, and have read other parts of
Aristotle. He is an astonishing stiff author. It is the
seeing the superiority of others that convinces one how
much is to be, and must be done, to get any fame; and it is
this which spurs the emulation, and feeds that ' Atmosphere
of Learning ' which Sir Joshua Reynolds admirably describes
as ' floating ' round all public institutions, and which even the
idle often breathe in, and then wonder how they came by
it.'

I was surprised at Ramsay's [4] calling on me the other day,
though I owed him a call.

He seems to have been delighted with his call at
Kinnordy. He came into the gallery to hear me examined
yesterday, but had not patience to wait till my turn, which
a great many kind friends did; and the rush they made from
the top benches to the lower ones, when I was called up, so
completely deprived me of all confidence, that it was not till
after many exhortations and entreaties that I could raise
my voice enough for the masters to hear me.

As for the confidence and quickness which you were
speaking of, as one of the chief requisites of the Bar, I don't
know whether intercourse with the world will supply it, but
God knows I have little enough of it now in company, and I
was surprised at feeling so much unconcern as I did in the
Schools, but that was no criterion, since all things flowed
then *secundo flumini.* Lord Exmouth [5] paid a second visit
here last week : he was presented with a Doctor's degree the
first time, and the last with the freedom of the city. Our
Rector was honoured with his last visit, therefore we got a
good view of him.

We have had torrents of rain here these last two days.
I hope it is not so in N. B. Ramsay describes the tenants
at Bamff as being in as miserable a state as any of those

---

[4] Sir James Ramsay, Bart., of Bamff, Perthshire.
[5] Admiral Viscount Exmouth, who bombarded Algiers, 1816.

round Kirriemuir can be. I have not heard from Bartley since I have been up, but I shall write to-morrow, and believe me, my dear father, your affectionate son,

CHARLES LYELL.

To CHARLES LYELL, ESQ., *Bartley Lodge, Stony Cross, Hants.*

Yarmouth:[6] July 20, 1817.

My dear Father,—I got to London after not a very good night's rest, which I attribute more to the length of my legs than any other cause. I am sure I could sleep soundly if they could be shortened, but as I have no wish to undergo an operation like that practised with so much success by Procrustes, I shall hope that some opposition coach will be established, deeper and longer, to tempt the long-legged. As soon as I arrived at the Gloucester Coffee House, I dressed, put my letters in the post office, took my place in the Norwich coach, and walked to Sowerby's.

When searching about the Row for his house, behold the very identical Ammonites Bucklandi[7] was lying on the steps! I went in and introduced myself, telling him by what means I had discovered his house. ' Ah,' said he, ' little I believe did they think at Oxford what advantage I should take of that joke. I hear Buckland was perfectly astonished when he read it.' I exclaimed involuntarily, ' Well he might be,' which he took in good part, laughing heartily.

I visited the cast of Phidias and (talking of things on a grand scale) the elephant at Exeter Change; also Bullock's Museum—got several commissions from Bullock to Mr. Hooker.[8] Saw the whole of Francillion's collection of foreign and British insects, the first in the world. Might

---

[6] On a visit to Mr. Dawson Turner, F.L.S., antiquarian and botanist.

[7] Remarkable for having frequently lost the inner whorls, which circumstance has given rise to its name, in honour of the Rev. W. Buckland, who, having found a large specimen, was induced by his ardour to carry it himself, although of considerable weight, and being on horseback it was not the less inconvenient. But the inner whorls being gone, so as to allow his head and shoulder to pass through, he placed it as a French horn is sometimes carried, above one shoulder and under the other, and thus rode with his friendly companions, who amused him by dubbing him an *Ammon Knight.*

[8] Sir William J. Hooker, F.R.S., F.L.S., author of many important botanical Works, and Director of the Royal Gardens at Kew.

have identified all I wished, but, as you may imagine, had
only time to write down the names of a few. The famous
Kangaroo Beetle, the only one ever taken, is a Scarabæus,
its hind legs of wonderful length and size, though I
confess I admired it as the father in Horace his pet child :

Hunc Varum distortis cruribus.—Sat.

Let those who wish to have an idea of the magnificence of
Nature, visit the elephant, those who wish to judge of her
varietas insatiabilis, see Francillion's collection. I saw Sir
James Smith [9] at Norwich and the Linnæan insects, with
some British added by Kirby. Saw the cathedral, and view
from the spire, and found an immense number of Belemnites,
Echinites, and bivalves in the chalk-pits near Norwich.
Set off by the afternoon coach and arrived here safe. All
well here, and desire their kind remembrances to you. Dr.
Arnold,[1] the Javanese traveller, is indeed impenetrable, but
you soon see he has much in him. The only subject
on which he launches out is on Fossil Remains, and then
only if you get him quite alone. He has a large collection
here, obtained from Norfolk and Suffolk, of Echini, Ammonites, and mostly Alcyonia found in flints. He has had
the good nature to go over them all with me. I have copied
for Buckland part of his paper, being a list of those which
are described, and shall copy the rest. A very large part are
nondescripts, which he thinks of publishing. A Mr. Wigge,
a botanist, who breakfasted here this morning, wished to
know particularly if you had discovered any new Junger-
mannia, and why I, being in the New Forest, where you have
found so many, did not also take up botany. He is a
character, but a very good-humoured, pleasant old man.
Mr. Turner surprises me as much as ever. He wrote
twenty-two (!) letters last night after he had wished us
good-night. It kept him up till two o'clock this morning, it
is true. A great many letters are constantly arriving from

---

[9] Sir James Edward Smith, botanist, founder of the Linnæan Society,
b. 1759, d. 1828.

[1] Dr. Joseph Arnold, F.L.S., a zealous naturalist, who accompanied Sir
Stamford Raffles, the governor, to Sumatra, and fell a sacrifice to his exertions
on their first tour into the interior, when they discovered the gigantic flower
of a new genus of plants, afterwards named *Rafflesia Arnoldi.*

travellers abroad, which he entertains us with by reading.
His smack scheme was a good one, after all. They always
know by signal, regularly kept up, of the arrival of Scotch
smacks in the roads, nine hours before they can come
opposite the town.

My dear father, your affectionate son,

CHARLES LYELL.

To CHARLES LYELL, ESQ., *Bartley Lodge, Stony Cross, Hants.*

Yarmouth : July 28, 1817.

My dear Father,—I received your letter yesterday, and
am not passing my time less pleasantly than you conjectured.
What I see going on every hour in this family makes me
ashamed of the most active day I ever spent even at Midhurst. Mrs. Turner has been etching with her daughters in
the parlour every morning this week at half-past six!
Harriet has as much talent as all the others united, and her
knowledge of Latin is astonishing. She has a more perfect
conception of Virgil than I had at fourteen, and earns a
shilling at least three times a week by doing her Latin
composition without a fault, and does all with energy and
good will.

Dr. Arnold returned on Saturday, after being away four
days. I was very glad of it, for as Mr. Turner has been
much employed in the bank, I have had time to examine
and consider the geological wonders of this country. The
Doctor says my conclusions are exactly like his, which
nobody ever knew he had made, and has become in consequence very communicative, and quite another person.
Yarmouth is a delta formed at the mouth of the Yare.
When first these sands rose, by the opposition of the sea-tide
and river, Norwich was a great seaport (as we find records
of), the violence of the tide being kept off by that bank, the
estuary filled up with 'fluviatile detritus.' The Yare then
wound through the present marshes, and entered the sea
north of Yarmouth (Mr. T. says 'No'). The reason that it
then turned off at right angles was, that the mouth being
stopped up by the sea, it was obliged to find a new course,
and the north river meeting it there, it flowed with it, south-
ward, and entered a little south of the town, then two miles

farther off, then four miles, at Gorleston, where the pier is, which you saw. All these ancient channels I found, and the Doctor confirmed them, though Mr. Turner laughs in spite of facts and tradition. The last movement of the river threw inland at least a mile of perpendicular sand, cliff 15 feet high, on which the village of Gorleston stands. The terrace or platform on the top of this cliff is on a level with the marsh land reaching to Norwich. A friend of Mr. T.'s told us yesterday, that thirty-five years ago he could stand by the river, and see the hulks of the ships over the Deens, which rise now six feet and more, too high on the sea side for such a prospect, and yet the sea has not been over them all that time. Dr. Arnold and I examined yesterday the pit which is dug out for the foundation of the Nelson monument, and found that the first bed of shingle is eight feet down. Now this was the last stratum brought by the sea, all since was driven up by wind, and kept there by the 'Rest-harrow'[2] and other plants. It is mere sand. Therefore, thirty-five years ago the Deens were nearly as low as the last stratum left by the sea; and as the wind would naturally have begun adding from the very first, it is clear that within fifty years the sea flowed over that part. This, even Mr. T. allows, is a strong argument in favour of the recency of the changes. Dr. Arnold surprised me by telling me that he thought that the Straits of Dover were formerly joined, and that the great current and tides of the North Sea being held back, the sea flowed higher over these parts than now. If he had thought a little more, he would have found no necessity for all this, for all those towns on this eastern coast which have no river god to stand their friend, have necessarily been losing in the same proportion as Yarmouth gains, viz., Cromer, Bakefield, Dunwich, Aldborough, &c. &c. With Dunwich I believe it is *Fuit Ilium.*

The Doctor told me that he has always thought that it was the meeting of the great north current with that of the English Channel that burst open the Straits of Dover. With this I was delighted, for he did not know that to the very same cause both Werner, Humboldt, Buckland, and others,

---

[2] *Ononis arvensis.*

as well as myself, have been attributing the existence of
Great Britain as to its insular and probably political situation, and by means of which it must for ever maintain the
former, and will, let us hope, the latter also for a very
long time. Had not the Yare been turned off, as I told you,
at right angles, it would have split into the *Septem ostia
Nili*, and Yarmouth would have now been an extensive
growing delta. All the acquisitions are now making south-
wards, of course.

A Mr. Edwards, an entomologist, dined here yesterday. I
have found some Coleoptera scarce at Bartley. I have been
learning Musäus's ' Hero and Leander,' a Greek poem which is
very beautiful. Mr. Turner has it at his fingers' ends. We
play chess every evening. I have played twice with Mr. T.,
and beat him once. Miss Elizabeth plays nearly as well as he
does. Mr. Turner has given me some plates of mosses for
you, I believe from Mr. Hooker.

Between Dr. Arnold's long catalogue of Norfolk fossils,
and a map which I think I shall be able to make of this
country, I flatter myself I shall compile some interesting
information for Buckland, who is quite of White's opinion.
' Local information, from actual observation, tends more to
promote Natural History and Science than all that is done
by the speculations and compilations of voluminous authors.'
I want to find chalk cliffs at Norwich, which must exist, I
am confident, and I think I shall have an opportunity of
going there this week.

Pray give my love to all, and believe me, my dear father,
your affectionate son,

CHARLES LYELL.

JOURNAL.

[Extracts from Journals in August and September, 1817, while travelling with his father to Forfarshire, and afterwards on a trip to Staffa with Sir James Ramsay, Bart., and Mr. Corbett, of Christ Church, Oxford.]

August 20, 1817.—The York coach carried us out of London early in the morning. The tunnel at Highgate was a new object, it was dug through the London clay almost entirely. We took a chaise to Helmsley in going out of York, and saw Lord Fitzwilliam's carriage and six, with six out-riders, and his son Lord Milton's carriage and four, with four outriders, both beautiful equipages. Lord F. runs a good many horses, and gained most of the plates this year. Lord Harewood was there ; his equipage exactly the same as Lord F.'s ; he runs no horses. Mr. Fawkes told my father that in the grand election contest between Lord F. and Lord H., to get their sons in for the county, when Wilberforce was brought in free of expense, it cost each of them 200,000*l.* Lord Lascelles was ousted, but the majority in favour of Lord Milton was very small. Lord H. immediately raised his rents 10,000*l.* a year.

At a small hamlet near this we saw an *Ulmus montana,* on a steep 'brae,' the most beautiful study for a landscape that can be imagined. A sort of stone work has been built under it, evidently to prevent the chance of such a treasure falling by the wind. We changed horses at Helmsley, and went through Duncombe Park. The old castle at the entrance is a fine ruin.

> And Helmsley, once proud Buckingham's delight,
> Slides to a scrivener, and a city knight—

alluding to this castle, and the present proprietor of Duncombe Park. This was not the first entrance to this beautiful seat which we had passed to-day : the first was a large arch with this inscription—' To the memory of Lord Nelson, and the unparalleled achievements of the British Navy.' In ascending the Black Hambleton Hills, we passed up a very deep narrow glen, without a river (unfortunately for the Huttonians), and in descending the same hills we had a beautiful view of Whitestone Cliff, and Gormire, a tarn on the top of a conical hill. From thence we saw the whole valley into which we were journeying.

I walked to Sowerby, a small village near Thirsk, and learnt from an old stonemason there, of the name of Dukes, that the lake of Gormire, which the common people think has no bottom, was fathomed once, and its depth was seven fathoms. The bottom is probably sandstone, as the whole hill is quarried for that rock. There are so many springs, that the crater overflows even after long droughts. Grouse are found on the Black Hambleton range, and the Duke of Rutland shoots there generally every year. These hills consist of magnesian limestone. The whole face of Whitestone Cliff fell down about twenty years ago, and left it in its present perpendicular state. It was dusky when we approached Newcastle, and the collieries made a magnificent spectacle. The heaps of loose small coal catch fire accidentally, they say, on account of the pyrites, which is highly probable. We slept at Newcastle.

August 24.—We dined at Dr. Headlam's,[3] and they showed us Sir Humphry Davy's newly-invented safety lamp. The dispute concerning the priority of invention (or rather the doubt entertained, for Sir Humphry has not condescended to notice it) is warmly espoused by Dr. Headlam in favour of Stephenson.[4] A service of plate has been voted to Sir H., which is to cost 15,000*l.* at least. There are large ironworks near Newcastle, and lead-mines. The workers of the latter are very destructive poachers of the muir-fowl of the hills ; for having powder allowed them to blow the stones with, and being able to cast shot themselves, they have ammunition put into their hands. There are several steamboats on the Tyne. The *Eagle* packet carried 500 people by way of experiment : 250 it accommodates pleasantly.

August 25.—We got into a coach to go to Edinburgh, where we called on Jamieson in George's Square. He could not show us the museum and collection, as they are packed

---

[3] A cousin of his mother's.   [4] The eminent engineer.

up, on account of the college being at present building. The
Regent's Bridge, which is nearly finished, will make a grand
entrance to the town. We walked on the Calton Hill, and
admired the new walks. There are a greater number of striking views from different points of this town than any place
I ever saw in nature, or painting. The new prison is built
with great spirit, and from the bridge forms a good support
to the Nelson monument.

August 27.—We dined at Mr. Mason's.[5] He told us
that a few days before, Sir H. Davy had given him an
account of an artificial island formed by a Cornish miner in
the sea, to get at a mine (of either lead or copper), which
was truly wonderful. Mr. Mason, speaking of the Crown
Church at Edinburgh,[6] told us that the pinnacles of Gothic
architecture are not mere ornament, but indispensable, for
unless an arch is loaded, it flies at the top, its first tendency
being to fly at the bottom. This of course it cannot do in
ground arches, but in the arches of the Crown it was necessary to add those ornaments, which I think disfigure it
much, and make the Crown of Edinburgh much inferior to
that of Newcastle. The next disposition is to fly laterally
when secured at top and bottom : an additional weight therefore is added to the centre. In the Roman semicircular
arch the shape was perpendicular. The Grecians had none,
which is very extraordinary.

August 28.—We had engaged places in the Dundee coach,
and set off early. Leaving Kinross, we came to a most beautiful glen, called Glen Farg, well wooded, and with a small
river passing through it, which tumbled over the rocks the
whole way. We could get no horses at Dundee, and were
obliged to sleep there. I walked to the top of Dundee Law,
where there seems to have been an old fortification, perhaps
Pictish. We travelled over the Siedlaw Hills : the rock was
whin[7] chiefly.

September 1.—Kinnordy. I walked up towards Glen
Prosen ; the banks of the Prosen consist of whin wherever
they are rocky; took several *Phalænæ.* The rocky glen

---

[5] Laing Mason, of Lindertis, Forfarshire.　　　　[6] St. Giles's Church.

[7] Whinstone, a Scotch provincial term for greenstone and other hard trap
rocks.

formed by the Carrity is through whin. Took *P. napi* and
polychloros; Sylpha atrata, and a new one, *S. stercolarius,
fimelarius*, and several *Phalœnœ*. Afterwards I examined
the Kirriemuir quarry, an enormous rock of red sandstone,
with a little *terra ponderosa* or sulphate of Barytes in it.
We found this same earth afterwards in greater plenty
in Mr. Mason's quarry at Reedie.

September 3.—We called at Cortachie and saw Lord
Airlie. I rode to Forfar and examined the slate quarries.
They told me they sometimes found in the slates the impressions of plants and animals, but they showed me no satisfactory specimens. I then rode four miles farther on the
old Brechin road, to some slate quarries, where I obtained
some crystallised sulphate of Barytes, if I am right in supposing it to be such. I have not seen the *terra ponderosa*
in sandstone, whin, and slate. The vein of it in the quarry
two miles from Forfar, on the Dundee road, is very considerable.

September 6.—Rode to Finhaven over the bridge of
Shielhill, through Tannadice. At the foot of the castle
hill is the old ruin of the castle; at the summit a large
vitrified fort, length 160 yards, breadth 40, as near as I
could measure by stepping; height of mound or wall, 10
feet. The western side of the hill is a considerably lower
level, and at this part there is an excavation 30 feet deep or
more.

TRIP TO STAFFA.

September 14, 1817.—Set out on a tour to Staffa with Sir
James Ramsay and Corbett (of Christ Church, Oxford). Passed
through Blairgowrie, and following the Tay through Athol,
we arrived at Aberfeldy, and saw the Falls of Moness in all
their beauty : the accompanying wood and enormous cliff
are only surpassed by the fall itself. Saw Schiehallion in
the distance. Slept at Kenmore.

September 15.—Rode by the side of Loch Tay. We stopped
at Killin, the town on the western end of Loch Tay, where
the rivers Dochart and Lochy, uniting, form the loch. Lord
Breadalbane's grounds occupy both sides of the lake : they

are well wooded—oak, birch, larch, hazel, alder, and ash.
Ben Lawers we could hardly see for the mist. Visited the
Castle of Finlarig, which our learned host, a Cameron, informed us signified etymologically Larig the abode, Fin of
Fingal, and Killin itself was Kill the churchyard or burying-
ground, In of Fin or Fingal. I was surprised to find that
the people here spoke English so infinitely better than those
in Angus, but they told me it was taught them at school as
Latin was, Erse being so exclusively their native tongue.
They showed us the grave of Fingal, and MacNab's burying-
ground, with the Dochart roaring around us.

September 17.—We set off for Tyndrum, through Glen
Dochart, and after sleeping at Dalmally we proceeded along
Loch Awe, passing under Cruachan. After this, we had the
finest mountain view I had ever seen. On our right was Ben
Cruachan, some way off Ben Lawers, and the prospect ter-
minated by a fine conical hill. After riding past Bonaw we
saw a monument to Lord Nelson, which a person might feel
more proud of than of all the columns and pillars erected
to his memory by committees of counties and towns. An
immense mass of unhewn granite, of a beautiful red colour,
has been erected by the Lorn furnace-workers, the people
employed in the iron-works. They found it in the valley,
and have brought it to a small hill, from the summit of
which it forms a conspicuous object. It is about twelve feet
high. The country from thence to Oban is very singular.
You pass by high cliffs of whin, rising suddenly here and
there, yet you have never to mount any, but seem going
through the cracks in a torrent of lava. One view from
here of Cruachan with a double peak was very fine. We
arrived at Oban by the middle of the day, and engaged a
small cutter. We passed Kerrera and came near Lismore.
The next morning we came into Aros Bay, the hills of Mull
on the west, and the Morven on the east. Aros is a collection of two or three small houses and the ruins of an old
castle. We landed and walked to the head of Loch-na-Kael,
took a boat there, and sailed down Loch-na-Kael with a
prosperous wind, and had a fine view of Ulva, with its
ranges of basaltic steps, like Salisbury Crags, Edinburgh.
I believe we counted thirteen. We arrived at Staffa in the

evening, and the wind being northerly we were able to enter the cave, with the width of which I was much disappointed. When the boat was in they could nearly touch each side at once with their oars. The height is magnificent, and two or three broken lines of columns at the bottom on each side form a superb base to the pillars. The roof, hung with the broken heads of pillars, is likewise grand.

There was a slight agitation and swell of the sea, which prevented us from going in very far. The wave which entered sucked us in, and the boat sunk at the same time several feet. We then heard the wave creeping up towards the end of the cave, till it struck against it, we in the mean time remaining motionless : then the wave returned, and gently heaved us up, and carried us to the mouth of the cave. This was very fine, but after two or three trials we rowed away for fear of an accident. The pillars in the Boats' Cave are higher and finer than those in Fingal's, but a striking part of the entrance of Fingal's is that the pillars are ranged in a fine rounded swell on each side, instead of being straight in the manner of a wall. With the beauty of Buachaille and the cliff we were astonished and delighted. The Clam-shell Cave is also curious.

Instead of returning to the inn at Ulva, which we had intended, we went on to Icolmkill that evening, as Mr. McDonald of Ulva, to whom we had letters of introduction, was not at home. It was late when we landed at Iona, but the landlord of the inn and his family had not returned from shearing. We were obliged therefore to go into 'But,' in as miserable a specimen of a shieling as can be imagined, a conical hut open at top, with a large fire on the ground in the middle of the floor. Round the walls above us were immense stores of dried fish, tied two by two, and slung over a rope : it appeared a winter's store. The whole was dirty in the extreme, and *lacrymoso non sine fumo*, as may be supposed. When the landlord returned, and we were introduced into 'Ben,' we were but little better off, the smoke, our great enemy, being as dreadful as ever.

To avoid this, and relieve my eyes, I walked out. It was moonlight, and the finest aurora-borealis I ever witnessed was lighting up the east. The ruins of the abbey appeared to great advantage, but they are but poor after all. Sir James got a bed at the schoolmaster's, and Corbett and I slept at the inn.

September 20.—The children crowded round us in the morning, with Icolmkill pebbles to sell. They are Serpentine, which they find on the beach, very unctuous to the touch, and yellow.

The schoolmaster, McLean, showed us the antiquities, and told us the customary story of the interment of St. Oran, &c., and pointed out the tombs of the kings of Ireland, Norway, and Scotland. There were two fine crosses standing seven or eight feet high, of which there were once 360 standing. They were all thrown down at the time of the Reformation. The Coltsfoot [8] flourishes most magnificently over the tombs, and hides much of the rubbish and treasures. The ruins of the church look like so many gable ends of houses, set up rank and file, and one of the inducements to go directly from Staffa to Iona might be to contemplate the triumph of the architecture of nature over that of art.

McLean informed us that the inhabitants now on the island, who amount to 450, only hear divine service four times a year, and not that often, as the roughness of the weather sometimes prevents the Presbyterian clergyman from coming over from Mull, for Iona belongs to a parish in Mull. He preaches in the open air, and in English, which part only of his congregation understand. What a change from the state of the island when the monks had service morning and evening. The schoolmaster, it is true, reads prayers every Sunday, but, as he confesses, no one scarcely but his scholars will attend. He teaches them English only, which they speak fairly. The islanders seem to exist by their fishing. Their only traffic is kelp, and a few black cattle which they rear. We saw some black cattle embarked, which was a very amusing and ridiculous scene. A rope is tied to the horn of the animal, and he is driven to the seaside. At first they make violent resistance, and the driver, while they are kicking and prancing, yields with the rope, and lets them exhaust themselves, as an angler does a large trout. They show great dexterity in always keeping in front

[8] *Tussilago farfara.*

of the animal. The beasts are then pulled into the sea, into which both men and women walk with as much unconcern as if they were by nature amphibious. The fore-legs of the animal are first lifted up by one person on each side, and then the hind. We saw one lassie seize hold of the tail, and help to lift him in that way.

About half-past nine, after having paid an enormous bill for our bad entertainment, we set off in our Aros boat, determining to get back to Oban by passing along the south of Mull. The immense multitude of small granite rocks which we passed through was very entertaining. This south-west end of Mull is a coast of red granite, and in McLeod's Bay it takes a decided columnar shape, the pillars being four-sided, whereas most of those at Staffa which we examined were pentagonal, none square. There is a large creek in this bay, with the pillars on each side about 150 feet high, and broken into regular steps like the basalt. Between these two cliffs, which face one another, and are only divided by a large rent two or three feet wide, a large rocky mass is supported. We sailed on with a good breeze, the granite structure in the coast still continuing. The small islands were like so many Buachailles, the coast like so many churchyards, pillars standing up with their square sides, like tombs. The last point to which this reached, they called Ardtornish, about four miles from McLeod's Bay.

CORRESPONDENCE.

Exeter College : October 21, 1817.

My dear Father,—I find so few of my old acquaintance come up this term, that I could almost fancy it a new college. It will be less difficult, therefore, to acquire what Paley so strongly recommended to his pupils, ' courage to be alone.' Ramsay tells me there are four new noblemen and eight gentlemen commoners this term. We have only one of the latter—the son of a man of consequence in Ceylon.

He is a copper-coloured black, under the care of Lord
Bathurst. Our Government educated his elder brother, as
they are doing this one, first under a private tutor, then at
Trinity Hall, Cambridge. They say, but I believe without any
foundation, that he is a nephew of the King of Candy. His
manner is very pleasant, and he talks admirable English.
He is to take orders, they think at present, and then follow
his brother to his native country. His name is M. de
Saram.

I saw Dr. Arnold in London, since which I received an
unexpected letter from him at Portsmouth, chiefly to inform
me that he had just discovered a paper written in the last
number of the 'Geological Transactions' by Buckland, on
flints owing their formation and shapes entirely to animals.
The pear-shaped flints we saw with Mr. Taylor at Norwich
are touched upon in it. Now that the time of parting is
come, the Doctor seems almost to shrink from the thankless
and dangerous task before him.[9]

Dr. Williams tells me that Buckland will not be in
Oxford till late this term, if at all. He has not been on the
Continent this long vacation, but making several 'home
circuits.'

When I was at Muddiford[1] the other day, I found the
cliff composed of what I think must be decidedly the green
sand of Werner, which you remember Phillips says always
accompanies chalk. I found at Bullock's Mona Marble
Exhibition in Town some fine specimens of Serpentine among
the unpolished rocks, some of which are precisely the rock
at the *Bridge of Cortachie*, with that green, shining, glu-
tinous matter on the outside, but none of course with the
porcelain appearance, which must be owing I think to heat.
I spent the morning at the British Museum, studying the
collection of minerals, a very great and splendid treat.

I received a letter this morning from Mr. Turner, sending
me an etching of Captain Manby and Talma, much good
advice, and an invitation to spend my Christmas vacation at
Yarmouth. I seem to have been at Yarmouth a day or two
ago, so very near did we pass within the roads. The sail

[9] Going with Sir Stamford Raffles to Sumatra.
[1] A bathing-place near Christchurch, Hants.

up the Thames, at the rate of twelve knots an hour, with
generally 300 vessels near us, moving rapidly in all directions, was delightful beyond description. A side wind shows
off every vessel in the most picturesque view.

> One side on high, one buried in the deep,
> Her deck to meet the ocean, slanting steep,
> Beneath the swelling sails her mast bowed low,
> She mocks in speed the following breakers flow.
> Back from her bows the broken waves are cast,
> And far behind her in the moving waste
> A level track declares her well-sped way,
> Where swiftly circling sports the eddying spray.

I admire my mother's picture [2] very much—a strong and
pleasing likeness, nor should I have discovered its unfinished
state at the usual distance of viewing it. Believe me, my
dear father, your affectionate son,

CHARLES LYELL.

To CHARLES LYELL, ESQ., KINNORDY, FORFARSHIRE.

Exeter College: November 11, 1817.

My dear Father,—I have been sitting in the schools for
seven hours this morning, a ceremony which I must undergo
again before I am qualified to go up myself. The day was
almost entirely taken up with Baring,[3] a gentleman
commoner of Christ Church, son of an East India Baring of
immense property. His examination will last perhaps three
days more, as he goes up for a first class in *both.* He went
on to-day in a splendid manner. He is a young genteel-
looking man, and has mixed with society in Christ Church
pretty freely since his residence. He was examined to-day in
Aristotle's *Ethics, Rhetoric,* Thucydides, Herodotus, Polybius,
Xenophon, Livy, Euripides, Sophocles, Æschylus, Virgil,
besides *Divinity.* The examinations are expected to be unusually severe, and it is feared there will be many before the
end of the term.

I had a letter this morning from Guilding, from the island
of St. Vincent, favoured by the Hon. W. Edwards, H.M.S.
' Hydra,' now at Portsmouth, who was to have delivered it to

---

[2] By Thomas Phillips, R.A.
[3] Son of Sir Francis Baring, and afterwards first Lord Northbrook.

me here, and been lionised by me over Oxford, but is detained from doing so at present. Guilding had procured then (17 September) 700 entirely new species of insects, one sphinx measuring eight inches from tip to tip of the wing, which he means to present to the Linnæan Society, with other rarities. He has got duplicates of most things for me, and is to send them shortly. He has also collected birds, fishes, &c. &c.

I have also received another letter from Dr. Arnold. They encountered a dreadful storm in the Downs, and lost their best bow anchor, put back into the Thames again, and afterwards were driven back by another gale off Scilly to port. He wrote from Kelston in Cornwall. He was making a tour, with Sir Stamford and Lady Raffles, through the mining district, and gives a wonderful account of that interesting country.

Oxford is of all places the most barren of news. Rather than send an empty letter so far, I will give you the only verses I have found time to make lately, which are but the beginning of a subject.

LINES ON STAFFA.

Ere yet the glowing bards of Eastern tale
Had peopled fairy worlds with beings bright,
Roamed o'er the palace and enchanted vale,
And dreamed a heavenly vision of Delight,
And told of realms rich with unborrowed Light,
On which the needless sunbeams never fell,
Whose noon of splendour never knew the night,
Illumed by lamps that burnt unquenchable,
And dazzling hung in air, upheld by magic spell:

All these and more, with which their wizard strain
Led far away deluded Fancy's child,
Till he would turn on Nature's self again,
And deem her charms a desert bleak and wild,
Himself from visionary heavens exiled;
While yet unheard that strain, the Time had been
When Nature's hand as if in sport she toiled
To build e'en more than could the thoughts of man,
Amid the Ocean vast, had framed a fairy scene.

For she had found a lone and rocky isle,
And at her voice a thousand pillars tall,

> She bade uprising lift the massy pile,
> And far within she carved a stately hall
> Against whose sides the entering waves did fall,
> While to their roar the roof gave echo loud—
> And she had hid each column's pedestal
> Beneath the depths unseen of Ocean's flood,
> While towered their heads on high, amid the passing cloud.
>
> And she had fashioned with an artist's pride
> The dark black rock where hung the sparkling foam,
> And many a step along its sculptured side
> Had hewn, as if to tempt some foot to roam,
> Some favoured foot of Mortal yet to come.
> She bade no shapes of Terror there abound,
> That pillar'd hall no guardian dragon's home,
> But Ocean rolled his mighty waves around,
> To guard from vulgar gaze her fair enchanted ground.

Whatever you may think of the poetry, you will agree with me in regretting that Werner should have died without the knowledge of this geological discovery concerning the origin and formation of basalt.

I am perfectly well, and believe me, my dear father, your affectionate son,

CHARLES LYELL.

CHAPTER III.

JOURNAL.

[In 1818 he made a tour in France, Switzerland, and Italy, with his father, mother, and two eldest sisters, and kept a close journal, from which the following extracts are made.]

June 8.—We left Dover and embarked on board the ' King George ' packet, and in three hours were landed on the pier at Calais, and the carriage and luggage having been taken to the custom-house, we went to L'Hôtel Dessin. In the morning I mounted the tower, where all the town and sea can be viewed, and then went into the church, the first Catholic one I had ever seen. A painting attributed to Vandyke was shown me, with many others, and images of saints. It was twelve before we left Calais. Four miserable nags were fixed to the carriage, tied on with tackle which any Hampshire wood-stealer would have been ashamed to see round his Forest colts. But the appearance was a small part of the evil: every two miles something (as might have been expected) went wrong. The strong fortifications round each town are of course one of the most striking novelties to an Englishman. Those of Boulogne presented a very different appearance from Calais, the town being divided into *la haute et basse ville,* the former only being enclosed by walls, and commanding the lower part. English inscriptions over almost every shop pointed out how many of our countrymen were residing there. The different châteaux which we passed between Boulogne and Samer were inhabited chiefly by English officers, and when we arrived at the inn at Samer, we heard a British bugle sound, to rest the horse-artillery, who are quartered there. It seemed that the English had taken quiet possession, both military and civil, of this part of the country.

June 10.—Early in the morning I was roused by the drums and fifes of 400 French soldiers, which passed through Samer on their way from Montreuil to Boulogne. They looked much out of feather, partly from their march perhaps, but partly, as I understood from our soldiers, from their national dirtiness. The scene was much changed when the British horse-artillery were drawn up in the market-place. There were 170 men of these, most of whom wore Waterloo medals, all admirably mounted. Six cannons, each drawn by eight fine horses, and six ammunition waggons, each with four horses, added much to the spectacle. When we had climbed the hill out of Samer, on our way to Montreuil, it was a beautiful sight to look back upon them following in separate corps, each detached body having with them one cannon and one ammunition waggon. They were coming up from the town to exercise near the wood of Montreuil. Going on through a flat chalk country, we passed through the Forest of Crecy, and came on to Abbeville, and having dined, we proceeded to Amiens, and were six hours in going five posts. Clermont was the next fortified town through which we passed. The citadel is on a very commanding rise. Here for the first time we left the chalk (?), which we had been on every step from Calais, and which we entered again before we got to Paris. The country changed with the soil, but not much for the better (with regard to scenery). We slept at Chantilly.

June 12.—In the morning we visited the magnificent stables which belonged to the Palace of the Prince of Condé, destroyed at the Revolution. The exterior is in grand style and good taste. The apartment in the inside which includes the length of the whole building is 620 feet long, and accommodated for some months 320 horses of the British cavalry, when part of our army were quartered at Chantilly lately. As we drew near Paris small towns began to thicken, as in the neighbourhood of the British capital. St. Denis was the principal place. Here we visited the church, which is very fine. The choir contracts at the east end, much in the style of Canterbury. The heights of Montmartre soon made their appearance after we left St. Denis, and vineyards became common. The road was a straight avenue, with a row of trees on each side all the way, till we entered Paris by the Rue de Clichy, then down Rue Mont Blanc and Rue de la Paix, into the Place Vendôme. Here we saw the splendid column made of the melted cannons taken by Napoleon in his different victories.

We dined at the Hôtel Meurice, Rue St. Honoré, and in the evening drove to Tivoli, a kind of Vauxhall. It was not a full night for these gardens. We tried by way of experiment *les montagnes,* and the whole party descended in these sledges.

Sunday, June 14.—After service at the English chapel, went to the Tuileries. After Mass, Louis came out on the balcony, and was received with what I thought a very cold *Vive le Roi!* We dined at Vérey's, the *restaurateur* in the Palais Royal. The views from his windows of the square filled with people, the avenues of trees, and the flowers, all kept in the highest order, and the fountain in the middle, formed a cheerful scene.

In the evening we drove up the Champs Elysées to the Beaujean Français, something like the gardens of Tivoli, which we had before seen, but on a larger scale, and more crowded, Sunday being the gala-day in France. Different groups were spread over the gardens, some round a band of music, in another place round a conjuror; in another, shooting at a mask with an air-gun was the amusement. Above all towered the Montagnes Français, down which I descended in a car. A balloon went up from these gardens, and the evening closed with fireworks.

June 15.—We visited the Jardin des Plantes, which occupied the greatest part of the morning. The Museum,

particularly the room of fossil remains, amused me greatly.
I lamented not being able to hear Cuvier lecture, as he is
now in England. The living animals are allowed each a fine
space to range in. Martin the bear would not be persuaded
to climb his pole. Some time since he had a fall, by which
he broke his ankle, or some joint, a sad misfortune to many
Englishmen, to whom the sights of Paris must now appear
comparatively a blank. We next visited the church of St.
Germain, the screen of which is worthy of notice. Capt.
Ogilvy [1] called on us early, and informed us that the water-
works of Versailles were to play that day in honour of the
Duke of Wellington, and invited us to dine with him at his
house in St. Germains, the day after. We lost no time in
profiting by this intelligence, but set off in a *calèche.* The
garden view of the Palace of Versailles is very grand, but it
appears to me too large even for a palace, at least in any
situation besides the capital. Many thousand people were
collected round the large basin of the Dragon,· where the
Duke of Wellington was to pass. He was dining with some
French marshals at the Trianon. At their arrival, with a
small body of guards, forty-two immense columns of water
rose out of the basin, the great Dragon, with all the dolphins,
&c. &c., began to spout out streams of water, and the effect
was beautiful.

June 17.—We departed in good time to pay Capt. Ogilvy
a visit at St. Germains. The distant view of Paris from this
road is interesting. As Malmaison was on this road, we
stopped to see it. The house might be that of any country
gentleman in England, but derives its interest from having
been so long the residence of Josephine, and so often that of
Napoleon. There are no pictures of value there, but the
great gallery is a fine room, and the full-length picture of
Josephine is said to be a good likeness. But I was most
delighted in seeing the garden and grounds laid out in the
best taste, for though Versailles is a magnificent specimen of
art, I confess the sight of winding walks, instead of straight
alleys; trees spreading their branches naturally around,
instead of being cut and tortured into shape; rose-bushes
growing on the ground, instead of being grafted, according

[1] The Hon. William Ogilvy, brother of Lord Airlie.

to the French fashion, and looking like flowers bound on
the top of a hop-pole, gave me more of the real charm of a
garden than I thought I was ever to experience in France.
I understand that the Trianon at Versailles is laid out in
this manner, and that the French look upon both as singular
curiosities, and have taste enough to admire them. A tulip
tree at Malmaison was the first I ever saw in flower. That
Englishman who, when he first beheld this tree, and mar-
velled at the industry of the gardener who had clipped every
leaf so evenly with his scissors, would meet in France with
enough real instances of labour employed in perverting
nature. We saw some very vivid lightning on our return to
Paris, but escaped a storm which fell round us. A few
glowworms on the bank were the first I had seen in France.

June 18.—In the morning we paid a visit to the Louvre,
where, besides the paintings, we had a view of Louis XVIII.
setting out for St. Cloud, from the Tuileries. We saw
him well, as we looked down into his open carriage, which
was drawn by eight horses. We went in the evening to the
Opéra Comique. This was the first theatre we had been at
in Paris, for besides the heat of the weather, Talma and all
the first-rate actors and actresses were away from Paris, at
different country theatres. I saw little to admire in the
house, and was not Frenchman enough to follow the very
rapid pronunciation of French comedians; but the orchestra
was good, and music is an universal language. It was the
composition of Grétry.

June 19.—We went to the Jardin des Plantes, to see the
collection of comparative anatomy, which is very beautiful,
and might tempt anyone who had the opportunity of staying in Paris to take up ardently the study of anatomy.
There are a vast number of exquisite models, coloured to the
life, descriptive of the anatomical structure of the human
frame; besides skeletons of a vast number of quadrupeds,
birds, fishes, serpents, &c. &c. I afterwards visited the
Panorama of London in the Boulevard des Capucins. We
dined at Beauvilliard's, a *restaurateur* in the Rue de Richelieu,
and afterwards went to the Théâtre Français and heard
Corneille's play of 'Le Menteur.' I followed the actors
with the help of a book well enough to enjoy it, and laugh

heartily, but often while I was cutting a leaf with my knife, and turning over, I found to my astonishment that they had got down half the next page. ' Le Menteur ' is more of a farce than a comedy.

June 20—I went in the morning to the Bibliothèque du Roi, and called for the first volume of Cuvier's work on Fossil Remains, which was immediately brought, as was a French Dictionary and map of France which I asked for. They seem even to take pleasure in waiting on you. The library is of great use, there were six or seven tables full of students, most of them writing. How different in every respect to the Bodleian at Oxford! except in books, in the number of which I suppose it cannot surpass the Bodleian. I read Cuvier's paper on the Geology of the Country round Paris (the chalk basin of Paris).

June 21.—We went this morning to hear Mass performed in Notre Dame. Service was going on in several different parts of the cathedral at the same time. I could neither understand a word they said, nor admire a word they sung; there was no music in the chant, and it was unaccompanied by the organ. A single instrument, a bass-viol, I believe, now and then joined its bass notes with their voices. We next visited the model of the colossal elephant which is to be erected over the canal near the ruins of the ancient Bastille. The idea of this is said to have been entirely Napoleon's. Water is to be conveyed into the animal, and it is to spout it out at its trunk, and thus serve as a gigantic fountain. The taste of this is much disputed, and what the effect will be when the whole is of bronze, and when it stands as a public monument in this city, I will not pretend to antici- pate ; but the idea is unquestionably no mean one, and the proportions of the model are so good, though above seventy feet in height, that the front view is wonderfully striking. It is done over with plaster of Paris, and has quite the ap- pearance of stone.

We then went to the cemetery of Père la Chaise, a large burying-ground in the same quarter. It would be well worth visiting were it only for the extensive view of Paris which presents itself from this rise. But it has sufficient interest in itself. The immense number of tombstones and monuments, the order and neatness of the walks, and of each little plot round the sepulchres, which are filled with beautiful, and sometimes valuable plants. It would seem here that the cypresses were not—as Byron says,—

> The only constant mourners of the dead,

for we observed wreaths of flowers lately placed on the tombs of some who had departed a long time since. We went in the evening to Les Jardins Ruggieri, which I thought the best of the kind we had visited. But its being Sunday probably added much to the crowds which were there. Sunday in Paris is celebrated like most of the old Roman sacred rites, by dancing as well as prayer. *Le saut de Niagara* is the most conspicuous object in this garden. It is like the other *montagnes*, except that before the car descends it is suspended in the air on a kind of drawbridge, which, like the cross board of a see-saw, is balanced on its centre and loose at the extremities. The declivity down which you are shot is infinitely more rapid than any other in Paris, but very delightful, not only to Parisians. Some splendid fireworks finished the evening.

June 22.—Went to the Jardin des Plantes, where I again looked over Cuvier's lecture-room, filled with fossil remains, among which are three glorious relics of a former world, which have added several new genera to the Mammalia. We also walked over the greenhouses, and admired many fine tropical plants. We had a great treat afterwards at the Opera. ' Le Devin de Village ' was first performed, interesting not only from its real merit, but as a literary curiosity, since both the poetry and the music are by Rousseau. ' Le Rossignol ' followed, and then the ' Carnival of Venice,' a pantomime, both excellent in their way. The whole was got up with great splendour and magnificence, but a filthy curtain which was let down between the acts could not fail to remind us of that total want of *convenance* which one has so often heard remarked as characteristic of the French. It was like the filthy Stygian pool which flows perpetually down their finest streets.

June 23.—I spent all the day in the Louvre, and became more than ever convinced of how much there is there worthy

of admiring and studying for as many weeks as we had
hours to give to it.

June 24.—We drove to the Royal Observatory, and then
to the Church or Abbey of Val de Grace, but on entering
were astonished to find, instead of the signs of a consecrated
place, nothing but enormous piles of blankets and sacks of
straw and other things. By a remarkable conversion, it has
now become a military storehouse and hospital. I got
back to the Louvre in time to spend one hour there before
its closing. There I met Sir James Ramsay and Mr. Malcolm, two Oxford friends, who had just arrived from London.
In the evening we attended the Opera. The piece was ' Les
Danaïdes.' Nothing very striking in the music, still less in
the poetry, but all good, and the scenes magnificent, particularly the concluding one of the infernal regions. There
were a great number of soldiers about the door, as there are
wherever you turn in Paris. They keep excellent order, but
bayonets are not very pleasant things to see in a crowd, and
the very authoritative tone in which they order about gentlemen's servants and coachmen cannot at all suit an Englishman's feelings.

June 25.—I spent again nearly all the morning in the
statue gallery of the Louvre. I was particularly struck with
the Demosthenes on the right as you enter the Salle de la
Paix, with Venus Genetrix, and many of the Roman emperors, Trajan, Augustus, and a bust of Caracalla. I was
much delighted also with Jason stooping to untie his sandal,
but more than all with the Fighting Gladiator.

June 26.—We left Paris on our way to Switzerland, in
time to reach Fontainebleau, where we slept. In the Forest
of Fontainebleau a (limestone) rock bursts up in huge
masses through the soil, and presents a very singular appearance, not in any way in character with so tame and flat a
country. The timber here is magnificent, of many species
and great extent.

June 27.—The exterior of the Palace of Fontainebleau
is perfectly inelegant, not to say frightful, and the gardens,
if intended for the French taste, very inferior to those of the
Tuileries or Versailles, and if meant for English, they are
nothing compared to Malmaison. In fact they seem an

absurd attempt to combine these two contrary styles.
Garden grounds in France openly profess to display the
skill and perfection of art, and have certainly the advantage
of show and splendour ; but in England they are laid out on
the principle of *Artis esse celare Artem,* and to make these
two opposites meet in the smallest degree is evidently impracticable. The rooms in the Palace are many of them
fine, and the furniture princely. And though the apartments in which the Pope was confined by Napoleon are in
themselves nothing peculiar, and still less the small table on
which Buonaparte signed his abdication before going to Elba,
yet they all serve to remind us of events of History.

June 28.—Through our whole journey from Auxerre to
Maison-Neuve there was no single point of interesting
scenery, and my only pleasure was from the few passing
remarks I was able to make on the geology of this country.
It is chalk as far as St. Bris, between which and Vermanton it becomes more like a white limestone, and after
leaving Vermanton I observed stratification unlike chalk.
At Lucy le Bois it remained much the same, and between
this place and Avallon we came upon a blue limestone
abounding in Ammonites, some as large as the crown of my
hat, some considerably larger, with many gryphites and other
shells, continuing to Rouvray. Here the fields are enclosed
with hedges, and the country exactly resembled some parts of
England. I was astonished at the small village of Rouvray
to see the streets paved and the houses built of granite. On
our way to Maison-Neuve we found the road cut deep
through a red porphyritic granite. Yet there was no change
in the scenery, as is usually the case upon entering a portion
of primitive country.

June 29.—Upon leaving Maison-Neuve we came again
upon the limestone we had left at Rouvray, containing
Cornu Ammonis and other fossils in abundance. We had to
mount a steep limestone hill on leaving Vitteaux. Every here
and there were large stones of the colour of ruddle,[2] containing apparently the same fossils as the limestone. At
Dijon we slept, the largest place we had been in since
Paris.

[2] Or reddle, red iron-chalk.

We passed through Dôle, a pretty town, between this and Vaudray. I saw gravel strewed in large hillocks over the plain, as it is in Scotland. In some small villages near Poligny we were surprised at the change of the buildings, for some were roofed with wooden tiles, and the roofs of some projected a great way over the side of the house, in the Swiss style.

July 1.—The Jura chain rises abruptly from the flat plain. Poligny lies under the mountains, being built on the slope. We had to thank Napoleon for the noble road which carried us by an easy ascent up the precipitous rise of these mountains. Large limestone cliffs presented themselves, the stratification very horizontal. On looking down from the height, we had an extensive view of the plains of Burgundy which we had left. The limestone in the Jura I conclude is of a different age from what we passed through before Dijon, for the latter abounded in organic remains, whereas I could not discover one fossil in the Jura. By the roadside I picked up many beautiful petrifications, which must be forming daily here, where the water is charged plentifully with lime. The minute veins in the leaves of many of these petrifactions were preserved with exactness. I was surprised to see one limestone cliff, which had been cut through to make the road, spotted with dark flints, something like moderate-sized chalk flints, but not in layers as in chalk. Between Champagnole and Maison-Neuve the steep rocks covered with fir-trees reminded us of Dunkeld, and the situation of Morez, where we slept, was at the bottom of a deep valley, where the cliffs and steepness of the rocks on each side, and the large bed of a river at the bottom, formed a romantic scene.

From Morez we ascended to Les Rousses, the frontier town. Here our passports, &c., were examined. We had now climbed to nearly the summit of the Jura. La Dôle was full in view, a mountain that reaches more than 5,000 feet above the level of the sea. It was not many hundred feet above us. On this, and on a few other mountains, small patches of snow were pointed out to us. Now that we had gained the height of the loftiest mountains in Scotland, we could not help reflecting with astonishment on the contrast

between the Grampians and the Jura. Instead of a cold
piercing wind, we were here glad of an umbrella to screen
us from the heat of the sun; instead of a desolate waste, the
dwelling of moor-game alone, here were towns and houses
scattered about, and cattle; in place of a boundless expanse
of heath, here were tall forests of fir, with the Alpine rose
and a thousand beautiful flowers. The *Veratrum* in particular,
and *Gentiana lutea*, with their broad leaves, covered the
ground in some parts, and gave it, to me, the appearance of
garden ground. In descending the Jura from Lavatay to
Gex, we had a most magnificent view of a vast extent of
country. Below us the Lake of Geneva and the Pays de
Vaud, before us the Savoy Alps towering up to the clouds,
and, in spite of their great distance and the height on which
we stood, extended in a long line before us like an army of
giants, Mont Blanc rising high above all in the middle as
their chief. We saw the Dent de Midi to the left, shooting
up his two remarkable peaks, with many more of extraordinary and picturesque forms. We went to Secheron,
and visited Geneva, about half a mile from our hotel. I was
rather disappointed in the town itself, but the appearance of
the people I thought a great improvement on France. The large
spreading straw bonnets are very becoming to the women,
and a pretty girl seems not such a rarity in a shop here as in
Paris. We crossed a bridge over the Rhone, which flows
through the town, coming out of the lake with a rapid
current, of a most peculiar deep blue colour. We called on
Professor De Candolle, but found he was gone to Paris.

July 4.—We departed from Secheron early on an expedition to the Valley of Chamouni. Geneva we found busy
and lively in all parts, it being market-day. A few hundred
yards carried us out of the precincts of this republic, and
upon entering Savoy, at the small village of Carouge, we
underwent the farce of having our baggage examined. We
soon passed Mont Salève on our right, and the Voirons on
our left, the two nearest hills to Geneva, both calcareous, as
is the Môle, a picturesque mountain on all sides, and very
verdant up to the top. We passed the Menoge, a small
torrent which winds along a deep ravine, apparently the
gorge of old of a vast body of water. Here I should have

expected, as in approaching the Primitive country in Scotland, to see large micaceous boulders of the old rocks, but there were scarcely any of the kind. Whether they are buried in the soil, or whether the ruin of the Môle (a large part of which seems evidently to have been torn away) has covered everything. The scenery towards Bonneville, where we dined, and after it, improved at every step. We had been following the Arve for some distance, and after passing through the town of Cluse, we turned with it up a deep valley, or rather an enormous ravine, where the rocks rose to our left to a very precipitous height, and quite perpendicular. Le Brison on the right was a very picturesque mountain.

Le Cascade d'Arpenaz, a fall of 800 feet perpendicular, was looking very beautiful on our left. There was not much water, but the wind carried that little quite away in smoke, before it descended half way, so that farther down, where the scattered spray fell, and collected on a projecting portion of the cliff, a new cascade seemed to spring forth. Our ride was truly romantic the whole way to St. Martin, where we slept. Mont Blanc had been concealed by clouds the whole day, but this evening we enjoyed a full view of him. The setting sun rested on his summit long after it had ceased to light up the valley and neighbouring mountains. Though the outline from here is beautiful, I confess I was disappointed in the near resemblance which it bears to a cloud, for the snow naturally takes the same tints as white vapour from the sun. L'Aiguille de Varens, a calcareous crag which overhangs the village of St. Martin, is of immense height and very grand, especially when half buried in clouds, and the steep rock appearing through them, cut off from its base.

July 5.—Leaving St. Martin, we wound round the Aiguille de Varens, the base of which mountain is partly of slate, partly of schist. The Cascade de Chêne is pretty, but the Chute d'Arve we saw from such a height, that it seemed only a river roaring over a rocky bed. The view from the Pont Pelissier is very delightful. Here we passed the Arve, and soon came in view of the glaciers. After passing the small Glacier of Taconay, the Glacier de Bosson, the most picturesque of all, came in view. It has advanced farther than any of them in a direct line across the valley, and particularly in the last three years. A fir wood on the farther side marks the spot to which it used to extend before that time. Marching on with its enormous bulk, it has trodden down the tallest pines with as much ease as an elephant could the herbage of a meadow. Some trunks still are seen projecting from the rock of ice, all the heads being embodied in this mass, which shoots out at the top into tall pyramids and pinnacles of ice, of beautiful shapes and of a very pure white, which is finely set off by a background of dark fir. This lower part of the glacier is urged on by an enormous precipice of ice above, descending from the upper parts of the mountain, and which increases in weight every year. It has been pressed on not only through the forest, but over some cultivated fields, which are utterly lost. The poor woman whose land has been thus destroyed, came to present us a petition signed by the minister. They have placed a small wooden cross at the bottom, with the intention, no doubt, of arresting the progress of the evil which threatens to overwhelm many houses immediately before it, and indeed all human power can do nothing else towards saving them.

The Glacier de Bosson marches almost in a straight line, though inclining a little down the valley. There is a declivity before it, all the way to the opposite side, which I think it must reach in less than twenty years, even were it to advance at half the pace of the late three extraordinary winters. Should it ever do so, it must cause a hundredfold the calamity that has just befallen the Valley of Bagne and Martigny, since it would flood the valley above, by stopping the Arve, and then deluge it below when it broke loose. The mischief it has already caused, and the much greater evils which it threatens, must raise in every one a feeling of horror, in spite of the admiration excited by contemplating this picturesque phenomenon. When you walk over the ice you find the air very cold, though a burning sun was warming the valley when we saw it, and causing the vegetation to flourish with luxuriance up to the very foot of the glacier. One of the choicest of our garden shrubs, the Rhododendron, was in flower there.

July 6.—I took a guide early in the morning, and leaving
the inn at Chamouni, began to mount l'Aiguille de Brevent,
a mountain which rises on the opposite of the valley, and
affords the nearest general view of Mont Blanc. It rises
steeply, nearly 6,000 feet above the valley, and, as may be
supposed, requires some labour to mount, but it well deserves
it. Planpraz, a point about two-thirds of the way up, affords
a grand view of the summit, and all the points of Mont
Blanc, but at every step you ascend, he still rises, till at the
top of the Brevent you may consider yourself raised opposite
his middle height, the most advantageous station of course.
My guide, Joseph Marie Coutet, I found very intelligent.
He was the son of one of those who attended Saussure in his
famous ascent, had travelled much as a soldier under Napoleon, but was not in the Russian campaign, to which a vast
number of the inhabitants of Chamouni were taken, and few
ever returned.   Yet Buonaparte's policy has augmented the
population of Chamouni. By never requiring married men to
go, many were driven to this alternative, and as we are told,
and might have expected, many very unhappy marriages
have been the result. We crossed some large patches of
snow in mounting, and found the top of the mountain quite
covered, except the highest crag, on which I sat down to eat
my dinner. It was the height of luxury in such warm
weather to be able to cool my wine by plunging the bottle
in snow. The view from this point is truly magnificent.
Mont Blanc full in front, with his four largest glaciers descending down his sides—of Bosson, Mer de Glace, Argentière, and Tour; the villages of Servoz and Tour at each
extremity of the valley, and Chamouni at our feet, and the
river Arve. Behind us rose the snowy summit of Buet,
with the Aiguille de Varens, and many others, the Jura in
the distance.

In going down the Brevent, my guide chose out as carefully the tracts of snow as he had avoided them in ascending.
He had provided me with a pike, tipped with an iron nail,
which he used himself. He directed me to lean back on
this with one hand, as he did himself on his, and then lock
my other arm firm in his. We thus slid down together with
amazing rapidity, and precipitous places, that required half

an hour's labour in mounting, were conquered in two or
three minutes. It reminded me much of the Parisian amusement of *les montagnes*. This, however, lasted no farther
down than Planpraz, which we soon reached, and there I
found my father and the ladies, and we descended together
to Chamouni. The Rhododendron reddened some of the
cliffs, with many other flowers.

July 7.—Having got my breakfast early, I set out a
little before six o'clock, with Marie Coutet as a guide, to
visit Le Jardin on Mont Blanc. We ascended the Montanvert in less than two hours. I rested a few minutes at Blair's
Hut, to look over some of the small collections of specimens
of the different rocks of Mont Blanc. The Mer de Glace is
by no means a picturesque object from this point. We then
had to scramble by a difficult path which leads along the
cliff which overhangs the Mer de Glace. The whole of this
steep side, covered with the Rhododendron in full flower, was
a wonderfully beautiful sight. It was now growing hot, and
I was glad to refresh myself with a second breakfast, and
some cool water mixed with my wine at the fountain.
Hanging from the vaulted roof of the high hollow in the
rock, from the bottom of which the fountain bursts, is the
Ranunculus glacialis, a rare plant. It was in full flower, and
a very beautiful and conspicuous ornament, with its white
blossom, to the black rock above. It is impossible to climb
up to this. A few steps from hence brought us to the place
where we were to embark on the Mer de Glace. The first
part of this is the most formidable and hazardous of all, and
causes many a traveller to relinquish the enterprise. The
ice is traversed in every direction by rents and deep fissures,
over which you must leap. I could scarcely summon confidence at first, not from any fear of the depth of the chasm,
but the unsafe appearance of the landing-place on the opposite side, which, though in fact hard frozen, looks exactly
like soft snow, a resemblance which the multitude of small
holes in it full of water serves to confirm. In one place a
narrow bridge of frozen snow, not above a foot or two deep,
leads over a chasm of great depth. This we were obliged to
pass.

We had to follow the course of this glacier nine miles up,

but the six last are comparatively easy. I did not find the
air so much cooled by the ice as the day before, on the
Glacier de Bosson, but this might have arisen from the contrast not being so great with the air I had just left as in the
bottom of the valley. I saw, however, that even in July,
this large body of ice creates frost enough in the night to
encrust over the water which fills many of the chasms.
Thus there are but a few months out of twelve that can
melt the glaciers at all, and in these not twelve hours perhaps out of twenty-four, and even these twelve hours must
in the hottest month be partly occupied in doing over again
the work which each night undoes. We had proceeded, I
believe, more than half way, before we arrived at what may
be considered the river Arveiron, yet unborn. All the meltings of the upper part of the glacier from the summit of
Mont Blanc, collect and form a torrent, which confines itself
in as correct a bed down the ice as any river does in a plain ;
and rushing along with great swiftness, it suddenly comes to
a deep round well in the ice, just large enough to receive it.
It plunges down this and appears no more, till under the
name of the Arveiron in the Valley of Chamouni. This
chasm is called ' Le Grand Moulin.' As you stand on the
opposite side to where the torrent enters, you feel a current
of air blow on your face, caused by the water. We saw a
rainbow on many sides of this, a phenomenon which I am
sorry I did not pay more attention to at the time. It is
also wonderful, I think, that since the well is perpendicular,
the side against which the water is thrown with great
violence, instead of being worn away and hollowed, seems
rather to project.

While we were here, an avalanche broke loose from a
rock which overhung the Mer de Glace, and came down
with a noise like thunder. Many butterflies were flying
over the glacier. Every here and there huge boulders of
white granite covered the ice. Two in particular, my guide
led me to, each of which had one side covered with crystals
of quartz, some very perfect and richly charged and coloured
with green chlorite. The large space that was covered with
these made a splendid show as the sun shone full upon
them.

After a walk of nine miles over the ice, it was a wonderful transition, on arriving at the foot of the Couvercle, to
sit down on a plot of grass so richly strewed with our blue
garden violet that the air was really perfumed.   The large
Veratrum, blue gentian, and 'Forget-me-not' were in plenty
there.   Here I took a repast, and seeing my guide fast
asleep, I could not resist following his example.   After
about half an hour he awoke me, and we climbed up a long
and steep precipice to the summit of the Couvercle, a projecting rock, which forms part of the lofty Aiguille du
Moine.   From the foot of this rock you see a valley of ice
extend up to the summit of Mont Blanc : around you are all
his shapeliest points, ending in beautiful pinnacles of pointed
rock.   The picture is in short entirely composed of the sublime and terrible—bare rock, ice, snow, and sky, with a dead
silence round.   The Couvercle is the very seat which the
imagination of Virgil chose for the poet Orpheus, when
abandoned to despair.

Solus Hyperboreas glacies, Tanaimque nivalem, &c.

Before descending from the Couvercle, I was glad to
take a second dinner, for the mountain air has certainly a
wonderful effect on the appetite.   I observed below me, on
the side of the Mer de Glace, a double moraine remarkably
regular.   The only way in which I can conceive this to have
happened, is by the gradual decomposition of two cliffs
above, of different altitudes.   The fragments from the higher
would naturally be thrown out farther, and form the exterior
line.   The granite sand of the moraine near the head of the
glacier contains, as might be expected, much mica, but at
the source of the Arveiron, and foot of the Glacier de Bosson,
it contains none, but is a white powder and as soft as pounce,
consisting probably of quartz and feltzpar, and mostly of the
latter.   It is this which gives the Arve its milk-white colour,
and not, as one often hears said, mere melted snow, than
which nothing is more clear, as I was often glad to find when
thirsty.

The Arve is much of the colour of streams and ponds
on chalk in England, than which nothing can be more un-
picturesque, unless perhaps the torrents which rush from

l'Aiguille de Varens when the snow melts in the spring, which my guide assured me are as black as ink, from the slate. We had two miles and a half still to walk before we could reach Le Jardin, a fatiguing addition to the march, since we were up to our knees in snow at almost every step. This little oasis, a small plot of ground only a few hundred yards in circumference, and always surrounded by a glacier, is the highest elevation at which vegetation exists in Europe, but this is not the only reason which renders it worth visiting. You stand on considerably higher ground here than at the Couvercle, and the highest part of Mont Blanc seems to have risen much.

I took a third dinner at this place, and after writing my name on a slip of paper, to put into the wine-bottle there, which is full of names, began to retrace my steps to the Couvercle. I was surprised here at the number of insects, but my guide assured me they were surprisingly diminished since the year 1815. It is only strange they are not completely annihilated at this elevation, since their number has been so sensibly affected all over Europe by the late extraordinary seasons. Many would attribute this to the scarcity of entomologists near the summit of Mont Blanc, but the fact is, I suppose, there are no birds. A flight of hungry swallows would destroy in one evening more than an entomologist in a year.

On the snow which we had to pass between the Couvercle and Le Jardin, I observed that rose-coloured tint which is so difficult to account for, but the snow was not harder in those places, as I had heard affirmed.

I followed my guide with very rapid steps on returning over the Mer de Glace, for I began to be afraid of night overtaking us. The chasms are much easier to leap in descending, since you are always on the higher side, and for the same reason you can leap many which before caused a great detour. We stopped only a few minutes at the Fountain, where, after some pelting, a stone brought down two flowers from *Ranunculus glacialis*. At Blair's Hut I found two English travellers, looking over the collections of rocks. They informed me that my father and the ladies had just left the Hut, and were gone to visit the source of the Arveiron. I followed them there, by descending the contrary side of the Montanvert to what I had climbed in the morning. This lengthens the return a little. The precipice here is very steep.

There is an album at the Montanvert, which I did not look into, but one of our party discovered in it a stanza which ran, I believe, nearly thus—

Mont Blanc is the monarch of mountains,
They crowned him long ago,
Enthroned in ice, with robes of clouds,
And diadem of snow—

which, however far from perfection, contains more real poetry than I thought could be found in all the albums of Europe. The blue vault out of which the Arveiron bursts full-grown at its birth is well worth visiting. While we were there, a fine avalanche fell, which we saw and heard to great advantage. The sands of the Arveiron are delicately white and fine, but the grains of gold which they say are found in it must be transferred, like those of the Pactolus, to the poet's verse, before they will serve as ornaments to the scene. The rest of our party rode from thence to Chamouni in chars-à-bancs, but I, though much fatigued, preferred finishing this short distance on foot. The shortest way from Chamouni to Le Jardin and back they reckon forty-eight miles, eighteen being over the Mer de Glace, and five over the Glacier du Jardin. Too much for one day.

July 8.—The number of visitors to the Valley of Chamouni last year they told us was 1,400, 1,000 of whom were English. The town was perfectly inundated while we were there, fifty arrived in one day. We determined to-day on a visit to the Col de Balme, from which there is by far the finest view down the Valley of Chamouni. Chars-à-bancs carried us some way, and mules nearly all the rest, which I was very glad of, as I was very tired from my preceding day's expedition.

The Mer de Glace is advancing terribly every year, but fortunately is going down the valley.[3] The Glacier d'Argentière is perhaps more threatening, from its nearness to the Arve. In the valley I took no less than seven specimens

[3] The Mer de Glace has of late years much diminished.

of that rare insect *Papilio Apollo.* The rock of the Col de
Balme is a brown ligneous slate, with some veins of white
quartz intersecting it : the appearance is very curious. On
the top was the richest carpet of turf I ever saw, spangled
with thousands of the deep blue gentian, red trefoil, and other
mountain flowers, yet exceedingly level. Farther down
quantities of the *Cnicus spinosissimus* made a magnificent
show. There is a small lake on the top of the Col de Balme,
where we were ; but there is still a higher point, with an iron
cross at the top, near which M. Eloher perished by his rash-
ness. To this we did not ascend.

July 9.—Left Chamouni on our return to Geneva. The
clouds entirely screened the distant mountains, and there
were some partial but very heavy showers, which the covering of our chars-à-bancs but ill defended. About Servoz it
is extraordinary how abundant the wild barberry grows. At
St. Martin we found our carriage again, and dismissed the
chars-à-bancs, and Chamouni guides. The latter are a most
intelligent set, but the people of this valley are in general
frightfully plain, owing chiefly to the goitres, to which the
women are more liable than men. At St. Martin there are
few in comparison with the great number at Sallanche, on
the opposite side of the river. We left Geneva, to commence
as it were anew a tour through Switzerland. Coasting along
the lake, we passed Coppet, and slept at Rolle, a neat little
town on the lake.

July 12.—We entered the Pays de Vaud, and arrived at
Lausanne, and walked down to Ouchy, the Piræus of Lausanne, and went out on the lake in a boat. We remarked the
blueness of the water, which cannot be owing to the reflec-
tion from the sky, since it was very cloudy ; nor is it exclusively where the water is deep, for I observed near the shore
that the weeds, which when pulled up were perfectly green, were
b!ue when seen through the water. The town of Lausanne
is uninteresting, except to us from having been so long the
residence of Gibbon. From Iverdun to Neuchâtel is a
pleasant drive, but the road at first is execrable along the
shingle of the lake, which is of a very sea-green cerulean
colour, and very different from the blueness of Geneva. In
Neuchâtel there were certainly more good gentlemen's

houses than we had seen in any place of the same size on
the Continent. One of Monsieur Pourtalis, in his time the
greatest merchant in Europe, whose family still lives here.
One of Monsieur Pourri, also a very wealthy merchant. The
latter built the large town-hall here, the former the hospital.

July 15.—After leaving Neuchâtel, we soon approached
the end of the lake, and went through St. Blaise, a large
village, and saw next the Lac de Bienne, in the middle
of which is Rousseau's Island, where he resided so long, and
by it the 'Isle des Lapins,' which it was his fancy to people
with rabbits. We were obliged to have six horses to mount
a long and steep hill, from which we enjoyed again a grand
view of the distant Alps. A fine mist between us and the
base of the chain had exactly the appearance of a second
atmosphere, and the gigantic summits of the Alps seemed
cut off and unconnected with the earth. Here we took leave
of the Alps for a time, and entered a very romantic valley,
fine forests of fir on all sides, and high limestone cliffs, the
stratification much disturbed, but answering in general one
side with the other in their inclination. We passed under
Pierre Pertuis, a curious excavated rock, but more celebrated
perhaps than it deserves.

July 17.—Went to Basle by the side of the river Birs, up
a long valley, in which there was a great variety and abundance of wood. After dinner we walked to the cathedral, and
climbed the spire, from which we saw into France, Germany,
and Switzerland. I rather liked the effect of the painted
tiles on the roof of the cathedral, which were laid in the
mosaic style.

July 18.—Holbein was a native of this place, and there
are in the Bibliothèque Publique many of his drawings, and
nine small, exquisitely finished paintings in the same frame,
representing 'The Passion.' They have also here a great
treasure, an edition of the 'Laus Stultitiæ' of Erasmus, with
pencil vignettes drawn on almost every page by Holbein, and
which show a great deal of genius. The collection of fossil
remains, taken chiefly from the limestone of the Jura, is rich.
I am much struck with some which I believe to have been
Tulip Alcionia. They have also a large *Monoculus*, which
they consider unique.

We next went to the gardens of Mr. Fisher, banker, which are laid out with taste, and look down upon the Rhine. In the rock-work and arbours were some beautiful specimens of Tuff. This modern formation, which abounds here from the large quantities of limestone in the Jura, is found very useful in building under water, since from its porous nature it hardens every year by deposition. It is also inestimable from its great lightness in the higher storeys of their houses, which are chiefly built of wood.

After a walk round the small botanic garden here, we proceeded on our way to Schaffhausen as far as Stein, a small village on the Rhine, where we slept. The first red sandstone I had seen since I had been on the Continent was in the cathedral of Basle, which is built of it, but I thought I observed it to-day composing the banks of the Rhine.

July 19.—Before the windows of our inn at Stein, on the opposite side of the Rhine, was Säckingen, a considerable town, which, with the two towers of its cathedral, its long bridge finished with a tower, the broad river, and the harvest going on in the fields, formed a pleasing picture. From this place we went on to Lauffenburg, and again crossed the Rhine into Germany. A short distance below the bridge, the Rhine is confined between some rocks, so as to appear only a third of its size. It rushes with amazing violence, the greenness of its colour, the quantities of foam, and the large waves and whirlpools which it causes below these rocks, give it a more exact resemblance to the sea, during a brisk gale, than anything I ever saw inland.

July 20.—When we had approached within a mile and a half of Schaffhausen, we sent on the carriage, and walked down to Lauffen, to see the Fall of the Rhine. This approach is very favourable, and we got many picturesque views of the Fall before we arrived at the Castle of Lauffen. In a room at the top of this, we had a fine front view of this magnificent object. Afterwards they darkened the room, and threw a beautiful landscape of the Fall on a white piece of canvas suspended from the ceiling, by means of a cameraobscura. It was a complete picture, with the addition of motion. The water tumbling over, the clouds of foam, and the agitation of the water in the large basin below, appeared

beautiful on this flat surface. We passed over afterwards
to the top gallery, from which you look down on the Fall,
and see admirably the contrast of the river before and after
it has been hurled down the precipice. We then descended
to the lower gallery, where you are well repaid for the wetting
you get from the spray. Your eye meets the Fall midway,
and you can form a just idea of the body of water which is
carried down. The rock is limestone, and I found some
fossil shells in it. The masses which break the Fall so
picturesquely, are much worn away on the side which is
opposed to the current : one is perforated. They must in the
course of time gradually waste away. They are happily
clothed with underwood. I observed. as the sun shone on
the foam, it took very much the rose-coloured tint so remarkable on the snow in the Alps. In our walk back to
Schaffhausen, the heat was very oppressive. In consequence
of the fineness of the summer, they look forward to as rich
a vintage as 1811, the memorable year of the comet. The
waiter at the Couronne assured us that at their inn alone
they had, in the course of last year, 745 English, not including other foreigners. From Schaffhausen we went to
Eglisau, a town on the Rhine, where we slept.

Between Eglisau and Zürich we passed through a very
rich corn country. They cultivate much the poppy here for
medicine. The flowers are single but large, chiefly white,
but sometimes a good proportion of red and pink intermixed.
When in full blow, and with a field of ripe corn on each side,
it is the gayest crop imaginable. We arrived early at
Zürich, and drove to the Hôtel de l'Épée, a celebrated inn,
but which I found partook more than any of a fault too
common in Switzerland. They have their stables and cow-
houses under the same roof, and the unavoidable consequences may be conceived, till they can fall in with a man
as able—as 'Hercules to cleanse a stable.'⁴

The next morning we paid a visit to the library, in which,
besides many curious fossils, they showed us a literary curiosity interesting to an Englishman, the Latin letters of Lady
Jane Grey to some learned man of Zürich, in which were
many Greek and Hebrew quotations. From Zürich we pro-

⁴ *Hudibras.*

ceeded to climb Mount Albis, from the summit of which we enjoyed a most beautiful and picturesque view of both ends of the Lake of Zürich. On the other side, the Alps were obscured by clouds, but we saw the Righi and the Hackenburg, a very remarkable broken mountain. After dining at the inn near the top of Mount Albis, we passed a small lake which they call the Türler See, and then went down to Zug. The latter part of the descent was almost dangerous for so heavy a carriage as ours.

July 23.—The rest of the party having decided upon going straight from Zug to Lucerne, I engaged a guide to conduct me to the Righi. I landed at Arth, a small place in the centre of Schwytz, where the costume amused me much, for the women wear an exact imitation of the crest of the hoopoo on their heads, a high double frill reaching from the back of the head to the forehead, very ridiculous, and I should think useless.

At the inn an English traveller, a Mr. Hope, who was just sending forward his servant to the top of the Righi, politely offered to engage a room for me at the same time. I then made my guide conduct me to the Valley of Goldau, that I might have a new view of the Rossberg, a mountain from which a large part fell twelve years ago, and buried a village. The guide pointed out very evident marks of a more ancient *éboulement,* which had extended considerably farther than the last, but of which they do not know the date. The recent pile stands exactly over the former one, yet, wonderful to say, a church and a large inn have been erected already over it. A large stripe marks the passage down which the fragments rolled, from the top to the bottom of the mountain. They are chiefly of pudding-stone, and the best explanation I could imagine, after hearing the different accounts of the people there, is that a vast quantity of water, after long rains, had collected at the juncture of the pudding-stone and a stratum of argillaceous earth, about half-way down the mountain, which at length gained strength sufficient to wash out part of the clay, and let down the pudding-stone above. The great difficulty is, that all the accounts of the people there seem to agree in stating that large masses were blown up, as if by volcanic force, perpendicularly in the air, which some attempt to explain by the force of water, others by air, confined.

It was a scorching day, but I proceeded immediately to mount the Righi. About every hundred yards was a high crucifix, with a neat little painting at the top. Numbers of travellers were climbing at the same time.

At Righi, a place about half-way up, where there is an inn, I met with a party of Englishmen, Sir John Wrottesley and his three sons, the oldest of whom is at Oxford, whom I know by sight only, and as intimate with some friends of mine. Having entered into conversation, we ascended by a very fatiguing march to the Staffel, where we dined together. There were a great many fountains of excellent fresh water, which relieved us much the whole way. The mountain appeared composed entirely of pudding-stone, and some very fine cliffs of this rock presented themselves to us in a very singular manner during our ascent. It is a very practicable mule road the whole way, but very laborious.

The view towards the Alps was in part obscured by clouds, and had opened so gradually to us at every step, when we cast our eyes back, that we seemed to see nothing more on this side on arriving at the top, but on looking down the precipice on the western side, which was free from all clouds and mist, a view opened to us which astonished me more agreeably than anything of the kind I had ever seen before. It gave me precisely the idea of being raised to a great height, and looking down upon a whole world at once. Numberless small lakes, rivers, and towns, could be distinctly taken in at the same time by the eye. The large lake of Lucerne, at our feet, happened to have the sun full upon it, and was in a blaze, which at such a depth below us had a singular effect. The lake of Zug also made a great feature in this enormous map. Mount Pilatus seemed of great height, even viewed from this, and the Altorf end of the lake of Lucerne was very remarkable, however much it lost by the distance. As the sun descended, it totally left the lake of Lucerne, and filled with a blaze of splendour the Sempach, which is much more distant, but so considerable a sheet of water, that it became a very striking feature.

We staid on this point till sunset, and then determined

upon mounting to the Culm to sleep, which we found very full of travellers. Fortunately my room was secured, and Mr. Hope came to say he would give also his room to our party, for he should go down the mountain again. In these two rooms were four beds, which was all the landlord said he could spare for our party of five. In this little hut, 4,350 feet above the level of the sea, this Bedlam of languages, this chaos of confusion, were shut up nearly forty travellers, with many guides, and the servants of the inn. Many French, Germans, English, and Swiss, were supping together in the *salle-à-manger,* which was in a few minutes to be a barrack bedroom.

As none of the people of the inn would attend to us, we took possession of two rooms, which contained five beds, but our host turned us out immediately, and said if we were discontented with the accommodation we might turn out of doors, for he had no scarcity of inmates, and for his part he should have thought us well off with half the number of beds, ' since what objection could we have to sleep two in the same bed ? ' On this dilemma, Mr. Johnson, an Englishman, came to say there were two beds in his room, one of which he should be most happy to give us. We went with him, but found the people taking out this extra bed, and in the meantime some ladies took possession of our single-bedded room. Thus were we reduced to one room with three beds ! The landlord, however, soon gave us back our right, and we made up another bed for ourselves on the floor, by pillaging a pillow from one room, a sheet from another, &c. I must not forget, however, the assistance which Mr. Reichenbach afforded us, by coming, when we were nearly in bed, with his bedgown and nightcap on, to beg we would accept *son couverture,* of which he had no want ! His large fat figure, which almost out-Falstaffed Falstaff, appeared to such advantage in his nocturnal habiliments, that we could not resist a laugh of admiration, which he very good-humouredly joined in. The eldest of the Mr. Wrottesleys having laid down on this shakedown, I got into a small bed, one of those which our worthy host thought a handsome provision for two of us. I was immediately attacked by a legion of fleas, who seemed to consider me as private property. There was too much noise for anyone to get much sleep, and so thin were the partitions, that we heard every one snore distinctly at the distance of three rooms.

July 24.—Before three o'clock, and long before the rising of the sun, nearly forty miserable half-dressed and shivering wretches were to be seen crowding round the crucifix on the highest point of the Righi-Culm. Scarcely a word was spoken, and a stranger might have mistaken them for so many pilgrims, and admired the fortitude with which they bore up against their evident distress. Not a single cloud or mist prevented our seeing the outline of the Alps, and I suppose it must have been our inconvenience (for we were wet through in our feet, as well as chilled by the air) that made our party all agree that the view was less grand by sunrise than sunset. After breakfast we descended to Küssnacht. This side is much more precipitous than the one we had before taken to mount from Arth. A short way above William Tell's chapel is a most delightful view of the lake of Lucerne, Mount Pilatus on the left, and the Stanzerhorn on the right. Went from Küssnacht to Lucerne in a boat with Sir John Wrottesley and his party, and here I rejoined our family.

CHAPTER IV.

JULY–SEPTEMBER, 1818.

MEYRINGEN—THE GIESBACH—GLACIER OF THE RHONE—VIEW FROM THE
GRIMSEL—THEORY OF PINK SNOW EXPLAINED—GRINDELWALD GLACIER
—LA BELLE BATELIÈRE—AVALANCHE FROM THE EIGER—BERNE—DR.
WITTENBACH—THE VALLAIS—DESOLATION FROM THE BURSTING OF THE
DRANSE—VAL DE BAGNE—THE BRIDGE OF LOURTIER—CONVENT OF ST.
BERNARD—ANCIENT TEMPLE OF JUPITER—BRIEG—THE SIMPLON—LAGO
MAGGIORE—MILAN—VERONA—VENICE—BOLOGNA—FLORENCE.

JOURNAL (*continued*).

July 25.—Went in a boat from Lucerne to Alpnach;
many magnificent views of Mount Pilatus. The rain
detained us some time at Alpnach, and the ladies were
obliged at last to set out in a char-à-bancs before it quite
stopped, in order to reach Sarnen before night. I preferred
walking, and went with my guide by a short cut, which
brought me there half an hour before the carriage arrived.
We crossed a bridge of wood 500 feet long, but only two or
three feet broad, over a torrent which has astonishing power in
the spring, but which then, in spite of the showers that
morning, had not a drop of water in it. Sarnen is a miserably poor place, yet nevertheless calls itself the capital of the
canton of Unterwald. Here they are all Catholics, and
Nicholas de Flue is their guardian saint, who for many
hundred years has saved their wooden walls from flame.
His tomb is not far off, at Sachseln, which we were fortunate
in reaching on Sunday, since many pilgrims were offering up
their prayers there, some getting down below the floor of the
chapel, to pray under his body. The chapel was full of
curious offerings, and the variety of the costume of people
come from different parts was entertaining. Both the
lakes of Sarnen and Lungern are picturesque. At Lungern
we dined, and then began to mount the Brunig, a pass

which the ladies found rather nervous on mules, and which I preferred walking. The scenery was very romantic the whole way, till at last we descended into the valley of Hasli to Meyringen.

July 27.—Went from Meyringen to Brientz. We were rowed, and rather rapidly, by two women and a boy. This is a common custom with the women in Switzerland, arising probably from the men all staying for some months in the mountains with their herds. Having landed on the opposite side, we were conducted up to see the different falls of the Giesbach, one above the other, which present numberless views, all extremely picturesque. The body of water is considerable, and very white, and the accompaniments of wood and water beautiful. We returned immediately to Meyringen, and were imprisoned by the rain the first time since our arrival on the Continent. In the morning I visited, in spite of the rain, the Alpbach, a cascade which falls from a considerable height only a few hundred yards from the town. Its colour is a Stygian blackness, and we had heard it all night pitching over, with a horrible clatter, large stones from the top to the bottom. This torrent some years ago nearly washed away the town of Meyringen, after which the inhabitants with great labour built an immense thick stone wall, reaching nearly from the Fall to the Aare, to conduct it in a straight course to the river. Nothing but this saved them on the Saturday before we arrived. The clergyman of the parish had exhorted them, in a sermon on Sunday, to keep a strong guard on the wall night and day, and when the torrent attacked any particular part, to throw branches of fir and other things to divert it from the wall. This guard we found still actively employed, for the river was constantly changing its direction, from the large mass of matter which was moving in its bed. This mischief happens after long droughts, when the sun has had great power on the slate rocks above, and caused them to crack and decompose. None but those who have witnessed it can conceive the strength of water charged with this black slate powder. The rocks seemed actually to float in it, and a trifling depth of it was pushing heavy bodies along.

July 29.—Our party went from Meyringen up the valley

of Hasli to Handeck, to see the Fall of the Aare, the distance
of six leagues, a very fair mule road, but which I preferred
walking. The Fall of the Aare is truly grand, the height
great, and the body of water considerable. Down the same
precipice another river of inferior size, but of a better colour,
falls over at right angles, and meeting the Aare in the middle
of its fall, causes a magnificent conflict of water, with clouds
of foam. Two bright rainbows were spanning and connecting with a sort of fairy arch this tremendous rent in the
rocks. The Fall is well shown from below, and from the
precipices which overhang it, where you are obliged to lie
down full length and look over. Here you are placed between
the two rivers, the smallest of which was a large body of
water when we saw it, from the late rains. Finding here
three German-Swiss gentlemen who were going up to the
Grimsel, I determined to walk there with them, and see the
glacier of the Rhone, while my father and the ladies
returned to Meyringen. I had two leagues farther to climb
to the Hospice, the scenery becoming at every step more
savage. Several times the path crossed large streams by
bridges of snow, the remains of spring avalanches, which
last all the year, coming down the beds of these torrents,
and in many places bridging over the Aare. We passed
some extraordinary large bare planks of granite rock above our
track, the appearance of which I could not account for. A
thick mist afterwards came, which prevented us almost from
seeing one another, and I was heartily glad to reach the
Hospice, though a dreadfully rough reception one meets with.

July 30.—My feet had suffered so much from my walk of
eight leagues over rocky ground the day before, that I left
with regret even the wretched bed they gave me at the
Grimsel, to undertake eleven leagues more to see the glacier
and return to Meyringen. I got, however, on the march
with my German friends by five o'clock, though the fog was
as thick as it had been the evening before. We could only
see three or four feet before us, but when we were on the top
of the Grimsel, the guide gave us notice it was going to clear
up. We none of us believed him, but it grew more light
rapidly every minute, and suddenly the mist which was
round us rose above the summit of the Finster Aarhorn, the

loftiest of all this chain of Alps, and is of a very picturesque
shape as seen from here.   The rest of the vapour fell down
upon the little lake which is near the Hospice, and seemed
to cross the valley like a compact white body, as a glacier
might.   The whole of this ' drawing the curtain ' did not, I
think, take above ten minutes, and was one of the most
beautiful phenomena I had witnessed in Switzerland.   Soon
afterwards we came to a small lake, above which was a steep
slanting precipice of glazed snow, a pass which, without
pikes, would have been difficult.   Here we remarked that
bright rose colour on some patches which is so extraordinary.
As the guide I had taken from the Hospice could talk
nothing but German, I requested my companions to ask each
of our guides their opinion on the subject, not with any
expectation of their clearing up a difficulty which Saussure
could not, but in order to judge of their intelligence.   *Their*
man, whom they described as *un bon garçon, mais un peu bête,*
seriously declared his belief that wine had been spilt over the
snow.   *My* guide, after having joined in our laugh at this
solution, said the cause was evident enough, viz., that when
the sun had melted the snow for five or six years, it invariably turned of this colour.   We soon came in sight of the
glacier of the Rhone, and I was disappointed to see that it
was inferior to every one of those of Chamouni in beauty,
and the phenomenon comparatively nothing, since instead of
the rich cultivation of that at the foot of the Mer de Glace,
there were only some poor pastures.   I was almost sorry
afterwards I had undertaken the task of descending to the
source of the Rhone, since it was a precipice which costs
much labour to go down and up again.   My German friends
having gone on to cross the Furka, I returned to the Grimsel
to breakfast, and as my guide was of no use, speaking not a
word of French, I found my way back to the Fall of the Aare
alone, having dismissed him at the Hospice.   Here I found
Mr. Johnson, whom I had before met on the Righi, and I
returned with him to Meyringen.

July 31.—At the inn at Meyringen they keep a boy who
is almost an idiot to black the shoes. He can speak but two or
three words, and understands scarcely anything but signs.   I
tried to let him know, by pointing to my watch, that I wished

to be called at half-past three, and, to my astonishment, he came to a minute. I started by five, determined to pass the two Scheidecks before evening, which is fourteen leagues. I had performed the preceding day what the Swiss call eleven *good* leagues, or nearly twelve, and the day before that eight. When I boasted of this to my guide, he observed coolly, or almost I thought with a sneer, 'Oui, c'est assez pour un Monsieur.' A just reproof, and which might come home to many a bragging pedestrian, since, let these gentlemen do what they will, the peasant will follow them as far, carrying their bags, and without the same mental stimulus to assist him. At the foot of the Scheideck I went to see the *lower* fall of the Reichenbach only. In this I saw nothing remarkably fine. At the châlet, which is six miles on, I stopped to breakfast. Many small avalanches fell while I was here, with a noise like that of distant thunder, from the Wetterhorn, which rises here very grandly. A precipice of snow near its summit presents a curious furrowed appearance, from avalanches which have rolled down there. We descended into the valley of Grindelwald, and I went a little out of my way to view the glacier. It is less picturesque, but very unlike those of Chamouni. I had here, as at the Rossberg, another attestation of 'how little distant dangers seem' to most men. The peasant who lives here told me that the glacier advances rapidly, but has not near reached an ancient boundary, for that he had built his châlet on an ancient moraine. If the next three winters are in any way as severe as the last three his habitation will unquestionably form a part of a new moraine. At Grindelwald, where I dined, I saw *La belle Elizabeth*, formerly *La belle batelière* on the lake of Brientz, now the wife of the landlord of the 'Chamouni,' who has the reputation of using her very ill. I bought some trifles of her, to entitle me to write my name in her album, where I found those of hundreds of my countrymen. A remark of Southey's I thought curious. 'R. S. recognises in Elizabeth a striking resemblance to La Fornarina of Raffael.' She has not the least affectation, either in her manner or her dress, and deserves, I think, her great fame for beauty. I did not lose much time in the valley of Grindelwald, with which I was rather disappointed, but climbed the Wengern

Alp, or lesser Scheideck, under a most burning sun. From
the top of this the two Eigers are seen to great advantage.
The great Eiger rises at once from its base to its summit,
one huge and tremendous precipice, far above the height of
perpetual snow, but so perpendicular that little can rest upon
it. From one of the ledges of snow at about its middle height,
I saw a fine avalanche fall over a precipice down to a parallel
ledge below. A loud crash announced to us the detaching
of the mass, and the leisurely way in which it appeared to
our eyes to descend this seemingly short precipice, gave me
a vast idea of the height of the whole Eiger. The mass had
reached its destination, and a great deal of the cataract of
snow which it brought with it had followed, before the report
reached us, which was like that of a loud clap of thunder.
I was prevented by the clouds from seeing the Jungfrau, but
got a fine view into the valley of Lauterbrunnen in descending. There I found the inn so full, that they could not give
me a bed after my long march. I was obliged to lie down
in the *salle*, with several other travellers, where I got no
sleep, not from the want of a bed, for I was tired enough to
have slept anywhere, but from the incessant noises. At
Thun we met our carriage, and went to Berne through a well-
wooded corn country, often much resembling England.
Berne was the most regular built town we had been in.

I called with my father on Dr. Wittenbach,[1] with whom
we were much pleased. He walked with us on the ramparts,
where we got a superb view of the Alps, lighted up by the
setting sun. Among other things the Doctor mentioned his
having been much enraged with Cox, the tourist, who in his
Fauna of Switzerland included the white bear as an in-
habitant of the Alps! and as if on Dr. W.'s authority.

In the exhibition at Berne are some good water-colour
drawings of scenes in Switzerland, most of them for sale,
and dear enough. Here we fell in with Dr. Wittenbach
again, and he gave us a most entertaining lecture on the
geology and mineralogy of Switzerland. He has a pretty
large collection of specimens from many distant parts, as
well as the Alps, which he himself has examined carefully. He
invited us to bring our whole party to drink tea with him in

[1] Eminent Swiss naturalist.

the evening in a public promenade, which is much the fashion at Berne. We were glad to accept the invitation, and were fortunate in having a fine evening, with all the most distant Alps perfectly clear.

August 7.—We went on to Payerne, through a rich culti- vated country. Before descending to this town we saw in the distance the lake of Morat and the opening of the vale of Travers in the Jura. We also passed a small lake before Vevay, containing only a few acres of water. The descent to that place along the side of the lake was most beautiful.

August 8.—From Vevay we went along the shore of the lake, which is much more picturesque here than at the Geneva end. We passed the Château of Clarens to our left, immortalised by the Héloïse of Rousseau, and then came to the Château de Chillon, which Lord Byron has lately made equally celebrated. Here we were conducted down into the dungeon in which the prisoners were confined. Leaving Villeneuve to our right, we passed on to Bex. From the inn we engaged a char-à-bancs which carried us up a valley, branching from Le Vallais, down which we looked full upon the Dent de Midi, a magnificent mountain. The excavations which have been made in hopes of discovering more salt-springs are on a very large scale: galleries of immense length have been cut through the rock. We were not content with walking through the lower one of these, but went up to the higher gallery. The great length, narrowness, and the wet of this, and the wind which blew out our lamps constantly, made it a fatiguing expedition, and we gained little enough by it. Each held a lamp, and the guide one, and all put on some ragged surtouts, to preserve us from the dropping of the water, which gave the party a grotesque appearance in the subterranean chambers which we arrived at in some parts of the gallery. After returning to see the House of Graduation, where the salt is extracted from the weaker spring by a very ingenious contrivance, we got into our char-à-bancs and proceeded towards Bex. From this we went on to St. Maurice, a narrow gorge through which the Rhone passes, and where Le Vallais properly begins, though the valley of the Rhone certainly reaches naturally to the lake of Geneva. There is a large cliff at the right of St.

Maurice, with very horizontal lines of stratification, some of which form terraces. On one of these is placed a house very high up, which seems stuck on to the perpendicular preci- pice. We went on the Martigny road to the Pisse-vache, a justly celebrated waterfall. Its singularity depends on the distance which it is thrown from the rock where it shoots over at the top, and an appearance of falling rockets which the little jets of water make when they are parted from the main body of the Fall.

After this a scene of lamentable desolation began,[2]— fields of corn under water still, and further on entirely covered with sand. The road was in some places under water, and the carriage was obliged to pass through very deep places. At last we left the old road altogether, and went over a tract of sand, under which had once been hedges and fertile fields. The trees alone remained bursting up through it, and at the foot of each of them was a heap of dried trunks and branches of trees and planks of houses which the current had brought down and lodged there. There was lying by the side of the road one striking instance of the great force which the flood had had. A large pine of full eighteen inches diameter, and proportionable height, had floated down and been stopped by a pollard poplar. The water had rushed with such violence against the two ends of the pine, that it had cracked in the middle and been twisted round the poplar. The tree was green when snapped, and therefore still held together. The diameter was eighteen inches at the part which rested against the poplar where it was broken.

Our inn at Martigny, the post-house, presented a sad scene. The first storey had been filled with mud, sand, rubbish, planks of houses, and boughs of trees. All this had been thrown out into the court. Everything I saw here increased my curiosity to ascend farther up the valley which had suffered by this calamity, and to reach the glacier next day. I found it was necessary to get at least to Sembranchier before night. It was late, but I took a mule and a guide

---

[2] The Dranse, having been blocked up by the Glacier de Bagne, formed a lake, which burst on June 16 of this year, sweeping away villages, and devastat- ing the fruitful and beautiful valley,

with me, and crossed Mont Chemin. When we got to Sembranchier it was dark as midnight, and a storm of thunder and lightning as terrible as I had ever witnessed. Nevertheless my guide advised me to press on, as there was no rain, to the next place, where there was a better inn, and as we had too much for the next day. It was three miles farther. I determined to walk, and let the guide drive the mule. A little beggar-boy, a child of twelve years old, who I suppose had lost his way and was terrified at the storm, seized hold of the mule's tail, and in spite of the guide's threats clung fast to it, till we got to Bagne, allowing himself to be dragged over stones and through two or three large brooks.

August 10.—The house which I slept at, the residence of President Gard, the chief magistrate for the commonalty of Bagnes, had once been a monastery, and if the accounts of the peasants of this valley can be believed, the French have done the country much good by abolishing it. The monks appear to have been a selfish, luxurious, and despotic body. There are 3,500 souls in the valley of Bagnes; all speak French. Mons. Gard condescends to wait on his guests in character of waiter. He seems popular with the common people, and was chosen, they told me, by the French, as being the only one who could read and write among their 3,500 inhabitants.

At every step as we advanced my guide pointed out some devastations of the Dranse, which I should not have remarked. Many châlets, indeed most of them that stood in the way, have been so completely buried, that a stranger might pass all up this valley, and see no greater signs of desolation than the wide channels Alpine rivers so often present. I had not rode, however, nine miles, before I had feeling assurance of the destructions of the Dranse. The bridge of Lourtier had been carried away two nights before, and only a single plank of the new bridge as yet thrown over. The mule I was of course obliged to leave, and the crossing so furious a river by a single plank would have been dangerous to a nervous person. About eighteen people were employed in erecting this wooden bridge, and I was much amused with the simplicity of their operations. They first cut down three of the largest firs they can find near the

river, square them, and then send two or three men by some
bridge, whether up or down the torrent, across to the other side.
A rope is thrown over to them, fastened to the end of one of
the trees. By this means the first plank forms a communication, after which two more are soon got to the side of it.
They then merely fasten cross boards on these three trees, and
thus over a river fifty feet wide in the smallest part they
can drive hundreds of cattle every day. Large masses of
micaceous schist rock have been finely laid open to the day
by the late catastrophe. Near the cascade of Lavanche,
where the Dranse falls down from a considerable height, the
soil has been laid open to a great depth, and it is easy to see
that it consists of rounded boulders of an immense number
of different primitive rocks, which do not seem to belong to
this neighbourhood. The scenery of this valley is very
beautiful, like that of Hasli. Owing to the annihilation of
the old road, our path was very intricate and difficult, and
we had to walk five or six leagues before we arrived at the
head of the valley. Here there is a châlet, where six men
live who have in charge eighty-four cows and twelve pigs,
belonging to perhaps nearly as many proprietors in the
valley, who pay them for herding and milking their animals.
They make almost all of it into cheese. It is wonderful that
six men should be equal to this labour. The heat of the sun
was very oppressive, and when I saw an immense fire in the
châlet for boiling the milk, I of course felt no inclination to
enter. Having got some milk from them, and made a
dinner on the provisions I had brought with me, I walked
a short way up, which brought us in front of the glacier of
Getroz. I was much gratified here by seeing at the first
glance exactly how everything must have happened. My
guide informed me that this spot was formerly called Mont
Mauvoisin, from the near approach of the two perpendicular
cliffs on each side of the valley. Opposite to us was Mont
Pleureur, a tremendous precipice, over the brow of which
peeps the glacier of Getroz, which increased so much in the
last severe winter, that it thrust over immense masses of
ice. These falling with the snow in the spring, completely
blocked up the narrow chasm into which the valley is here
contracted. An immense lake was soon collected from the

stoppage of so considerable a river as the Dranse. Every
exertion appears to have been made, and two-thirds of the
lake successfully drained out. It was the remaining third
which had force to lay waste so much land. The remains
of this immense avalanche are diminishing greatly every day,
but are still a grand object. The perpendicular precipice of
ice must be in the middle (to speak within bounds) 150 feet,
and the slanting line of snow from the top of this to the
side of Mont Pleureur as much again. Between this precipice of ice on the side of Mont Pleureur and Mauvoisin, the
cliff on the opposite side of the valley, there is little more
than room for the river to pass, and very considerable danger,
as every one here believes, of the same calamities recurring
next year. There is, however, great hopes, from the heat of
the present summer, that the glacier will retire, and the river
force its passage as of old, under the spring avalanches. In
looking up the valley I saw the glacier of Chermontane, which
appears very large. Though there were such numbers of
houses destroyed by the flood of June 16, yet only twenty-
five persons lost their lives, and those were mostly at Martigny. The engineer had given all due warning, but the
people of Martigny thought themselves secure, from a tele-
graphic communication which they agreed upon with some
people at the top of the valley. Fires were to have been
lighted as beacons, but those on the two first hills by some
mistake were never lighted, and if a shepherd on Mont
Chemin, on hearing the roar of the torrent, had not set fire
to his beacon, hundreds would have perished. In returning
home I stopped to eat some whortleberries, which grow on
the side of these mountains of greater height and perfection
than I ever saw elsewhere. In this patois they call them
Loutres, and use them for a dye.

My guide pointed out to me a dead snake like an adder.
Some beetles were feeding on it, which I was amused to
find were the *Silpha vespillo* which I have often observed on
dead snakes in England. The abundance of grasshoppers
and their different notes is really to a stranger a characteristic part of Switzerland's rural curiosities. Some green
ones are of an enormous size, and, like Anacreon's, sing
chiefly in trees. There is a grey one with scarlet wings

(the colour of the under-wings of *Phalœna Caja*) which makes
a curious cackling noise as it flies.

Most of these Grylli, I observed, make their noise by
pressing the thigh against their side, but some keep their
thighs immovable, and make it with the upper part of their
wings and wing-cases. Neither of these organs would make
a very poetical figure in Anacreon's elegant ode, as instruments of sound, however much Aristophanes might have
enjoyed paying this animal, which the Athenian loved so
much, the compliment of having 'a good thigh for singing.'

The bridge of Lourtier, which I had crossed by one plank
in the morning, I found nearly finished. Here I mounted
my mule, and rode to near Lourtier. The river we saw had
increased much by the heat of the day, which had melted
much snow. Farther on, we found the road I had passed
in the morning was washed away by the swelling torrent.
My guide was obliged to ascend the mountains with my
mule, and make a detour to Lourtier. I scrambled on, and
joined him there. Here, to our further disappointment, we
found the bridge we had before crossed in the act of being
swept away. The Dranse had brought down rocks which
had raised its bed twelve feet or more, just above the bridge.
Over these a cascade was pouring, which fell with fury upon
the bridge, and had already broke much of it up.

A herd of goats belonging to the different villagers,
which I had seen driven over this in the morning, were thus
prevented from returning, and many poor wretches went
in consequence to bed without their supper, for which they
depend solely on their goats. Some men were dismissed to
the mountains, to milk them, with a few pails only. The
greatest part they said they should be obliged to waste on
the rocks, rather than let the goats go two days without
being milked, which is very injurious to them.

Since it was now impossible to advance farther, I got a
lodging in a miserable châlet belonging to the Frères Mi-
chaud, who have lost greatly by the late deluge. They were
before reckoned wealthy.

August 11.—I left my hard bed of straw very early, and
proceeded to the President Gard's, at five miles distance. We

had to scramble over a difficult mountain path, for the main road on the opposite side of the river was closed from us by the top of the bridge at Lourtier. Having breakfasted here, I went on to Sembranchier. At the hamlet of Matignet, my guide pointed out the place where a barn had been carried away. The widow who had built it, finished it that spring on June 11, and had an inscription carved on one of the boards: ' The widow Anne M. Fillet built this in spite of the lake of Malvoisin.' Five days after came the flood, and this very board with the inscription was picked up at Martigny, nine miles off, and is, I understand, still shown at one of the inns.

In the valley of Sembranchier I was able to ride my mule much pleasanter. There is nothing a Swiss peasant points out with such exultation as a field of corn, nothing is to their eyes so picturesque. The scenery, in fact, of this valley is much hurt by the extent of cultivation, and is in every way inferior to that of the valley of Bagnes. At Liddes I met my mother and sisters returning from St. Pierre. They had not been able to accomplish their visit to St. Bernard, from their fatigue, and at St. Pierre I met my father returning from St. Bernard. When about half an hour's walk from the convent I was overtaken by a most violent hail storm, which drenched me completely and chilled me. I was very glad to accept the offer of some clothes from the monks, who came out most courteously to invite me in. I found three English gentlemen there, and a party of Piedmontese gentlemen and ladies. A very handsome supper was served up at seven o'clock, and the monks made all feel themselves heartily welcome. The worst part of this magnificent institution appears to be the unwholesomeness of the situation. From its elevation the snow rests on it so long, and the walls are made so damp by the meltings in spring, that they are all subject to rheumatism, and it is only the youngest who can remain there at all. They have another convent at Martigny, which receives all the older monks for the remainder of their lives. Those, however, at the Great St. Bernard, seem to be a happy active set, and acquire, of course, a great deal of information, from conversing with travellers from all parts of Europe.

In the time of the Romans there stood a temple of Jupiter where the convent now is, and a number of antiquities have been very recently dug up there. It is the opinion of the monks that Hannibal sent part of his army over here, for (supposing him to have crossed the Rhone at Lyons) he would have fallen in with a Roman station at Vevay, and would from thence of course have been able to learn this communication with Italy. The convent is very far from a handsome pile, but the scenery round it is wild and savage, and reminded me much of the Grimsel, for there is a lake here also. The St. Bernard dogs, so much celebrated for saving people in the snow, are fine handsome animals.

August 12.—Having taken leave of the monks early, I set off in a thick mist, which prevented our seeing anything, in company with the three Englishmen, whom I found very gentlemanlike men. A rock of mica schist which overhangs Sembranchier is decomposing, and peeling in so alarming a manner, that they fear, when the rain follows this excessive heat, that the town will be demolished by an *éboulement* similar to that of the Rossberg. The guides assured us there were enormous cracks traversing parts of the mountain, which had only appeared within a few weeks. We returned to Martigny.

August 13.—From Martigny we proceeded up Le Vallais, in the scenery of which I was much disappointed. We passed Sion; the castle here on a hill is a striking object; when we came opposite to Leuk, I left the carriage, which proceeded on to Turtman with the rest of the party. I had about nine miles to walk from Leuk to the baths where I slept, with the intention of rising very early next morning, and climbing the Gemmi. My walk was in dead silence, for the guide I procured at Leuk, to carry my *sac de nuit*, could speak nothing but German. I found it, however, by no means tedious, for this small valley is beautifully wooded with fir, and the moonlight was exceedingly bright. The Gemmi, from which I expected to enjoy a glorious view of all the highest Alps, was covered with clouds, and rain had been falling on it. I determined, therefore, to give up the expense and trouble of this arduous enterprise, and content myself with seeing the baths. The company at these was not particularly numerous at this season, but it is a curious sight in the large room. I visited some of the smaller hot springs, and a small waterfall about half an hour's walk higher up than the village. On my return I found my father, who had rode up there from Turtman, to which place we returned together, and joined the ladies, and went on in the carriage to Brieg. We passed at Visp the opening of a valley, down which comes an immense river from Monte Rosa and meets the Rhone. The barberry-bush, which grows abundantly wild here, is a beautiful shrub.

August 15.—The first view we had of Brieg was singular, and rather picturesque, from the multitude of towers. A Jesuits' college is just re-establishing itself here. From Brieg we began to mount the Simplon, and were obliged to take eight horses throughout the ascent, which is, however, quite unnecessary. In mounting one enjoys extensive views back into Le Vallais, but nothing very grand for Switzerland. The descent, on the contrary, on the Italian side, was something new, and very magnificent. It is one of those valleys or deep ravines so common on the sides of smaller mountains in Scotland, but magnified to a size which makes it worthy of Monte Rosa. This road is a wonderful work of Napoleon's, and shows the skill of the French engineers. The labour of the galleries, which are blown with gunpowder through granite, must have been immense. This road now belongs to the government of Le Vallais, and must be valuable to them, since so many of their people are thus kept in employ, and paid by the numerous travellers, chiefly English. We paid, I think, 26 shillings at the barrier. The road cannot be in better order. We left Le Vallais at Isella, and entered Piedmont, and went on to Domo d'Ossola. The difference of the Italian side of the Alps is as remarkable as that of the southern side of a garden wall. Houses and fruit-trees reach to an extraordinary height up the mountains, and the crops of tall maize and hemp make a great show in the valley. Domo d'Ossola is prettily situated. Here you take leave of the mountains, which are by no means finished by the descent from the Simplon. We never saw the very summit of Monte Rosa, because of the clouds, but from what we did, I should conceive it not to be a grand mountain for its enormous height.

August 16.—We were fortunate in entering Italy when the great heat had ceased, which we learned had been excessive everywhere. We embarked on the Lago Maggiore at Feriola, and were rowed to Arona, having sent round the carriage, visited the Borromean Islands, which are so celebrated, that all are disappointed with them. They would, however, be pretty enough, if the buildings on them were not so frightful. The large ugly palace on Isola Bella is so out of proportion, that the little islet looks as if it would sink to the bottom with its weight. In the interior, however, it is a handsome nobleman's place. In the gardens here we saw a fine American aloe in flower, and two bays of extraordinary size. They both spring from the same root, are 105 feet in height, being nearly equal each, and 10 feet in circumference each, measured 5 feet from the ground. Arona we found a delightful little place. Near this we saw at a distance the colossal statue of San Carlo Borromeo. The Italian lakes are much more cheerful than those of Switzerland, from the quantity of houses, wood, and boats, which latter are sadly deficient on the Swiss lakes.

August 17.—Went on to Sesto Calende, which stands at the end of the Lago Maggiore. Here the carriage crossed the Ticino, in a ferry boat, and we entered Lombardy. From thence to Milan is a richly cultivated country. The Indian corn, of which the quantity is extraordinary, stands from ten to twelve feet high. The hemp is very tall, and the vines, which are festooned on low trees (chiefly mulberries I believe), form elegant partitions to the fields. The flat on which Milan stands is so dead, that you see scarcely anything of the town till you are in it. The avenues round it, made by the French, are planted chiefly with the *Bignonia catalpa*, a singular and elegant tree. The lines of pavement in the streets of Milan, for the wheels of carriages to run on, is a refinement which was new to us, and deserves imitation. It must save carriages amazingly, for they glide along with delightful ease and swiftness, and make no noise. We went to the Hotel d'Italie.

August 19.—Went to the Brera, once a convent, now

converted into a picture gallery. A valuable painting of Raphael in his first style, and of Perugino his master. Attached to this is a small botanic garden, which has some good things in it. We then went to the amphitheatre which Buonaparte built, capable of containing 50,000 spectators, in which was exhibited chariot races, &c., and water could be introduced for the Naumachia. The Austrian soldiers now use it to learn to swim in.

The famous fresco of Leonardo da Vinci, ' The Last Supper,' is still fine, much as it has suffered. The French made a stable of the room of the old convent which contains it, and the damp would shortly have destroyed it, had not Beauharnais heard of it. The triumphal arch of the Simplon is the finest work of Napoleon's I have seen. It is of beautiful white marble, and the sculpture and designs of the alto-relievos are fine. It stands, unfortunately, unfinished, though, had it not been so, the Austrians might perhaps have pulled it to the ground, since some fine friezes which were prepared related to Buonaparte's victories over them. The Emperor of Austria can feel no inclination to go on with this trophy, and the expense would be great, since what is prepared it would be unnatural and unpolite to erect.

There is in a street of Milan a fine ancient colonnade of fluted pillars, said to have belonged to a palace of Nero's. There is another triumphal arch of Buonaparte's to commemorate the battle of Marengo, which is not remarkable. The Austrians have erased his inscription, and in its stead have inscribed ' Paci Populorum Sospitæ.'

August 20.—We drove about the town of Milan over the delightful pavement of these streets. There are no very fine squares in the town, except the opening opposite the cathedral. The white marble loses much of its effect, from the darkness of the bottom part, which has been long built, and beautiful as each of the numberless sculptured pinnacles are in themselves, the whole building is not grand. When you stand on the roof, the *coup d'œil* of the ranks of pinnacles, each bearing a statue, is very beautiful. From the top of the tower we had a fine view of the town, and the vast plain of Lombardy bounded by the Alps and the Apennines. The view of the mountains was not clear.

We then went to the Mint. The merchants have just struck off many thousand gold coins with Napoleon's head, dated 1814. The Ambrosian Library, which we visited, contains the Virgil of Petrarch, which it is said he wasted his time in writing himself.

We went to the Opera in the evening, which is a much larger theatre than that in London, but much inferior in beauty. The dancing I thought superior to the music.

August 21.—We paid a visit to the Observatory; the air was clear, and Monte Rosa looked grand from his enormous height, though his outline was not good. It is a very near resemblance to the Blumlis Alp, as seen from Thun.

I was interested in seeing through a telescope one of the spots on the sun. In the evening we went on to Brescia, through a richly cultivated plain, where irrigation was carried to beautiful perfection. Streams often crossed the road, which passed at right angles over others by small aqueducts.

August 22.—Brescia is an immense place. We visited several churches, in which are some pictures by the old masters; Titian, and others.

We afterwards went on to Verona. The vines are beautifully festooned on the trees, and the grapes purple. The amphitheatre here was the grandest Roman antiquity I had ever seen. The marble of which the innumerable tiers of steps are built is uninjured. Fortunately, one small portion of the outer *façade* still stands to point out its ancient height and magnificence. This enormous building serves now for a stage to mountebanks and conjurors, who when we entered were amusing a crowd, which seemed actually lost in one of the corners of this immense oval. Such sports may at first sight appear a degradation to this glorious relic, but what were those which once drew a Roman audience there?

I must not forget that between Brescia and Verona we passed the beautiful lake of Garda, Virgil's Benacus, and the fortress of Peschiera. We only skirted the lake, but could we have gone up it, the scenery promised to be beautiful. It is not much inferior in size to any lake in Italy. There is a large Roman arch of triumph by which we entered Verona. The antiquities are the principal magnificence of this large place. The Cathedral has nothing to recommend it but an Assumption of the Virgin by Titian. From Verona we went on through a country resembling a rich garden. The hedges even by the road-side sometimes festooned with vines, and besides the Indian corn, another enormous grass of similar growth and height, and even more handsome, is much grown. It yields a grain which they give to animals, and make brushes of the heads. They call it Sciorgo rosso (*Holcus Sorghum*[3]). We stopped some hours at Vicenza to admire the beautiful architecture displayed in the palaces and public buildings built by Palladio. Vicenza was his birthplace, and we were shown the house which he once inhabited. He has raised here to his memory a most interesting classic treasure: a theatre, taken from plans of Vitruvius, which may be considered as an exact model of what the ancients used for their theatrical exhibitions. Though a small model of an ancient theatre, it would be considered a good-sized modern one, and strikes one immediately as extremely elegant and chaste in proportion and decorations.

In some of the churches here are pictures by masters of the Venetian School : Paul Veronese, Bassano, and Bellini. We afterwards went on to Padua. Passed some fields of rice, which before I had never heard was grown in Italy. It grew dark before we reached Padua, and I was delighted at seeing several (*Fulgoreæ*,) fire-flies, resting on the willows by the side of our road ; some were on the wing. Their light almost rivalled that of our glow-worm, but only appears at intervals as they fly, the wings I suppose covering it each time they strike.

August 24.—At Padua we walked to the Hall of Justice, an enormous but not handsome room. An inscription here reminded us that the great historian Livy was born here (Patavinæ). Here we met again Sir J. Wrottesley and his sons, who were going on to Venice. They told us that the day before, a courier on his way from Milan to Venice with 40,000 francs had been robbed near Brescia. A soldier who accompanied him as a guard was shot dead by the robbers, and he himself was nearly beaten to death. From Padua

[3] Millet.

we travelled on to the sea-side and embarked at Fucina for
Venice, in a boat which makes part of the regular post.
The distant view we got from the lagunes, of Venice, made
our hopes of its magnificence fall considerably, as did the
entrance by a narrow street, but they were soon raised again
to their full, on entering the Grand Canal. We were rowed
up through crowds of gondolas to the Hôtel Grand
Bretagne.

We arrived in good time at Venice, and went out the
same evening in a gondola to St. Mark's Piazza. This beautiful square is not to my mind much ornamented by the
great square tower. Being of red brick and with a green
cap, and with a line of small apertures running unequally
down one side, the effect of it is not in harmony with the
beautiful regularity of the rest, however well in some pictures this is concealed and softened down by alterations of
the colours. The form of St. Mark's, which is a copy of
that of St. Sophia's at Constantinople, is very cheerful and
gay, consisting of one large dome surrounded by four smaller
ones. This closes beautifully one side of the piazza, and
opposite to it is a new palace which Buonaparte built, having
destroyed, to make room for it, a church and buildings which
were before a more picturesque finish to that end of the
square. This is a change much to be lamented.

In the bronze horses of Lysippus which stand in front of
the great dome of St. Mark's, being raised not too high to
have a fine effect from below, I confess I was somewhat disappointed. The heads are fine certainly. It is the fashion
now to complain that these four steeds are thrown away,
from being placed where they are not to be seen from their
height, but travellers have only to take the small trouble of
ascending by an easy staircase, and they may see them near.
Le Place Petit St. Mark is also exceedingly beautiful, and
gay from the crowds of gondolas, and the fine width of the
Jews' Canal, and the churches and buildings which come in
view there. La Douane, St. Mark, and St. Giorgio were
conspicuous. There was a true Italian glow in the sky,
which gave a colour to the buildings highly favourable. But
if this scene is now so fine, what must it have been when
Venice could boast the empire of the sea, and saw at this

spot vessels from every country of Europe pour in commerce.

August 25.—From our windows we had a delightful view of the Rialto and the Great Canal. Early in the morning it was very beautiful to see the boat-loads of fruit, particularly of the most immense pumpkins, each single one almost too heavy to be lifted. The gondoliers stand universally as they row, as is the custom on the Italian lakes and some of the Swiss. This practice gives astonishing life to the scene, and the motion is remarkably elegant.

We went out in good time, and visited many of their splendid churches. In the Chiesa dei Gesuiti (Santa Maria del Rosario) the altar is ornamented with small pillars of lapis lazuli of great value, built by Marsari. The architecture is very chaste and elegant both within and without. We next saw the Chiesa del Redentore; then to the large church of Santa Maria della Salute. In this, three very fine specimens of Giordano's are to be seen, three large pictures representing the Nativity, Presentation, and Assumption of the Virgin. We then saw the church of San Giorgio Maggiore, and from thence to St. Mark's Piazza. Here we entered the Doge's Palace.

The room of the Grand Council must fill every one with vast ideas of the ancient power and magnificence of this Republic. Of great size, yet elegant proportion, the ceiling rich with gold and innumerable paintings by Giulio Romano. Statues down the middle of the apartment, and at the sides large and grand works of the two Bassanos. At the end where you enter is an enormous Paradise of Tintoret, containing a truly innumerable quantity of figures, yet so arranged as to produce a good effect. This is the largest painting in oils, they told us, in Italy. We entered afterwards St. Mark's Church; the whole ceiling there, of the five domes, is in mosaic, and executed after the designs of some of the first masters of the Venetian school. The pavement, which is done something in the same way, has large hollows in it in some parts, and is very uneven, the piles on which its whole weight is supported having given way here and there. We went up, after admiring the riches of the altar, to the gallery, where we saw close the horses of

Lysippus, and had a view down into the square. In this church are several trophies brought both from Jerusalem and Constantinople after the Crusades; pillars, statues, &c. In the evening went out to see the most wealthy of all the churches we had yet seen: Santa Theresa, built by eight noble Venetian families; there are in it six most magnificent altars, rich with rare marbles, precious stones, &c. &c.

On our way to this church we passed under the Rialto; it is an exceedingly fine arch, but the shops upon it certainly injure the effect much. All the fine things of Venice lie very commodiously near one another, at least the rapidity of the gondolas makes them appear so. It was growing late, but we determined on visiting the gardens which the French made; the only thing of the kind in Venice. They pulled down a church and several convents and other buildings, the ruins of which raised the soil higher than almost any other point in Venice; on this they planted avenues of trees, and made handsome walks, which are laid out certainly with taste, and in a good style, and look already very handsome.

In our way to this we passed Lord Byron's gondola, who was returning from the Lido, a small narrow strip of land which is near to the continent, and faces Venice. It is an island, and the only place near Venice where a ride can be obtained. On this Lord Byron has built stables, and keeps a stud of six horses. After breakfasting at ten o'clock, he is rowed in his gondola regularly almost every day to this island, where he rides six miles out and six back again. He then returns—is generally alone in these excursions. His residence was in one of the old palaces not far from our hotel.

August 26.—We visited a most superb room built by a fraternity of merchants in Venice, in which all the pictures were by Tintoret, many exceedingly beautiful. The effect of this hall is very striking, and impresses one with the riches which commerce can give the smallest state, and at the same time with considerable admiration, that the taste of men employed so much in business should have risen so much beyond what is usual in such persons.

After returning to our hotel and dining, we again took a gondola, and went to see 'The Martyrdom of St. Peter,'

Titian's *chef d'œuvre* [4] in the eyes of artists, at the church of St. Jean et Paul. As the evening was delightfully mild, we made them row us all round the large naval arsenal, from which once were launched such powerful fleets, and round the new garden, thus making a complete circuit of that side of Venice.

We spent almost the whole morning in the magnificent gallery of pictures belonging to Count Manfrini, in the Manfrini Palace. This it seems is the most splendid collection in the possession of any individual at this time in Italy, made by the late Count, since the French Revolution; a man who rose by farming tobacco at an immense extent, from beggarism to enormous affluence. He purchased a title, which is easily procured by money at Venice, and had the taste to select a most choice set of specimens of almost all the great masters. To study these would have required weeks of course, but the things which made the chief impression upon me during the few hours we were there, were a Queen of Cyprus by Titian, a very beautiful portrait, and a portrait of a man by Rembrandt, the most astonishing production I ever saw of his.

After spending the morning here we left Venice, and after crossing the water, went on to Padua. In this town we had before left much unseen. Everyone should visit the church of St. Antoine here. It is most extraordinary that of the sculptures in relievo in the chapel of the patron saint, in which the different miracles are represented, no less than four were *raising from the dead*; so that in this, the farthest stretch of miraculous power in suspending the laws of nature, they believe St. Anthony to have done more than is recorded of Christ himself.

The chapel of this saint is full of votive pictures which attest miracles performed since his death. While we were looking at one of the offerings, our cicerone, a man who belonged to the church, said, 'This, sir, is a miracle I saw with my own eyes; the child had a deformed leg, and upon the mother praying earnestly at the shrine of St. Anthony, it was restored to her whole as the other.'

From the manner in which the man related this, it ap-

4 Since destroyed by fire.

peared to me that he believed what he was saying, to such an extent does their superstition extend! The greater part however of the votive tablets relate not to miracles, but providential escapes from imminent dangers. As all these are most religiously attributed to the guardian saint, it is easy to perceive how perpetually his influence is kept alive. But everyone who has looked over in a cursory manner the collections of votive tablets which in almost every church he meets with, must be struck at the vast proportion which *being run over by carriages* bears among the accidents. The careless and extremely swift pace which is so much the fashion here in driving through towns, and the great number of carriages, easily account for this.

The Botanic Garden here is worth visiting, as indeed must every Botanic Garden be in Italy, where our greenhouse and hot-house plants flourish as if in their native climate, though in winter here they are obliged to protect them most carefully from the severe cold. A fine sugarcane here was interesting, and an immense fan-palm tree, which the gardener informed me, to my great surprise, is wild in Spain; also fine Banyan trees, *Laurus camphora,* &c.

August 28.—From Padua we proceeded on to Ferrara, through a vast plain. The castle and fortifications of Monselice make a show in this flat country, standing on a small rising. Near Rovigo we crossed the Adige in a ferry boat, by which the carriage also was transported. This river, as indeed are most of those both small and great in this plain, is of a most formidable height above the plain. The deposition of earth which they bring from the mountains raising their beds continually, and the inhabitants being obliged to bank them in strongly for fear of a deluge.

In the plain of Lombardy there is a remarkable scarcity of houses between the great towns; indeed one never sees a building, and wonders where the farmer's necessary offices are in so richly cultivated a country. In return for this, towns of immense size are met with at every step. Ferrara is one of them, and of 22,000(?) inhabitants, yet covers more ground even than is necessary for this, from the porticoes on each side of nearly all the streets. We visited the Cathedral here, and afterwards the Library, in which is the tomb of Ariosto, who was a native of this town. The inscription consists of some common-place complimentary Latin verses.

From Ferrara to Bologna is a continuation of the same dead flat which we had been on ever since we entered Italy, but here we in a great measure lost its amazing fertility. We were delighted while dining at the inn with some street music, which was a great treat. The band consisted of seven blind men, who had all studied in the South of Italy, though they were originally from Bologna.   There were violins, violoncello, clarionet, &c. &c., and also a respectable female singer, wife to one of them, who relieved the instrumental performance now and then. It is their custom to play always on the arrival of any English at dinner every day.

Some of the finest works of the great masters of the school of Bologna have been lately restored to their native town from the Louvre. They do not mean to attempt to replace them in their former stations, but are putting them up in a large public room, in which I think they will stand certainly a better chance of being seen well than in most churches, where huge candles, &c. are posted before them. But we were in luck to arrive there before this new arrangement was quite finished, since most of these immense works were on the ground, resting against the wall, by which means we saw them at whatever distance we chose ; whereas I fear some few of them will by-and-by be lifted up much too high. In a grand picture of Domenichino's, ' The Massacre of the Christians by the Albigenses,' some of the female heads delighted me exceedingly, but the painter, to preserve their beauty, has certainly suppressed all the horror which they ought to be feeling.

August 30.—We hired a calèche and drove about the town to the different palaces. At Casa Ratta was the famous Sybil of Domenichino's, a spirited head. This is to be sold, as are many of the fine things in private collections. Some have been already bought at great prices by the English, and have left this town. There is great appearance of poverty in Casa Ratta.

August 31.—We left Bologna and began to mount the Apennines. The view back upon the town was not

so fine as one we had enjoyed the evening before from a
small rising on which stands the prison. It must however
always be a fine object, and the hanging (or leaning) towers
a most singular feature. As soon as we began to ascend,
they harnessed on four oxen before our four horses, which
was to us a perfectly new style. We crossed a river soon
after Bologna, which is very destructive sometimes, but in a
valley which we looked down into on our right hand during
most of the ascent, we saw the Reno, the most desolating
stream I ever saw, though the waste of land here is I have
no doubt owing to the crumbling and decomposing nature
of the sandstone of which the mountains here are composed,
and from which accident they give a very peculiar and
extraordinary character to the scenery on this side of the
Apennines. For the superstratum of the soil has been completely washed away, and no vegetation remains excepting
the trees, which are abundant; between these is seen the
white bare sand, and here and there a larger space perfectly
white, so that the surface as far as the horizon both ways is
chequered with green and white.

After mounting some miles, we had a very extensive
view back on the vast plain we had left, which seemed as
level as the ocean, and indeed is very nearly so. They say
moreover that from hence on a clear day one can discern
the sea in the distance, but this I should think without a
telescope must be always impossible. We saw on all sides
very extensive forests of sweet chestnuts, which from its
foliage, and the contrast in the colour of the fruit, which
was in profusion, was a very picturesque tree.

As soon as we left the sandstone, and came upon the
limestone hills, the curious character of the scenery ceased.
We entered Tuscany; the officers of the Douane, quieted by
a fee, let us pass unmolested, as they will everywhere in
Italy. We proceeded on to Coviliajo, a small town where
we slept, having passed a volcanic fire which proceeds from
a mountain a few miles from here, intending to visit it at
leisure on our return over these mountains.

September 1.—The rocks about Coviliajo are principally
basaltic, which agrees well with the volcanic fire in this
neighbourhood; indeed the steep whin cliffs which here and

there burst up strike the eye immediately as not belonging
to the tame limestone hills which form the chief character
of the chain.

Throughout the first part of the day the scenery was
wild and picturesque ; on descending the Florence side, it
became rich and fertile : the hills covered with the sweet
chestnut, vines, and quantities of olives ; the gardens full of
standard peaches, &c. ; the grounds about the numerous
country-seats abounding in the cypress and the *Pinus pinea*.
All these, with a clear sky glowing with heat, presented for
the first time to our eyes a characteristic Italian landscape.

The entrance to Florence is fine, from the public walks
and triumphal arch. We drove to Sneider's Hotel, which is
famed as being the first in Europe. Its situation and magnificence, joined to the real comfort in it, entitle it to its reputation. It is nearly monopolised by English families all the
year.

September 2.—The attacks made upon the rich purses of
the English, of which foreigners have certainly formed vast
ideas, are almost more numerous at Florence than at Paris.
The first thing in the morning, at breakfast, arrived a letter
celebrating both in verse and prose, both in English, Italian,
and French, 'the auspicious arrival of Sir Lyell and his
illustrious family.' After celebrating the wealth and naval
power of England, the poet hoped ' to share in the fruits of
the unbounded generosity of the English.' All the production was execrably bad, and we were sorry to find, when the
author called in person, that he was too old to be amusing
as a fool. We made him however recite to us the verses, of
the merit of which he entertained evidently very different
opinions from us.

After this son of Apollo, came a statuary with alabaster
figures ; then a sort of gentleman-beggar with elegant
bouquets of flowers, with which he begged to present the
ladies ; then a print-seller, which finished the morning's
assault, at least as far as we know, for we drove out to the
Pitti Palace, belonging to the Grand Duke of Tuscany. Here
there is a magnificent gallery of pictures. The first that
arrested our attention, were two large landscapes by Rubens,
of great beauty. The first they called ' Going to Market,'

the second ' a View in some Part of Spain,' in which the
bright gleams of sunshine and dark shade, so often seen in
nature, when in the day-time there is a high wind and
many clouds, was expressed very finely. There was a picture
of Michael Angelo, 'The three Fates,' which I did not think
anything surprising. The celebrated ' Vierge à la Chaise '
of Raphael has been newly framed, and a glass put over it.
We stopped some time to admire two other grand productions
of Raphael, the portrait of Pope Julius II., and St. Catharine
and Holy Family. The suite of rooms in this Palace, now
the residence of the present Grand Duke, brother to the
Emperor of Austria, is very princely, and in high preservation.

After satiating our eyes with pictures, we went to the
Great Gallery of Medicis, where we might study some of the
first statues of antiquity. In the Salle de Niobe, to begin
with herself, I thought the grief in her countenance, though
not perhaps equal to what it would be in nature, yet greater
than I ever saw in any statue or picture.

As a group, the effect is much lessened I thought by
Niobe's being colossal and her daughter of the natural size.
The statues of her sons and daughters which were round the
room are some inferior, but one or two of them exquisite.

From this room we went into the Tribune, an apartment
where all the most precious morsels both of painting and
sculpture are collected together. Here stands once more
the celebrated Venus de Medici, and it equalled all my expectations, which I am well aware it does to few. It appears
to me that in statuary those who know nothing of the art
are much better able to enter fully into the merits of the
first works in it than they can in painting.

[Here the journal ceases, for the note-book containing the
remainder is supposed to be one which was lost at Florence.]

CHAPTER V.

[In 1819 Mr. Lyell took the degree of B.A., obtaining a second
class in classical honours. He became M.A. in 1821. In March
1819 he became a Fellow of the Geological Society of London, and
also of the Linnæan Society.

After leaving Oxford he was entered at Lincoln's Inn, and resided
in London, and studied law in a special pleader's office (Pattison's). His
eyes, however, became very weak, and he was recommended to desist
from reading for a time, and to join his father in a tour abroad as far
as Rome in the autumn of 1820 ; he had then no time to devote to
geology, as they were busy seeing the towns, pictures, and natural
scenery.

In 1822 he made an excursion to Winchelsea for the purpose of
inspecting Romney Marsh, an extensive tract of land, from which the
sea has retired within comparatively recent times.]

CORRESPONDENCE.

To His Sister.

Rome : September 28, 1820.

Dear Marianne,— You will expect me to be very full in
writing to you from this city. I must tell you, however,
that there is only one thing which has gone very far beyond
my expectations, and that is the statues of the Vatican. I
had no idea that half so many exquisite productions of
Greece and Rome were still in existence. Mathews, in the

' Diary of an Invalid,' says that you must not look for beauty
in the remains of antiquity in Italy, for none of the ruins
here are so picturesque as those of our own Gothic abbeys in
England. This is most true; but he goes on to say, it is
from the great names they are connected with that they
become more interesting than ours. Here I must differ from
him, for if they contain so little now to admire in them, how
can they assist our imagination? I always had a grand idea
when reading the classics of what the Roman Forum and
Capitol once were, but now I shall always find great difficulty after seeing what it now is to raise again half so good
a picture in my mind. The Forum now is about half the
size of Grosvenor Square, the surface uneven from large
mounds of earth thrown up from excavations, and looking
like large dung-middens overgrown with weeds. Down the
centre is a double row of half-starved, shabby elm trees;
round it the celebrated arches of Septimus, Titus, and Constantine, each standing more than knee-deep in a large pit!
for the soil has risen everywhere at Rome in an almost un-
accountable manner, sometimes 40 feet from the ruins of
buildings shaken down by earthquakes, and, before an ancient
arch or pillar can be shown, a gulf must be made round it.
At almost every ancient relic I have had to say ' Amen ' to
Forsyth's exclamation, ' Those lying engravers! ' The
Colosseum, however, from its immensity defies exaggeration.
As for the seven hills of Rome they have been swallowed up
by earthquakes I suppose, for Ludgate Hill would match any
of them that remain. The Romans used brick for their
finest buildings, and then covered walls 20 or 30 feet thick
with a facing of stones. These stones have been stolen for
the palaces of modern Rome, and huge shapeless mountains
of brick and mortar remain everywhere with high-sounding
names, to give the lie to Piranesi's engravings, and to make
the poets and historians of ancient Rome appear almost
ridiculous. Yet, after all, there is still enough among all
this wreck, when it is carefully considered, to prove what
Rome once was, and how she surpassed all that now is in
architectural magnificence. The Belvedere Apollo, Michael
Angelo's Moses, Canova's studio, and the Dying Gladiator,
have more than equalled my expectations. We are to leave

this in a day or two, for the seeing Naples is out of the question. I should not go there at present if here alone. For when the late disturbances took place at Palermo, all the English were informed that they were prisoners. None were molested, and all are now free, but God knows how soon a fracas at Naples might end, and it would be most tantalising to be shut up in a lodging, and not even allowed to witness the row. They are making very free with their favourite weapon the stiletto, and six or seven leading men are got rid of every night by those of the four different parties. General Church would be torn to pieces but for a guard of soldiers, and a threat of our ambassador's, that if a hair of his head is touched England will avenge it. The English have fled nearly to a man. Every peasant has become a soldier to support their *political opinions* in Parliament, which meets on October 1. Add to this 400,000 Austrians are expected there, whom they hate, and who surely can have nothing to do there, even if invited by one party. These Germans must in their way pass through the Pope's territories, where they are so much detested that they will not, I expect, add much to their tranquillity. All may blow over, but were we to wait till we saw the event, the Alps would become impassable first, and it is my father's intention to recross them before we shall have any chance of rivalling Hannibal. Mount Leonessa, the highest Apennine seen from here, was completely covered with snow on Saturday last, when we took a grand view of ancient and modern Rome from the tower of the Capitol. This snow has now melted, but it proves how early winter may come in the higher regions.

My love to all. Your affectionate brother,

CHARLES LYELL.

To GIDEON MANTELL, ESQ.[1]

29 Norfolk Street, Strand : February 8, 1822.

My dear Sir,—Though I was detained late in the country, I found you had not yet arrived with your book, but I still thought I should see you at our Annual Geological dinner.

[1] Dr. Mantell, F.R.S., author of *Medals of Creation, Wonders of Geology*, &c.

I now begin to despair, or rather to hope that your professional practice has increased so much beyond your expectation, that you find it impossible to leave Lewes. Of the collection which came quite safe in ten days, I can only say that I trust among so many valuable and interesting specimens you have not robbed your own of any which it might have been more advantageous for science that you had retained. The Professors of Cambridge and Oxford were present at our dinner, and Buckland was called upon to explain the vast quantities of bones which he found in the summer, in a cave at Kirkdale in Yorkshire, of which he had a large bagful with him : innumerable jaws of hyænas, teeth of elephant, rhinoceros, &c., unmineralised like those in the limestone caves in Germany full of bears. He produced some light balls or pellets, which he said he brought to town at first doubting what they could be. Dr. Wollaston[2] (I think) first pronounced they were like some calculi sometimes found in some species of Canis. Upon being taken to Exeter Change by Dr. Fitton,[3] the man there recognised the production, and exclaimed, ' Ah, that is the dung of our hyæna ! ' On analysing it they find it composed of carbonate and phosphate of lime, the same as hyæna's dung, which being an animal, it seems, of an ossiphagous appetite, has always its dung proportionately more ossified than any other. Buckland, in his usual style, enlarged on the marvel with such a strange mixture of the humorous and the serious, that we could none of us discern how far he believed himself what he said.

The researches I made at Christmas were, I am afraid, sufficient to prove that I must give up my hopes of discovering the Isle of Wight fresh-water formations in our part of Hants, but the shells are not yet come to town which will enable Webster to decide.

Believe me most truly yours,

CHARLES LYELL.

[2] Dr. William Hyde Wollaston, natural philosopher, b. 1766, d. 1828.
[3] Dr. W. H. Fitton, b. 1780, d. 1861, an early member of the Geological Society, contributed many papers, and clearly defined the position and character of the Green-sands as separated by the Gault.

To GIDEON MANTELL, ESQ.

29 Norfolk Street : April 11, 1822.

I fear you may think I have but little motive for troubling you with a letter, since it is no other than to inform
you that I am to-day setting out for Winchelsea, where I
shall pass nearly a week with a friend, who will leave me
ample time for geologising if 1 can find a field for such
operations, and for myself I have little doubt of this, since
the country is so entirely *terra-incognita* to me ; but can you
not also make me useful to you in ascertaining any fact or
procuring any specimens ?   In several short expeditions
which I have made into Surrey 1 have examined very carefully the junction of the chalk with the beds below, about
Reigate, Dorking, and Guildford.   I have brought specimens
from what Webster would have us call the ferruginous sand,
exactly like some brought from the green-sand of other parts,
which certainly creates a great difficulty, but still there
seem to me only partial beds in that which we should
decidedly call iron sand in any other place, and I still think
that if we were determined to name beds by analogy to
those of which Webster first drew the line in the Isle of
Wight, we should be obliged to pronounce that *the green-
sand* is wholly wanting at Guildford, and Greenough seems
clearly to entertain great doubts on this head himself, but
when I have worked more eastwards in Kent I shall be
better able to form an opinion.

Since you were here, there came up from Stonesfield to
the Geological Society a most enormous bone of some great
unknown animal, and Clift pronounced it new, but finds it
belongs to the same animal as one which some time since
was sent from the neighbourhood of *Cuckfield* !   Would it
not form a strange addition to the wonderful coincidence
already discovered by you, between your beds and those of
Stonesfield, if an immense new animal should at the same
time be found in each ?   This animal, they say, must be as
large as an elephant, but I cannot learn of what kind they
conjecture.

Yours very truly,

CHARLES LYELL.

To His Father.

Winchelsea : April 16, 1822.

My dear Father,—I had always intended to appropriate
the first rainy day to the transmitting to Bartley a narrative
of my adventures in these distant parts, but the weather has
proved so fortunate that this is the first opportunity which
has occurred. My journey through Kent by Sevenoaks,
Tunbridge, and Rye was so stormy that I saw little of the
scenery, but I should think the view from Madam's Court
Hill near Sevenoaks must surpass all the rest. This is a
most picturesque and delightful spot, and the sea-beach and
the marshes, the cliffs and fine old ruins, have entertained
me so much, that I could stay a much longer time here
alone, and not feel a moment's *ennui*. Among other things
I have found the Hastings and Battle limestone rock, with
its peculiar fossils in the cliff of that isolated hill on which
this borough stands, which I believe is the most eastern-
most point in which it has been seen, and I hope will settle
Mantell's doubts about its position, as I have here found in
the perpendicular cliff the bed of sandstone which is above
and below it. The destructive incursions of the sea at
different periods all along the coast afford very copious and
curious matter of observation as well as of speculation. I
took advantage yesterday of a light cart, the only conveyance about here, to go through a great part of that wonderful
place Romney Marsh, now richly studded with ewes and
lambs. It set me down within two and a half miles of Lydd,
where I called on my fellow-collegian Denne. I did not
find him at home, but on my way back I met him. He
holds in his own hands 900 acres of the marsh! on which he
observed I might easily conceive how much capital he might
soon sink in these times, and much he has sunk. All the
old families of Kent nearly have had or have some property
here, but have cut the marsh as soon as their wealth would
permit them. Denne is without society, but as they have
elected him every year head magistrate of their jurisdiction,
and as he is justice of the county, he says he has employment enough to render him independent of it. Perhaps you
would think 900 acres might do so alone. Their house is

just in the town and is good enough for any gentleman, but
the back-door opening into the farmyard betrays the pre-
decessor to have been the farmer turned gentleman, not the
gentleman turned farmer. How short and direct is the
road through Eton and Oxford from the grazier on Romney
Marsh to the fine gentleman? or, to speak plainly, to the real
gentleman in ideas, manners, and information. His uncle,
Mr. Cobb, has just let a good deal of the marsh, *a few
months since,* at 3*l.*, 4*l.*, and not a little at 5*l.* 10*s.* per acre !
but the latter in small lots, and near the town, and then
thought extravagantly high.

The sudden suppression of all smuggling at Lydd, Rye,
this town and numerous others, threw more than half the
population on the poor rates, and has made bankrupts of the
tenants on many a large estate, and quite ruined some of
their ports. Either the having permitted such license for
ages, or the sudden stopping of an ancient source of gain, is
unpardonable ; for there is all the misery felt here which
can possibly follow the total failure of a great manufactory.
The harbour at Rye is filling fast, and will be soon added in
spite of their exertions to the marsh, by the same causes
which formed the latter ; and though these are so evident,
the people here, and old histories, and even my friend Denne,
believe the idle story of its total formation during one storm.
Had the sea retired suddenly it would have left nothing but
shingle as at Winchelsea, and had our ancestors been as
greedy in walling out the sea as this generation before it
has half coated the shingle, there would be no acre worth
1*l.* per annum in all. I like the lodging here, and the old
woman, who has only 3*s.* 6*d. a week* of her own, and 6*d. wages*
to support *herself and a daughter* of near seventeen, which,
though lodging, coal, and a small garden are found her, is a bare
existence, and does not keep off the ague which is brought
by the sea-breeze from Romney Marsh, and though it does
not injure Lydd or Romney, does great havoc here and at
Rye. This old woman, and she tells me others here, cannot
afford to drink a drop of beer, while those who are not too
proud to go to the poor rates live in comparative luxury.

Ever yours,
CHARLES LYELL.

To Gideon Mantell, Esq.

29 Norfolk Street: April 19, 1822.

My dear Sir,—Although I did not leave Winchelsea until the 16th yet your letter did not reach me there but has followed me to town. Had I received it in time I might better have deserved the crown which you held up to my ambition, for I had no sooner begun my examination of the cliff there than I found myself in the midst of the limestone beds, with every character of strict identity with those of Battle and Hastings, particularly that gloss of rather the felspar appearance which you know marks it, and in which Webster was one day pointing out to me it closely resembles the crystallised part of the Fontainebleau rock—and no doubt it is owing to incipient crystallisation. The present town of Winchelsea is situated on a rock which projects like an island from the alluvial marsh-land which surrounds it, and which was once at no ancient date flooded by the sea; it presents a cliff more or less high towards every point of the compass. Close to the roadside, near the ancient gate which leads to Rye, is a quarry which affords a most satisfactory section of part of the rock. At the top is sand and sandstone, not much indurated nor highly ferruginous; then comes immediately under, and distinct from it, with not the least passage from the one to the other, a bed of the limestone, very hard and silicious, 5 *foot thick*; beneath this comes the sand again, of what thickness I could not ascertain. So much for the quarry which ends at the level of the road; but lower down the cliff appears another bed of the limestone, which must be far below the former, and proves therefore the alternation. The flat slabs of the limestone when weather-worn afford generally casts of innumerable small bivalves, of ( ) this size (perhaps cardiums), and so does the sandstone exactly the same. In a walk of four or five miles along the high cliffs between Winchelsea and Fairlight cliff, I could see none of the limestone in the ferruginous sand, but I would have made more particular search had I known you had not been there.

What little I could see of the Weald clay makes me

think it quite subordinate to the ferruginous sand, and to doubt more than ever its answering in position to the Petworth clay of Buckland's syllabus.

Buckland has received from the Yorkshire Cave the bones of the weazel, the rabbit, the pigeon, and I believe one other bird in a beautiful state of preservation, and which are being drawn for the R.S.

Do not fear any poachers on your Sussex preserve. I wish I could see so much activity as to give cause of fear. At our meeting this evening I will not fail to introduce your prospectus, and nothing will give me greater pleasure than to take the first opportunity of paying you a visit.

Yours truly,

CHARLES LYELL.

[In 1823 Mr. Lyell was elected Secretary of the Geological Society, and made excursions to Sussex and to the Isle of Wight, the results of which were not published till Dr. Mantell, in his 'Geology of the Isle of Wight,' 1847, gave extracts from Mr. Lyell's letters, dated June 1823, which showed that he had cleared up some of the obscurity in which the relations of the Lower Green-sand and Wealden strata were then involved. His first geological paper on the geology of some rivers near his native place in Forfarshire was read this year to the Society. About this time he made more than one visit to Paris, and became acquainted with Cuvier, Alexander Brongniart, Alexander Humboldt, Constant Prévost, &c.]

To GIDEON MANTELL, ESQ.

29 Norfolk Street : June 11, 1823.

My expedition to the Isle of Wight has very much confirmed the views which I before entertained of the geology of Sussex. Professor Buckland, who was of the party, though he did not altogether give in to a theory which he admitted was new to him, was still clearly surprised, when I pointed out to him the fact that there was everywhere in the Isle of Wight beds of sand below Webster's blue marl greener than those above. You are aware that Webster never found in his blue marl any fossils except two Ammonites. At Compton Chine, however, I found several small

Inocerami (*I. sulcatus*, I believe), and, although I was only
there five minutes, I saw so many fragments of shells in
this same blue marl that I am inclined to believe the identity of this bed with the blue marl of Folkestone and the
Gault might be made out by a further search for the
organic remains which it contains.

The section from Compton Chine to Brook is superb,
and we see there at one view the whole geology of your
part of the world, from the chalk with flints down to the
Battle beds, all within an hour's walk, and yet neither are
any of the beds absent, nor do I believe they are of less
thickness than with you. This is so beautiful a key, that
I should have been at a loss to conceive how so much
blundering could have arisen if I had not witnessed the
hurried manner in which Buckland galloped over the ground.
He would have entirely overlooked the Weald clay if I
had not taken him back to see it.

This clay, however, is only partly exposed, the softness
of it having caused a ruin of the cliff, just at the point
where the Petworth marble ought to be looked for. Soon
after this sandstone containing layers of limestone with
bivalves appears, then some mottled beds purple and white,
then Pyritous coal like that at Bex Hill I suppose. The
white sands of Winchelsea and Fairlight are magnificently
exposed, &c. &c. I staid a day longer than the rest of the
party, for the purpose of searching in the Weald clay of the
Culver cliffs section for Petworth marble, but the cliff is
there also in ruin; I found, however, a rounded block of it
two feet or more in diameter on the sea-shore, nearly opposite the Weald clay, specimens from which I have brought
to town, some of which are at your service, with many
specimens from the Fresh-water formations which I collected
for you, and will send you when I have more leisure to pack
them up. Almost the whole of the back of the Isle of
Wight is in such a dreadful state of ruin, that I believe it
has been the cause of much of that confusion which has
found its way into the heads of some of our geologists with
regard to your Sussex beds, and some of those very same
geologists whom I have heard ridicule De Luc for supposing he had there discovered chalk beneath the green-sand.

You will be very glad to hear that Mr. Warburton,[4] on whose accuracy above all men we may rely, says that above the Gault in Cambridgeshire green-sand occurs, as well as below it ; this upper green-sand, as well as the firestone of Reigate, he chooses to consider as part of the lower chalk. Yet both he and Dr. Fitton choose to call the beds at Beechy Head green-sand, supposing it to belong to that which is below the Gault. In this they are completely mistaken. I am going to Paris towards the end of this month ; if you wish to send any presents of books or anything to anyone I shall be very happy to be the bearer of them; indeed it will be of advantage to me, as affording me some introduction to the donees, and, as this visit of mine is principally to perfect myself in the language, I court everything which brings me in contact with Frenchmen. I like De Beaumont much. Yours very truly,

CHARLES LYELL.

To HIS FATHER.

London : June 21, 1823.

My dear Father,—I am fortified with most excellent letters of introduction from Buckland, Greenough, Grey Bennett, Professor Brochart, Dr. Fitton, Webbe and Lambert, &c.—to Humboldt,[5] Cuvier,[6] Brongniart, Cordier, Duvan, Prévost, Lefroy, Royer, &c. Buckland has also made me the bearer of presents of his new work to many of the above, and I have sets of specimens of my Scotch and Isle of Wight rocks to give away.

We had a very full meeting of the Geological last night; many foreigners at our club dinner, who were very entertaining. Professor Oersted, of Copenhagen, pronounced the following eulogium of our scientific dinners of which, as it was spoken in English, you may imagine the ludicrous effect. ' Your public dinners, gentlemen, I do love, they are a sort of

---

[4] Henry Warburton, one of the founders of the Geological Society of London in 1807.

[5] Alexander von Humboldt, eminent naturalist and philosopher, b. 1769, d. 1859.

[6] George Cuvier, comparative anatomist, b. 1769, d. 1832.

sacrament, in which you do beautifully blend the spiritual
and the corporeal !!'
A Mr. Underwood dined with us, who was so long a
détenu in France that he now resides there, preferring it to
England. He said, 'The Bourbons are becoming much
more arbitrary than before the Revolution; the prisons are
full of political offenders. A looker-on sometimes sees more
than those who are actually engaged, and my opinion of
the French people is, that they are much too *corrupt* for a
free Government, and much too enlightened for a despotic
one.'
I send you a specimen of Petworth or Sussex marble,
better polished than that which you have. With my love to
all, believe me, your affectionate Son,

CHARLES LYELL.

To HIS FATHER.

Hôtel Meurice, Paris: June 28, 1823.

My dear Father,—I left London Sunday, June 22, at nine
o'clock, in the 'Earl of Liverpool' steam-packet; crossed
to Calais *in* 11 *hours*! 120 miles! engines 80-horse power,
for 240 tons. Lord Lonsdale on board. As I was not sick
at all, he conversed with me much.
Sir Ralph Anstruther, a great friend of Sir J. Ramsay's,
was also a passenger, a young Scotch laird, has property in
Fifeshire and Caithness. I introduced myself as a friend of
Sir J., and he asked me to go with him to Abbeville in the
cabriolet of the diligence. The next day he took me in a
carriage which he hired to take him to Lisieux, near Caen,
some miles on the Neuchatel road. I then returned geologising to Abbeville, and spent the day in seeing the
manufactures of cloth, carpets, glass, &c. At the former
they have erected lately an *English* steam-engine of 18-horse
power, which enables them to dispense with 1,000 workmen.
The heads of these factories are very rich, and all *liberals*.
They seemed quite delighted to find an Englishman to whom
they might pour forth their abuse both of *their own* and *our*
ministers, which they did without sounding me as to my
politics. To the intrigues of our Cabinet they impute the

counter-revolution in Portugal. Since the peace they have
established two monasteries for friars, and two for nuns at
Abbeville, and they are now building another nunnery.
The next day I went on per diligence to Paris.

Ramsay was gone to Baden when I arrived, and I shall
certainly not go yet to the Esmenards, as there are two
Englishmen there who talk nothing but English. Judge
Richardson introduced me in town to his *marshal*, Bosanquet,
a barrister of Lincoln's Inn, who gave me a letter to Hunter,
a gentlemanlike young Englishman who has a place in our
foreign office here. He takes me to-day to a French family,
where I shall hear their language alone spoken, and where,
if I fancy it, I may *place* myself; if *not*, dine there for 3
francs when I like. Grey Bennett's friend, Duvan, called
this morning ; he is a very pleasant man, has been in office,
now lives on his own fortune ; his hobby is botany. He takes
me to Cuvier's *soirée* to-night, when I shall deliver my letters, &c.

To-morrow morning I am to attend Brongniart's [7] *levée*,
to whom I have four letters, and who among the English
geologists has the highest reputation both for knowledge and
agreeable manners of all the French *savans*. I have left
my letters and Buckland's book at Humboldt's (or, as his
porter calls him, Hoombowl), but did not see him. I called
on the Airlies yesterday, saw Lady A. and Mrs. Drummond ;
the former looking very well, and in high spirits. She says
there are not so many English here as there were, and the
price of lodgings and houses sensibly diminished.

All the French gentry attached to the Court whom they
see accuse all the English, both Whig and Tories, of being
Radicals and Revolutionists, and they are much piqued that
the Duc d'Orleans' *levées* are so much more splendidly
attended by English *noblesse* and gentry than the Court.

Your affectionate Son,

CHARLES LYELL.

---

[7] Alexander Brongniart, a distinguished chemist and mineralogist, b. 1770,
d. 1847.

To His Father.

My dear Father,—I have passed my time here as yet very much to my satisfaction, but I fancy I have not been advancing so much in my acquaintance with the language as with the *savans*, to whom my numerous letters gave me a most favourable introduction. My reception at Cuvier's last Saturday will make me feel myself at liberty to attend his *soirées* every week, and they are a great treat. He was very polite, and invited me to attend the Institute on Monday. There he introduced me to several geologists, and put me in an excellent place for hearing.

Miss Duvansel, daughter of Cuvier's present wife by a former marriage, is a young woman of most engaging manners, and very clever, talks English (as does Miss Cuvier) excellently ; she is pretty, and very lively. She is not a slight acquisition to the attraction of Cuvier's *levées*, and the more so as *he* is *by nature* reserved and uncommunicative in his conversation. We had a collation of fruit, &c.; the chief discussion was on Sir Walter Scott's last novel,[8] the scene of which is in France it seems. Cuvier is delighted with it, as were the ladies.

Humboldt addressed me, as Duvan had done, with, ' I have the honour of being familiar with your name, as your father has laboured with no small success in botany, particularly the Cryptogamæ. I suppose when he was in Paris I was absent, or of course I should have met him somewhere. But I hope if he comes again to be more fortunate.'

He was not a little interested in hearing me detail the critiques which our geologists have made on his last geological work, a work which would give him a rank in science if he had never published aught besides. He made me a present of his work, and I was surprised to find how much he has investigated the details of our English strata. I am going to him this morning with some specimens, to make him master of the last point which was cleared up in my Isle of Wight tour with Buckland, and afterwards he takes me and Sir J. Croft to the Observatory to show it to us. He appears to work hard at astronomy, and lives in a garret for the

---

[8] *Quentin Durward.*

sake of that study. The King of Prussia invited him to adorn his Court at the last Congress; thence he went to Vesuvius just after the grand eruption, and brought away much geological information on that head, which he was good enough to communicate to me. He speaks English well. I attend lectures at the Jardin du Roi, on mining, geology, chemistry, and zoology, all gratis! by the first men. The lectures both summer and winter are very numerous. I am sorry I am obliged to conclude, as I have much more to say. I dine every day at Madame Guien's, where we talk mostly French. They are friends of Hunter's.

My love to all, and believe me, your affectionate Son,

CHARLES LYELL.

To HIS FATHER.

Hôtel Meurice, Paris: July 8, 1823.

My dear Father,—Although I do not believe I have anything to say which would much endanger my being treated like Bowring with the Bastille, yet I write with more pleasure, as I know my letter will not be opened by the channel through which I send it.

The seals of many letters going to England, and still more of those into the interior of France, are broken, which Royer tells me ' is very *proper*, for it has been the means of detecting many enemies!'

Humboldt took me on Saturday, as he had promised, to the Observatory with Sir John Croft, who has a property in Yorkshire, was once in a diplomatic situation with Canning at Lisbon, from whom he brought letters to Humboldt. On our way Humboldt talked with great vivacity and force, of which I may perhaps give you some idea. 'No, Cuvier gives no lectures, and the reason I regret to say is, that he is still a Politician—no, you were mistaken, if you imagined that the ministry have reached a pitch of ultraism beyond him, I sent him back to his books. That time is yet to come. You observe that his *soirées* are mostly attended by English; the truth is, the French *savans* have in general cut him; his continual changing over to each new party that came into power at length disgusted almost all, and you know that it has

been long a charge against men of science, that they were
pliant tools in the hands of princes and ministers, and might
be turned which way they pleased. That such a man as Cuvier
should have given a sanction to such an accusation was felt by
all as a deep wound to the whole body. And what on earth was
Cuvier to gain by intermeddling with politics? If his
ambition could have reasonably flattered itself with the hope
that his talents might one day open to him the heights of
power, the attainment of *such an object* would have been an
excuse for his having *the weakness to desire it.* But this
he full well knew was impossible in France. First, because
he was a *Protestant,* and, secondly, because he was a man of
low family. You well know with what contempt the old
aristocracy of all countries are apt to regard all new men of
whatever abilities. *We* feel that but too much in Germany,
but *here* it is a principle of party now to carry such
prejudices to the utmost length. Cuvier's situation was a
proud one while he stood in the very foremost rank of men
of science in France, but when he betrayed the weakness of
coveting ribbons, crosses, titles, and Court favour, he fell
down to the lowest among his new competitors.' ' The Duke
d'Orleans has 320,000*l.* sterling per annum. He is liberal,
though he does not spend his income, for he puts by for his
numerous family. He is domestic, though he lives in
princely style. He likes the English, and deserves to be the
favourite, which he is, with them.' ' You cannot conceive,
gentlemen, how striking and ludicrous a feature it is in
Parisian society at present, that every other man one meets
is either minister or ex-minister. So frequent have been
the changes. They are scattered as thick as the leaves in
autumn, stratum above stratum, and before one set have
time to rot away, they are covered by another and another,
and on the last are sure soon to fall those which are now
blooming in full verdure above them. The instant a new
ministry is formed a body of sappers and miners is organised.
They work industriously night and day. They are more
religious, more constant at mass, more loyal, and, above all,
they know better how to ape exactly not only the ideas and
manners, but the very air and the expression, of their ancestors
of some centuries back. At last the minister, as Chateau-

briand and Vilelle for instance at this moment, find they are become heretics, Jacobins, infidels, revolutionists—in a word that they are supplanted by the very arts by which a few months ago they raised themselves to power.'

At the entrance of the Observatory we met Le Marquis de La Place,[9] a very fine-looking old gentleman. The famous mathematician Arago [1] did the honours of the Observatory, where we staid some hours. On Arago's pointing out from the top some newly-built monasteries, Humboldt observed, ' We have made a calculation, and find that there exist already more religious houses in Paris than before the Revolution.' This extraordinary statement did not seem to startle Arago, but Duvari, who, though a placeman, is far from violent in politics, assures me it cannot be true; he said, however, ' The assertion is so curious that I will inquire.' Humboldt said, ' These houses do not *yet* contain so many *inmates*.' After what I have given you concerning Cuvier, I ought in justice to add my own opinion, which is that he is more liberal and independent than I believe most Frenchmen are. He dares to speak and often with praise of Napoleon, and before Frenchmen, who might, and no doubt do, turn it against him. We must not forget that Baron Humboldt and he are the two great rivals in science, for La Place and the mathematicians do not come in contact with them. Humboldt's birth places him on the vantage ground, and Cuvier perhaps tries to compensate this by a little political power. As for his *ratting* so often, *defendit numerus.* What French politician could throw the first stone at him? Humboldt's family is noble and ancient in Germany. His elder brother a man now in great power there. His talents entitle him to regard with the contempt which he expresses, and I have no doubts feels, for *mere rank*, but we may say of him as Chateaubriand said of our English Peers, that he is well aware that while he gets too liberal, he is in no danger of losing the station and the advantages which his birth ensures for him. I dined with the Airlies to-day.

I hope I shall soon hear from Bartley.

My love to all, and believe me your affectionate Son,

CHARLES LYELL.

---

[9] The author of *Méchanique Celeste*, b. 1749, d. 1827.
[1] Francis Arago, astronomer, b. 1786, d. 1853.

To His Father.

Hôtel Meurice, Paris : July 13, 1823.

My dear Father,—Your letter of the 4th arrived on Friday. The rain, which has injured your hay, has fallen almost every day since my arrival here, and the Parisians talk of the weather as if it was the most serious misfortune. It has a great effect on their spirits, and no wonder Montesquieu, as a Frenchman, dwells so much on climate as the cause of national character. A lady said to me at dinner the other day, ' There is a revolution in the heavens, and the Duc d'Angoulême should be sent to quell it, for in truth he is too good for us here.'

I have learnt, however, from a quarter that leaves me in no doubt, that the Duke and all the Royal Family, except the Duchesse de Berri, were against the war, as well as the ministers. The Duke is decidedly hostile to that ' extreme right ' whose fanaticism carried everything before it. Almost all the less violent royalists are still in doubt whether it would not have been better to let alone the campaign, but as that is now impossible, they reason thus : ' We shall derive from the war in Spain two great advantages ; first, the cannon has been fired under the white flag, the army is now decidedly with us : this gives a consistency and weight to the Bourbons which they before wanted. Secondly, the Duc d'Angoulême, who is a man of more νοῦς than the English give him credit for, has acquitted himself with honour. He entered upon a dangerous expedition, and in which *he* least of all expected there would prove no severe fighting. The consideration and weight which he has acquired with all will be employed at home against that growing evil, the extreme right, which, if unchecked, must ere long occasion serious mischief in France.' My fear is that when eighty-five new deputies are added next year, the *ne plus ultra* Ultras will acquire as much additional force as they did this year. I am told the new French loan of Thursday last was equal to the greatest sum ever raised in England in one day during the last war. It has added to their debt 18,000,000*l.* sterling, and something more, which, with the 5,000,000*l.* with which they started, is no small expense even were the business over. In spite of the revolutionary sponge, their debt will shortly be no trifle. Next year the interest will be well on to 13,000,000*l.* sterling I suspect; and if ours was reduced to three per cent. (and I hear Robinson will make a great step towards this next year), ours would only be 26,000,000*l.*, or double the interest of the French. Mons. Corriglione, brother-in-law of the famous Carnot,[2] dined with us the other day. He has some landed property near Paris. He says the stagnation of commerce has depressed agriculture here in an unprecedented manner. Wool, which was at fifty sous when the army entered Spain, has fallen to twelve sous.

My fellow-collegian Rickards arrived in Paris yesterday, and persuaded me to give two guineas to the Greeks, which I do with great pleasure, and not the less so, though Dalby stands up for the Turks. He was a staunch Liberal when starting for the bar. Since he has decided on the Church, Rickards says he has grown too hot, even for Oxford, in Toryism. He abuses Canning and Co. for opposing the most sacred of wars against Spain, &c. &c. Tyndall, one of the young Englishmen in the house where I dine, has just received from the Greek Committee the appointment of surgeon, and a letter to Lord Byron, whom he sets out to join to-morrow. I was glad to hear the Cantabs have had a subscription for the Greeks. The Airlies I am sure will feel it no effort to settle in Angus, for they seem to talk and think of nothing else : they had intended to go to Switzerland and Italy, but all is given up. Maule's son,[3] it seems, has gained his action. Lord Airlie remarked, ' I am sorry for it *as a general rule,* for in entailed estates sons were but too independent of parents before.' I quite agree with him, and as no doubt Maule will appeal, I shall be glad to see the judgment reversed in the House of Lords. A Mr. Buchanan,

---

[2] Minister of War under the consulate of Napoleon, b. 1753, d. 1823, author of mathematical works.

[3] The Hon. Fox Maule (heir apparent to the Hon. William Maule of Panmure, who was in possession of an estate of 10,000*l,* a year), having 90*l.* of pay as an ensign in the army, and 100*l.* a year from his father, was found entitled by the Court of Sessions, to whom he applied, to an aliment of 800*l.* per annum. This decision was afterwards reversed by appeal before the House of Lords.

nephew of Mr. Ogilvy of Tannadice, who has met you, dined there, and said Gray of Carse is in ecstasies. He says, ' Aliments for ever ! I shall get my son to sue me, and we shall extract something from the creditors.' I remove to-morrow to 9 Rue Montabor.

Your affectionate son,

CHARLES LYELL.

CHAPTER VI.

JULY 1823—OCTOBER 1824.

ALEXANDER BRONGNIART — MANUFACTORY OF SÈVRES — M. PICHON — TERTIARY FORMATIONS AT MEUDON—M. CONSTANT PRÉVOST—MARSHAL SOULT'S GALLERY—POLITICAL DISCUSSION AT CUVIER'S—EXCURSION TO FONTAINEBLEAU WITH M. CONSTANT PRÉVOST—DIVISION OF PROPERTY— BARON DE FERUSSAC—HIS COSMOGONY—POLITICAL CHANGES IN FRANCE —VERSAILLES—GEOLOGICAL EXCURSION TO THE MARNE—CENTRALISATION OF PARIS—RETURN TO ENGLAND—REV. W. D. CONYBEARE — PLESIOSAURUS FROM LYME REGIS—CONSTANT PRÉVOST'S VISIT TO ENGLAND—KINNORDY—ACCOMPANIES DR. BUCKLAND TO ROSS-SHIRE.

CORRESPONDENCE.

To HIS FATHER.

9 Rue Montabor, Paris : July 20, 1823.

My dear Father,—Friday morning I entered, by a ticket from Duvan, the Garde Meuble. The new crown, which is made entirely of diamonds and amethysts, is the most elegant and tasteful, as well as costly piece of jewellery I have ever seen. Afterwards I went by invitation to Sèvres, where Brongniart showed me the whole process of the manufacture, and explained it mineralogically. I then dined with his family, and a delightful party. His sister, Madame Pichon, has been a celebrated beauty, and sat for Gerard's Psyche. Her husband was *Chargé d'Affaires* of France in the United States for five years. She speaks English, an l adores America to a degree that surprises me. Her husband wrote a work on politics, the object of which was to prove the possibility of introducing the English Constitution into France, and from his attacks upon Napoleon's tyranny was obliged to fly to England during the Hundred Days. Nevertheless, he was cast out lately from the Council of State, for

giving it as his opinion that the King could not authorise
the taking the veil in convents, without the Chamber re-
pealing the existing laws against it.   Cuvier, a Protestant,
was President of the Council, and gave his vote the other
way.   Since this time the Protestants of Paris, who before
looked up to him as a leader, have held him in abomination.
He also sanctioned the Missionaries.   Le Baron Coquebert
de Montret, Brongniart's father-in-law, was also there.   I
had read an essay which he wrote lately, to accompany a
geological map of d'Alloi's, showing the connection between
the geology and statistics of France.   He was for many
years from 1806 Consul-General of France at London—
thoroughly acquainted with England and many of our leading
men, and has spent much time in Ireland, the whole of
which except three counties he has been over.   His country
seat is near Sèvres on the banks of the Seine, at Bas Meudon,
and as there is exposed there a fine section of all the Tertiary formations of the Paris basin, from the chalk upwards,
he very politely asked me to come and breakfast with him
yesterday, which I gladly consented to.   He took me over
the whole ground himself, and gave me a delightful statistico-
geological lecture on the formations.   I was much gratified
at the analogy between this section and the same beds in the
Isle of Wight.   I made a collection of specimens, and after
declining an invitation to dinner returned to Paris.   I found
him occupied in making tables of the comparative population, increase of finances, &c., of Great Britain and France ;
and he was not aware, though he reads English works so
much, that Lowe had anticipated him.   I begged for Lowe
a copy of a memoir which he has just written for the Geographical Society here, on the comparative population of
Les Iles Britanniques and France, the object of which is to
check the fears of many Frenchmen, who oppose themselves
to every improvement here, on the ground of its occasioning
an increase of population, which idea they have chiefly derived from mistaking Malthus.

With my love to all, believe me your affectionate son,

CHARLES LYELL.

To His Father.

51 Rue Richelieu, Paris : July 28, 1823.

My dear Father,—I have now finally settled in Rue
Richelieu, No. 51—two very nice rooms for fifty francs a
month!! I went over to breakfast and dine last Wednesday
with Mons. Prévost,[1] a friend of Dr. Fitton's, at his country
seat at Montmorency, about a league the other side of St.
Denis, and met another young French geologist there. A
long walk which we took gave me a great knowledge of the
geology of the environs, which is very interesting. Prévost
has travelled much in Europe, is a man of good fortune, and
an excellent *Géologue.* He has come to Paris since, and I
dined with him yesterday at his house here, a pleasant party,
consisting chiefly of his wife's relations, who have a good
deal of land in the neighbourhood of Paris. I have promised
to spend the whole of next week in different geological excursions with him, chiefly in the neighbourhood of Fontainebleau, at his own mother's. We are to visit the millstone
quarries which supply nearly all Europe and North America.

At Cuvier's, on Saturday, the conversation was more
than usually interesting, chiefly on Napoleon. Cuvier
vouches for the truth of a great part of Las Casas' narrative
of events, &c., at many of which he was present here. He
has been all along very obliging to me, and the other day
he asked me to dine with him next Saturday, and said, ' I
shall assemble some of our geologists to meet you.' Sir J.
Croft and I meet often, and are very good friends. He says
the labour of those attached to embassies is often prodigious,
eighteen hours a day, and he has known many knocked up,
and himself nearly. It is pure drudgery, as, for instance,
adding up and copying the Ambassador's washerwoman's
bill! If my eyes were equal to half this, I would take to
the law in earnest, which would introduce me to society of
which I have an infinitely higher opinion, and to a description of labour which is even, when most dry, an exercise of
the reasoning powers. Sir John is a well-informed gentlemanlike man.

[1] M. Constant Prévost, an eminent French geologist.

I believe I am improving in the language, as much as can be expected from one who is obliged to abstain from writing exercises, and almost entirely from the theatres and reading French works.

My love to all, and believe me your affectionate son,

CHARLES LYELL.

To HIS FATHER.

Paris : August 3, 1823.

My dear Father,—I had intended to have written yesterday, when I had time for a long letter, but my eyes were so fatigued I was obliged to lay down my pen. By way of resting them, I escorted Lady Airlie to Marshal Soult's magnificent gallery of Murillos, of which master he has a collection of immense value and beauty. Afterwards we went to the Duc d'Orleans', where there are many by modern French artists, which are said to have merit, but for which we were perfectly spoilt by having feasted on the splendour of the Bolognese school. Madame Pichon invited me to a soirée on Friday, and introduced me to her husband. I believe I gave you a part of their history. When *Chargé d'Affaires* in America he opposed Jerome Buonaparte's marriage with Miss Patteson, for which Jerome got the Emperor to *disgrace* him. Two years afterwards Jerome was made King of Westphalia. He then wrote to Pichon that he had used him unjustly, and as compensation had appointed him Minister of his Treasury. The Pichons accordingly went to his court, and lived there in great style. He held lately some high place in the administration relating to Law Appointments, which he lost by opposing the Missionaries, a system which Madame says is driving as fast as possible all the French into two sects, *Bigots* and *Deists*, for they are making their religion too superstitious and absurd for persons of *education* in the nineteenth century.

Mons. P. seems clever, is very well informed, and communicative, and the house will be a great resource for me, as I have a general invitation, which I have no scruple in availing myself of in Paris, as they do not ever give tea at their evening conversations. He mentioned that the

Government have the nomination, on the average, of 1,200 law places of considerable emolument *annually !* The number of Judges (not counting Justices of Peace, who are also paid) is 2,500. These are for life, but there is a regular chain of promotion, for which they are all panting, which makes the whole machine a most formidable political weapon. The number of decorations of ex-ministers in the party reminded me of Humboldt's remarks. When I gave Madame Guien (the Protestant lady with whom I have been dining *en pension*) an account of this, she said, ' Faith, what between things spiritual and temporal, " c'est le règne des croix." '

At Cuvier's dinner-party yesterday there was a discussion as to what were the motives which induced Lord Castlereagh to kill himself, in which all the French assumed that it was a deliberate act. Upon some one observing that Lord C. had a surprising power, when he liked, of talking for hours without imparting information on subjects on which he did not choose to be clear, Cuvier said, ' I suppose then he was of Talleyrand's opinion, who declares " that speech was given to man, the better to enable him to conceal his thoughts." ' The Abbé Florini, an emigrant, who has been thirty years in England, and now here only on a visit, was amusingly violent, said our Press was an insupportable nuisance to ourselves, and to France, &c. After dinner he was arguing with me on the Spanish War, when Cuvier came up and said, ' Mr. Lyell, you must come with me into my study, for I want to show you a collection of English law books, which, as an English barrister, you will be amused at finding in my library. I interrupt your argument without ceremony, because I know better than you do the serious scrape from which I extricate you. Let me tell you, sir, that my good friend the Abbé (tapping the Abbé on the shoulder) is a man of old and noble family in France, besides this he is an Abbé, and moreover an *émigré*, and all these grafted on an English Tory. Is not this enough to form a compound which may support even all that we can here produce of Ultraism ? ' At this he laughed freely, and the Abbé joined with a good grace. What made the scene more amusing and extraordinary to me, and what I believe enabled the Abbé to bear with the joke, was that Cuvier is himself notoriously considered an Ultra, and I fear with too much reason, one of the most mischievous Ultras in the present *Conseil d'État*. There was a full *levée* in the evening, and he held forth on his new work which is just coming out, on Fossils of Cetacea. His reflections on the former state of this planet and the creation were so grand, and delivered so elegantly and so unaffectedly, that when he finished I could not help regretting that he had dabbled so much with what Sir J. Croft styles 'the dirty pool of politics.'—Your affectionate son,

CHARLES LYELL.

To HIS FATHER.

51 Rue Richelieu, Paris : August 10, 1823.

My dear Father,—My excursion into the country last week with Mons. Prévost proved a very pleasant and instructive one. Our route was by Mélun, Fontainebleau, Moret, Montereau. We had the satisfaction of discovering some new geological facts, and I gained a pretty correct knowledge of most of the formations of the environs and of the position of the *grès*, or sandstone of Fontainebleau. The forest is very varied, contains some fine timber, and a few views which are singularly wild and picturesque. We start next week on a second campaign. I slept two nights at Mons. Bevière's, near Mélun. Prévost's father-in-law is a very gentlemanlike old man. His father was rich, and Napoleon made him a Senator, which added 35,000 francs per annum to his fortune of 40,000 a year. Prévost has married a daughter of Turgot, the famous French advocate, who regained before his death part of the land which he lost at the Revolution, and left 900 acres, very rich, in the neighbourhood of Paris. His widow enjoys half for her life, which will be at her death equally shared by the son and three daughters, who possess already equal parts of the other moiety. Young Turgot, whom I have met at dinner, has married a young lady of Normandy, who will inherit a great fortune, and he hopes to become deputy of that province. He is an advocate, but does not practise except when he volunteers a defence in some political case, which he has done several times. He seems a clever but rather violent Liberal. *La loi commune* is, I should say, a favourite with the best informed and most candid men whom I have yet met with here. Even Baron Montret, who is fully aware of all that can be objected against it, confesses that he should hesitate much before he could give his vote against it. Prévost says the immense size of properties before the Revolution was a great evil, and that the division has increased the wealth and population of the nation, and the general ease and happiness. Mons. Pichon says that there are 12,000,000 *propriétaires* in France out of 30,000,000! This law was the subject of an animated discussion between several Frenchmen at Cuvier's last night. They mentioned, as a favourable result of it, the facility with which young ladies married, and alluded to the difficulty which those of a certain rank in England found in settling, and the numbers that never did, as no small evil. But without reference to the merits or demerits of the system, there is no institution of which the effects are more remarkable, or better worth studying in France at present; and if it lasts fifty years more, the contrast between England and France will be most curious, whichever nation it may be in favour of. Prévost has travelled much over France, and he says he has never known a single château of a certain size (about the size of Bartley) erected in modern times. In the provinces the demolition of the old ones is most rapid. Even near Paris he cited seven great châteaux which have been sold and pulled down *since the return of the King.* Perhaps the most remarkable of these (of all which I have made a list) is Montmorency. General Arrighi laid out 50,000*l.* sterling just before his death in having the walls and ceilings preserved. It was sold, and the stones of the house and the park walls burned into lime. The same law which brings large estates into the market makes it impossible to find a purchaser for them. Large companies of speculators are organised, who buy up these mansions for the materials, and the French are uncharitable enough to call them *les bandes noires,* from the analogy of their occupation to those adventurers of that name who bought up the churches and châteaux at the Revolution, and generally paid the purchase money by the lead alone with which they were covered. If a merchant gives a dowry to his daughter at her marriage, and dies bankrupt, or, as constantly happens, with his affairs in a less flourishing state than at the time of the wedding, she must bring her fortune back to be divided among the brothers and sisters. If *she* is dead, still her children must *refund*, and this is the cause of no little confusion and insecurity, and is no bad harvest for the lawyers. By the way, Cuvier mentioned last night that the number of *Étudiants en droit* at the college has exactly doubled since the peace. He also said, with regret, that the study of mathematics was going fast to decay, and explained why the smaller scale of their war department necessarily occasioned this. I have made acquaintance with Le Baron de Ferussac, a colonel on the staff of the French army, who went through Napoleon's Spanish campaigns. He is a lively young man of very pleasant manners, of an old family, and in the first society. He is engaged in publishing a work on the Molluscæ, and has studied much those branches which relate to Geology. He has the finest collection by far of land and freshwater shells in the world. I spent two mornings with him, and he is certainly the most brilliant builder of theories I have ever met with. I fought hard with him for Buckland's notions of the Diluvian formation, in which Ferussac is not orthodox. A cabinet geologist can account for everything much more easily than one who takes the field, and looks all the difficulties in the face; but he is very ingenious, and his knowledge of shells gives him a powerful clue. As you have almost said enough to provoke a geological letter from me, I will endeavour to state in a few words an outline of Ferussac's cosmogony. He says geology yields ample proofs, both that the ocean was at a much higher level than at present, and that the general temperature of the earth has decreased. That from chemical reasons the diminution of heat would naturally be attended by a subsidence of the waters. That from the organisation of the molluscæ, some can only live at great, others at slight depths. On the shores and shallows of the Mediterranean different species now exist from those in the deep parts, and that we may conclude totally different ones again dwell in the fathomless sea. In this manner he accounts for the oldest and lowest beds presenting us such totally different fossils from the superior ones, and the reason that as we ascend in the series the productions resemble more and more those which now live, is merely because the ocean, as it became shallower, contained animals more and more analogous to those which alone we have now access to. At length the chalk was deposited, and covered with its analogous formations almost all the continents we know anything about. Much land was then left bare, and freshwater formations began. The present system of valleys did not then exist, of which there is proof, and instead of the waters being carried off, as they are so beautifully at present, the whole was nearly covered with great lakes, of which we know thirteen, those of the London and Hants and Paris basins being three. All these thirteen present similar phenomena of alternating fresh and salt water formations, and mixtures of both. As the ocean was then still nearly on a level with the chalk, it often broke down a dyke, as in Holland, and deposited oysters, &c., once more. Analogous phenomena may now be seen in progress in the Caspian and other inland lakes, &c. I wish I had room to go on, but I find I shall be unable to tell you my plans. I am going with Prévost this week to see those parts of the Paris basin which I have not yet seen. In a month I think of leaving this. I went to a delightful soirée at Madame Pichon's on Friday. Humboldt was there. He told me that as soon as I have finished my tour with Prévost he shall get me to give him two mornings, and shall read over to me all he has written on the tertiary formations for his second edition, as I have now compared the Isle of Wight with Paris. Besides obliging him, I shall get a good *cram* by this.—Your affectionate son,

CHARLES LYELL.

To HIS FATHER.

51 Rue Richelieu, Paris : August 23, 1823.

My dear Father,—My plan at present is to leave Paris in a fortnight, or at farthest three weeks. Whenever I return

here I shall have the full advantage of the acquaintances I
have made.   Paris is as come-at-able as Holland, and more
so, because any season will do for it : indeed the winter would
be preferable to this.   If a future opportunity should occur
of going to Holland, there will always be many other tours
which I could exchange for it—but I only consider it as a
route home, and I have already taken most of the others.   I
am dining now with Ramsay's friends, the Esmenards, a
pleasant family.   There is only there at present a rich, young,
handsome, fashionable, and gentlemanlike American, of
Philadelphia, who speaks French admirably.   He has a box
at the opera, which would be a lounge for me, but my weak
eyes almost preclude my profiting by it.   He has travelled
all over the United States and Canada, and part of Europe ;
speaks Italian and German, has read some poets, &c., has
studied law, but does not intend to practise.   He amuses
me much.   He is an aristocratic republican, well informed,
thinks his own the *first* country in the world, and England
the *second*, and that the latter would rival, if not surpass his
own, if not blemished by the anomaly of a King and a Peerage.   I have seen Humboldt often since I last wrote.   He said
the last time, ' I have just read the last " Edinburgh " and
" Quarterly Reviews," and am glad to see in the *latter* two such
liberal articles as those on Greece and Spain.   To be sure,
some may say that the last pages of the latter article are at
variance with the first part, but still it convinces me that
the majority of your Tories are against the war.'   I asked
him how he thought the two Reviews had treated him ?   ' I
have no reason to complain, my political opinions are well
known, like all others, I have fared accordingly.   I was
roughly handled by the Quarterly, but perhaps still more
injured by the extravagant praises of the Edinburgh.   For
when they paid me such insidious compliments as to say I
was the first of *savans*, &c., they could not fail to make me
enemies.'   He told me what he thought of many of our
great public men on both sides, with whom he has had personal acquaintance.

    You will wonder that such simple questions as some in
your last letter are not easily answered here, for it is common for a well-informed Frenchman to reply : ' You will

wonder I do not know, but there is no use in learning, for
our fundamental laws are changed every few years, and if
not, they are violated with impunity.' That the royal prerogative is very undefined here, is, I think, proved by their
establishing 'monasteries' and 'taking the veil,' which is
in general believed to be unlawful. I have asked, 'Can the
king grant a *majorat* to an unlimited amount?' The answer
has been more than once, 'If he did, the tribunals, depend
upon it, would not overrule it.' The more I learn of the
institutions of this country, the more I am convinced that
their tendency to absolute monarchy is irresistible. Had we
gone to war for the Cortes, we should have, I suspect, thrown
the power into the hands of the opposite party, but that
power would have remained absolute. But as to the *majorats*, a duke's is 40,000 francs, the other titles less, but I
cannot say what as yet. A man's property is divided into
as many shares as he has children, and one more, which last
(varying of course according to the number of his family)
is at his own disposal. For example, if a duke has 340,000
francs per annum, and only two children, he can leave the
eldest 240,000 francs a year, viz., 100,000 francs for his
share, 40,000 francs for the *majorats*, and 100,000 francs
which is at the father's disposal.

You know that Buonaparte introduced a military uniform
into the Polytechnic schools, and the scholars were summoned to meals, lessons, &c., by beat of drum. When the
Bourbons were restored they abolished this, as inconsistent
with establishments intended chiefly for educating youths
for civil employments. Within this last three months the
drum has been reinstated for the clock, and the military
costume for the toga. A cousin of the Esmenards' dined
with them on Sunday in uniform. He is intended for a civil
department. Schools of *Enseignement Mutuel* were flourishing and multiplying in France. They are now discoun-
tenanced, and it is expected will soon be extinct. For these
the missionaries have founded 'Schools for Good Christians.'
On the high road in the Forest of Fontainebleau has lately
been erected an open chapel, with a figure of the Virgin, and
candles always burning. When we passed, many women and
children were kneeling and praying in it. The churches are

recovering their splendour in Paris, and the priests are most gorgeously apparelled. The congregations are amazingly increased, chiefly because the ministers and persons in power, forming the bulk of their aristocracy, have made church-going so fashionable, that it is becoming contrary to *bon ton* for your carriage not to be seen there. But unfortunately it is still a mark of party, and from my own knowledge I can answer for this fact, that some conscientiously religious Liberals abstain with their families altogether from church, because they have not courage enough to brave the imputation of being place-hunters. But besides their indirect influence, the Government have resorted to open compulsion. For instance, every student of medicine is compelled to produce a certificate of confession before he can take his degree. Fortune does not favour the growth of constitutional freedom in France. Religious feeling is in the opposite scale from that in which it weighed with our ancestors. And this feeling is not extinct as some suppose in France, nor can we ever expect it to be, as it is certainly natural to the constitution of the human mind. At our revolutionary period, the Court and the fashionable civilians were the sceptics, and often the open scoffers. The Puritans were then the church-goers, and Liberals. The Republican leaders in England were able to acquire power to check the Crown by precisely the same hypocrisy by which the modern French Ultras acquire unlimited force for the prerogative.

I have not time to say anything of my last excursion with Prévost to St. Germains, Poissy, Versailles, &c.

With my love to all at Bartley, believe me, your affectionate son,

CHARLES LYELL.

To HIS FATHER.

51 Rue Richelieu, Paris : August 28, 1823.

My dear Father,—I have called twice on Redouté,[2] without finding him.

As an opportunity now offers, I am desirous of seeing the Low Countries, though in autumn, for I think the chances

[2] Botanist, author of *Les Roses* and *Les Liliaces.*

of my seeing Paris again pleasantly are two to one greater
than my being able to make a tour in the former in Spring.
    Next Spring, for instance, I must hold myself if possible
disengaged in order to conduct Prévost to one of our great
geological head-quarters in the neighbourhood of Bristol. I
hope that Dr. Fitton will escort him to the Isle of Wight,
which I know pretty well, and which Fitton ought to visit
before he publishes his memoir on part of France.
    As I know nothing of the west of England, I shall do
myself as great a favour as Prévost. My last excursion with
him down the Seine to Triel, by land and by water, was very
pleasant, and marvellously economical, for we were three
long days, and travelled between twenty and thirty leagues
for fourteen shillings each, the whole expense. The aqueduct near Versailles, which crosses a very pretty valley, and
conducts part of the water which supplies the palace, is
really like the work of a Roman emperor, and well worth
visiting. The frescoes of Le Brun, in the palace of Versailles, are very fine, and the chapel magnificent. I certainly never entered elsewhere so kingly a mansion, or any
theatre so well adapted for displaying the splendour of a
great court. I have little doubt that in the next reign
France will (to her cost) see its ancient glory restored, for
as a Bourbon is the hero of every painting, it must naturally
be a favourite of the family. The revolutionary violence of
which it was the scene is considered by the Royalists as
having consecrated it, rather than rendered it an object of
aversion. As further inquiries have confirmed all the few
facts which I have given you in former letters, I am encouraged to send you a few more, which I have from as
good authority. When a minister has been dismissed here,
he has a pension of about 1,200*l.* per annum sterling, a
handsome fortune in France, not only from its purchasing
more here, but because it raises a man above the general
level than perhaps six times the sum would in England.
As the ministers are so often changed, these pensions bid
fair to form a large item in the expenditure. If a man votes
against the party in power, he is stripped of his pension.
This has been done, and if the moderate Royalists had given
a patriotic and conscientious vote against the Spanish war,

they would have lost their pensions. As ex-ministers must
always form the most able, efficient, and enlightened and
useful opposition in a free country, this stretch of power is
clearly unconstitutional, and, what is worse, is intended to
be such. When Napoleon set up his Senate, he gave every
senator 36,000 francs for life. As councillors they were as
independent (or ought to have been) as the judges. These
senators the Bourbons made peers, with the exception of
such as betrayed them during the Hundred Days. A large
portion of them were of course as little able to support their
dignity as were most of the noble *émigrés* who returned
with the present family. To both these classes pensions
were assigned, but not, as Buonaparte gave them, for life,
but *durante bene placito !*   Such a gift certainly neither
‘ blesses him who gives nor him who takes,’ for it so de-
graded the House of Peers in the eyes of the nation, that
since that time it has lost all weight and consideration.
The moderate and rational Royalists, who saw the error,
proposed as a remedy, either that these noble paupers should
be put on the footing of Napoleon’s senators, or that a sum
be raised by loan, which should be given with a *majorat*
attached, and that it should be given to the *émigrés* as a
compensation for their losses in the Revolution.   The meaning of this latter proposition was to annihilate the hopes of
those fanatics who are for laying hands on all the property
which they lost at the Revolution. The scheme was regarded favourably by a large part of the Liberals and de-
feated by the Ultras. The following facts will explain this
paradox. You may see every day among the *affiches* in
Paris, *Vente des biens Patrimoniels*. This term is used in
contradistinction to all those estates which depend on a
Revolutionary Title.   The latter, even in the height of
Napoleon’s power, bore an inferior price to the former, and
what is curious, the difference between the two varied then,
and does still, in different provinces according to the attachment or aversion to the new order of things. Any remu-
neration to those who lost at the Revolution would cure the
Revolutionary titles, and, by raising their value, favour a
large and powerful party among the Liberals who hold those
estates. This is one reason why the Ultras would not relish

it, but still more because their hopes would then vanish of recovering their lost lands. Their cry is ' All or Nothing!' and for this, as in the war with Spain, they would not scruple to hazard the crown, and risk another revolution.

August 30.—M. Pichon, *Conseiller d'État*, called on me yesterday, and left an invitation to dinner, but I returned so late from Humboldt's, with whom I spent the afternoon, that I could not go, but I attended a gay soirée there. I had some conversation with him, in which he gave me some entertaining accounts of the state of Parties here. Among other things he said—' Do not suppose that the law of *Equal Division* is the work of Revolution, it was the almost universal Law of France (the nobles only excepted), and is therefore now consecrated by the prejudices of those who love the old as well as the new order of things. The Roman Law, you know, favoured division, and that spread over half France. And the customary law in general favoured it also. I and many others *do not feel sure* that the Law of Primogeniture is not preferable.'

The following speech, which I heard a lady make the other day, is I think characteristic of the ideas which the French entertain, and the manner in which they speak of religion when they are unreserved. ' La Duchesse de Broglie is an amiable woman, very unlike her mother, Madame de Staël, in many points, particularly religion. She is in danger of becoming a complete Methodist, and more so I think since her late tour in England. She will have some influence, too, on that score, on the mind of my old friend the Duke, her husband. When I saw much of him formerly, he had not that turn. I do not mean that he was irreligious, but certainly at that time *he took a philosophical view of things.*'

I have promised Humboldt to pass the afternoon to-day in his study. His new edition serves as a famous lesson to me, in the comparison of England and the Continent. There are few heroes who lose so little by being approached as Humboldt. Of Cuvier this cannot be said.—My dear father, your affectionate son,

CHARLES LYELL.

To His Father.

Paris : September 7, 1823.

My dear Father,—To-morrow morning, by ten o'clock, I shall be off for Compiègne in a ' *Clerifere*,' with an order from Duvan, who is Inspector of Public Works, to see the Château. Duvan and I have parted great friends. I have paid at T. and Wurtz's for Hooker's ' Flora Scotica,' which is on its way, and he has given me by way of souvenir La Roche Jacqueline's ' War of La Vendée.' Your roses, &c. from Redonté, I have had packed up with my two boxes of geological treasures by a regular *emballeur*, and they will go from Calais by steam to the Geological Society, where they will be in safety. Your debt to Redonté, which is of three years' standing, I paid—*twenty-four Napoleons*. Although he has made a great deal of money, yet I learn that he has contrived to be poor, and he seemed not a little desirous of touching the coin.

If I had taken up my quarters with the Esmenards on my first arrival, I should have got on very far in French, and it will be a most desirable point to make for on any future occasion. Her husband, in Buonaparte's time, had the administration of the theatres here, a profitable employment, and one which gave them access to the first society, which Madame Esmenard still keeps up. She is very sensible, genteel, and well informed,—in bad circumstances, and rather melancholy, I think for her daughters' sake rather than her own. The eldest daughter paints portraits. Some miniatures also of hers are exquisitely finished. Several of Napoleon and Josephine she has sent to England, and they sold well. As her labours assist the family, she works hard and constantly. She is very amiable. The second is pretty, has reached great perfection in flower-painting, under Redonté, sings moderately, and plays the piano well, *in the French style*. She is unaffected and exceedingly pleasant. The youngest has none of these female accomplishments, and Nature has not favoured her either in form or face, nor I suspect with a contented disposition. She has read, however, the Lord knows how many works, in French, English, and Italian ; not to mention that she has dipped into Titus Livy, with a French translation, and (by way of supplement) began the Latin grammar last month, with an Irish teacher, resident in Paris. Whether the young lady has in truth 'drunk deep of the Pierian spring,' I have had no opportunities of ascertaining, but I can see that the liquor is getting fast into her head, and will, I fear, soon render her d—h disagreeable.

On Saturday week Cuvier introduced me to Professor Van Breda of Ghent, who not only drew out for me a plan of a tour in Holland, but gave me letters to all the Dutch universities, Ghent, Amsterdam, Haarlem, Leyden. He is a friend of Greenough's, who showed him attention in London. As I finish by Ostend, I shall be glad to receive a letter from you there, as well as Rotterdam. I have had full leisure since my stay in Paris to sow all the letters of introduction which I brought from England to French people.

I have not heard of Prévost since he went into the country, so I suppose his wife would not give him leave of absence. I therefore went on Wednesday alone, for a pleasant excursion of three days, to La Ferté-sous-Jonarre, a town situated on the Marne, in a beautiful country sixteen leagues N. of Paris, five beyond Meaux. There are the quarries which supply half Europe and N. America with millstones. They are drawn from the most recent of the Antediluvian Formation, yet in some parts almost resemble the quartz of the oldest granite. They have a simple machine of a lever and a bucket, to draw water out of the pits, by which a boy of twelve years does as much work, I believe, as four men in the marl pits at Kinnordy. It costs but five francs erecting! I have taken a plan and measurement, and cannot but think it could be used with advantage in your loch. The Cathedral at Meaux is a beautiful specimen of the light style of Gothic. My love to all, and believe me your affectionate son,

CHARLES LYELL.

To His Father.

My dear Father,— Although I have not yet advanced far
on my route, yet as an opportunity offers, and it may be some
time before I have another, I will send you a few lines.

Madame Esmenard asked me one evening ' if I did not
think the French had less national prejudices, particularly
towards strangers, than the English.' I answered : ' Allow
me first to ask whether by English you do not mean those
Scotch, Irish, and English, who come from all parts and
countries, whereas by French you merely mean Parisians ? '
After a moment she said, ' I confess I do,' and she defended
this by assuring me that their ' provinces were so *behind,* so
prejudiced,' &c.

Frenchmen, judging from the best-informed and most
liberal whom I have seen, when they speak of the spirit,
state of 'feeling, opinion, manners, faults, and virtues of
France, merely mean by France, Paris. Mons. Pichon observed to me, ' If there is one thing more than another for
which I abhor the memory of Napoleon, it is that he levelled
the provinces, and the population of this country, by stripping
them of all the municipal rights which they had left : such
was his system. There did not remain any occasion on
which they could practise " self-government." Throughout
France they do not now choose a single alderman or mayor.
It has become almost unsafe to trust them with any power,
and much restriction which strikes an Englishman as despotic is most necessary.'

The effects of Paris being the centre of the system from
which all light and heat emanate, are some of them singular
enough. Their numerous judges, I before mentioned, are
made subservient by a chain of promotion. Sometimes the
President of a Tribunal where the salary is small is raised
to another where it is greater, but if it be of the same emolument, and a certain number of leagues *nearer Paris,* this is
regarded as a great step. Duvan assures me that there is
not a provincial town, not even Bordeaux, which in publishing, and adding to the stock of knowledge, is to be compared to many a small place in Germany of 3,000 inhabitants,

If a man is thought to display talent, he is hurried to Paris, as the only soil where it can be nourished or admired. There are no seats of education of as much consequence as our second-rate public schools out of Paris—*no* scientific establishments, with two or three paltry exceptions, as at Lyons, &c., which seem not to be so much respected as our New Philosophical Society of York, or the Geological of Cornwall. Although the population of this empire is one-third greater than that of the British Isles, there is no Oxford, no Cambridge, no Edinburgh, no Dublin, Glasgow, Aberdeen, St. Andrews, Hertford, Eton, &c. &c.—none of those numerous and often distant points which serve with us to circulate knowledge and the spirit of improvement through the system. Nor can I learn that there is any Gallery of Pictures or Statues, such as Wilton, Hamilton Palace, and a hundred others far and near, which with us enable many a provincial, who has not the time and means to travel to the great central monopoly, to gratify and improve his taste. Paris devours all. A Frenchman may well feel proud of his splendid capital, as worthy of an enlightened nation, and as one of 'the eyes of Europe.' But when we find that in knowledge, science, and in the splendour of wealth, all is a desert beyond, we cannot but regard her pre-eminence as invidious, and suspect that the spot where the court and chief political power has generally resided has acquired greatness at the expense of the other parts. It reminds me of that very original image to which Montesquieu, in his ' Esprit des Lois,' likens a despotic government, which he says resembles that custom of the savages of Louisiana, who, in order to get at the cocoa-nuts, first cut down the tree. I have met, however, with so many liberal thinking and moderate men of both parties in Paris, that I expect France will work out for herself a better order of things. There is always hope while the representative system remains, and liberty to publish their debates; and I hear that Labourdonay and many Ultras, who would tear the Charter in pieces if they could, set off as soon as the session ended, to labour to acquire influence in their provinces. This is well, and must raise up the provinces soon.

My love to all, and believe me your affectionate son,

CHARLES LYELL.

[M. Constant Prévost visited England in 1824, and Mr. Lyell made a tour with him from London to Bristol and the Land's End, returning by the southern coast to Hampshire, in the course of which they were entirely engaged in geological investigations. In the summer, Mr. Lyell made another geological tour with Dr. Buckland in Scotland, who accompanied him from Kinnordy to Aberdeen and Inverness, and they paid a visit to Sir George Mackenzie in Ross-shire, and after crossing the Grampians, visited Sir James Hall at Dunglass.]

To GIDEON MANTELL, ESQ.

29 Norfolk Street, London : February 17, 1824.

My dear Sir,—Your very obliging present of the 'Outlines of the Geology of Lewes' came to me almost as a reproach for having so long delayed sending you my letter on the Isle of Wight, which I ought to have returned to you long before. Your little volume is a very elegant illustration of your native town, and your contribution to it is an excellent proof how much the sphere of local interest is enlarged by geology. A few years since, a history of Lewes would scarcely have yielded any glimpse of information so far back as the seventh century. The Geological Antiquarian can safely rank his treasures of the youngest date as of an age of which the builders *of your old castle* had no traditionary knowledge.

W. D. Conybeare [3] is in town, and has been with us for some time. He is waiting for the arrival of the new Lyme Regis Plesiosaurus, of which he has an excellent drawing. The Duke of Buckingham has bought it, but it will be exhibited for some time at our rooms, 20 Bedford Street. It affords a great anatomical triumph to Conybeare, as most of his hypothetical restorations in his former memoirs turn out true to nature. The new animal is a very perfect skeleton, and a prodigy, for it has forty cervical vertebræ, whereas existing quadrupeds range from seven to nine, reptiles from three to nine. Aves reach no higher than twenty, the swan being the maximum. What a leap have we here, and how many links in the chain will geology have to supply.

Believe me, very truly yours,
CHARLES LYELL.

[3] Rev. William Conybeare, Dean of Llandaff, geologist, b. 1787, d. 1857

To His Father.

Bristol : June 6, 1824.

My dear Father,—I arrived here on Saturday with Mons. Prévost, after a very pleasant tour. We went to Gloucester and to Falfield, sixteen miles on the road to Bristol, where I called on Dr. Cooke at Tortworth Rectory. The Rector was not at home. By good fortune, however, his friend Mr. Weaver, a member of the Geological Society, who has made himself very eminent by his writings, was staying there, and keeping house. He received us with the greatest cordiality, made us breakfast and dine there for two days, and gave us more information than I had ever received in the same time. The Rectory is the most beautiful and picturesque spot in the world. It stands, like Airlie Castle (to descend from the sublime to the geological), on the Old Red-Sandstone Conglomerate, and on proceeding northwards all the beds which the Isla present are formed in succession, in particular the Cortachie slate, which contains organic remains here, proving it to belong to the transition-limestone series. The identity of these beds, so carefully studied here, throws such a light on Forfarshire, that my tour has been inestimable to me, were it merely for that, more particularly as Maculloch, not having compared this portion of English geology with Scotland, has no suspicion of it. The calcareous part, which unfortunately for us is wanting in Scotland, happens to be feeble at Tortworth, and this helped to make the analogy more striking to me. We came from Tortworth to Bristol, where Mr. Conybeare had been expecting us for three days. I do not expect to get farther west than Plymouth. Prévost's Norman paper attaches him to the Oolites chiefly, and makes him unwilling to approach the more ancient formations.

He is so good a naturalist, that I shall learn much from him. He is wonderfully struck with the beauty of our roads and stage coaches in England. This town looks dull, though I believe the commerce is thriving moderately.

With my love to all, believe me your affectionate son,

CHARLES LYELL.

To Gideon Mantell, Esq.

Bartley Lodge, Southampton : July 9, 1824.

My dear Sir,—The three letters which you have sent me, and which are still unanswered, arrived in town when I was absent on a tour with Mons. Constant Prévost, a French geologist whom you know from his works.

I am glad you are persevering in enriching your collection, which is becoming generally celebrated, and will soon be more known since the visit of our Oxford Professor, who was much struck with it.

My friend Prévost studied anatomy for three years, under Cuvier, and as he is intending to visit England again this summer, he will if possible call on you at Lewes. I gave him your address, and assured him that he would meet with a welcome from you. It would be of use to his geological work in Normandy to see your fossils, and I doubt not you will derive much light from his remarks on your osteological treasures.

Three weeks since a magnificent specimen of an Ichthyosaurus (tenuirostris?) was discovered at Lyme by the celebrated Mary Anning. It is about the size of the Plesiosaurus which you saw in town. M. Prévost took a drawing of it, which I have traced, and I send it you, that you may see it, as it will be long probably ere it is published. The sketch was taken by measurement, and, although rapidly, yet may be depended upon. While we were at Lyme we witnessed the discovery of a superb skeleton of Ichthyosaurus vulgaris, by Miss Anning. It was perfect, save the tail, which a cart-wheel had passed over. It was two feet long.

Believe me, yours very truly,

Charles Lyell.

To His Father.

Airlie Castle : August 10, 1824.

My dear Father,—I rode with Chambers [4] round the plantations. They have all received a check by the severe winter of 1822-23, which is evident. As the top of the

[4] The gardener at Kinnordy.

Castle Hill is so prominent, it would be worth while to take
out the few Scotch firs, and substitute healthy young larches,
which, if they do not grow so high, will still be lifted above
the rest of the wood. I went through the peat moss, the
draining of which proceeds rapidly. The number of wasps
in the garden is unprecedented. Chambers has accordingly
introduced ' commissioned bottles,' *i.e.* bottles made expressly
by his direction, eight dozen of which are emptied every two
days, full of dead wasps, but ' the plague,' he says, ' is
unabated.'

I am obliged to finish this letter before breakfast, and
have scarcely time to tell you of my ten days' tour with
Captain Ogilvy, which was most successful, and puts the
geology of the county in a new light, as little suspected by
Blackadder or Maculloch. The Siedlaws are what Werner
termed ' a saddle,' the stratification being this, the older
beds in the centre, flanked by younger ones ; but what is still
more interesting, the centre beds are the same as appear on
the Isla between Airlie Castle and the Mill of Craig, or
between Kinnordy and Catlaw.

The discovery of the New Red Sandstone explains many
anomalies which Dr. Fleming made by confounding it with
the Old. We went by Dundee, Arbroath, and Lunan Bay.
I drew the whole coast from boats, and we examined it
thoroughly, not stopped by rain a single day. Paid a visit
at Tannadice on our way back. Lord Airlie arrived here
last night, and is going to shooting quarters forthwith. Left
all well in Paris except Mrs. Drummond. He has given Glen
Moy [5] to Sir J. Ogilvy and me, but I shall begin with
Balentore.[5]

Believe me, your affectionate son,

CHARLES LYELL.

To HIS FATHER.

Kinnordy : September 6, 1824.

My dear Father,—I am much obliged to you for having
shown my friend Prévost an English race-course. On Tuesday morning I set out with Drummond, to the North Esk.[6]

[5] Shooting quarters.
[6] Thomas Drummond, botanist. Went to North America as assistant

Went by Hill of Careston, from which there is a glorious view of the Strath. I lodged at the shooting quarter, an inn of Maule's at Gannachy Bridge, close to the burn. Made myself master in four days of the grand succession of strata seen between Gannachy and the granite of Mount Battock, from the summit of which the view is very fine. We saw Aberdeen well, Loch na Garr, Ben Macdui, Cairngorum, &c. I improved myself much in geology, and Drummond also, who knew nothing, but was very anxious for a cram. The weather was delightful. I was to have gone to Airlie Castle yesterday, but begged off a day in order to arrange my concerns. I drive there to-day, and we are then to examine the Ericht from Blairgowrie upwards (perfectly unexplored ground), and then on Friday to dine together (the captain and I) at Pearsie, and stay a day or two there. I have received a most friendly letter from Buckland, in which he tells me that he has just finished his expedition to the Hebrides, and wishes me to accompany him on a visit to Sir George Mackenzie's at Cowl (in Ross-shire ?), and on a short tour, by Aberdeen, Inverness, down Loch Ness, Fort Augustus, &c., all to take less than *ten days.* He offers to make his time in some measure suit mine. I have chosen within two days of what he proposes, and have told him he must spend one day here, that I may show him the place, and also some specimens on which I have doubts, besides my map, &c. It will be I expect about the 15th of this month. I look forward to no small amusement in being ten days with him, when he is so full of new matter, as he must be after a visit to the Western Isles, so interesting and disputed a field for geological inquiry. On Saturday a pair of antlers were taken out of the marl two feet deep in the loch here quite perfect and of uncommon beauty. The frontal bone is entire, and supports them well. They have nine points or branches, and measure more than ten inches round in the thickest part. It will be a great ornament to the house. Some of the jaw and teeth were found also. The loch also produced me a fine dish of fish for dinner yesterday. There is a great sale of marl this year. Among the speculators

naturalist to Dr. Richardson, under Sir John Franklin, and afterwards collected plants in the southern United States, and died at Havannah in 1834.

who are investing their capital in buildings in the New
Town of Kerry, on the South Muir, is Mr. John Chambers,
forester at Kinnordy.

Your affectionate son,

CHARLES LYELL.

To His Sister.

Cowl, Ross-shire : September 26, 1824.

My dear Eleanor,—A fall of snow, which has surprised
the good people here in the midst of their harvest, gives me
an hour's leisure to write to you, in answer to your account
of the Salisbury gaieties, for which I have to thank you.
We arrived here yesterday morning from Inverness, and
shall probably quit it to-morrow for Brora in Sutherland-
shire, where my companion has an interesting point to
examine, otherwise we might both find no small entertainment here for a longer stay, as it is one of the most delightful houses I ever visited. Sir George Mackenzie took a
long walk with the Professor and me, both yesterday
and to-day. He is very gentlemanlike and intelligent, and
full of information. He was very desirous to show Mr.
Buckland a set of parallel roads on the small scale in a
valley here, and was evidently astonished with the rapidity
with which the Professor seized all the facts, and gave a most
excellent theory to account for them. They will make a
pretty figure in the second volume of the ' Reliquiæ Diluviæ,'
which Mr. Buckland is just going to publish. The country
here is very beautiful, the hill scenery picturesque, particularly at the present moment, when Ben Wyvis and the
higher hills are covered with snow. There are some of the
largest oaks near the house that I have yet seen in Scotland,
and the hills are clothed with birch to some height. The
house is new, and does credit to Sir George's taste. We
have of course been looking over our host's magnificent
collection of mineralogical treasures from Iceland. But his
present hobby is vitrified forts, on which he has advertised a
work, in which he is to attack Dr. Maculloch. Lady
M. is a very ladylike woman, and her two daughters
pleasant girls. They have four or five sons of all ages. We

are to have Lord Pitmilly, one of the Scotch judges, to dinner to-day, who is on the circuit. But I must now beg leave to treat you with a little retrospective history of my adventures. Mr. Buckland was so desirous of clearing up some puzzles which presented themselves on the banks of the Carrity near Kinnordy, that he agreed to see the Isla, and as this was found more than a day's work, we accepted Captain Ogilvy's pressing invitation, and dined and slept at Airlie Castle, and finished the Isla and Melgum next day, and after dining again with Captain O. returned to Kinnordy, and started next morning for Stonehaven. Saw Dunnotter Castle the same evening, and next day boated it to Aberdeen, and saw the termination of the Grampians in the sea cliffs. At Aberdeen we were in high luck, for Dr. Knight, Professor of Natural Philosophy, was an acquaintance of Mr. Buckland's, and invited us to go with him to an annual dinner, at which we saw the Principals of the two Universities, Dr. Jack and Dr. Brown, and all the Professors. The next morning we breakfasted at Dr. Knight's, then dined with Dr. Forbes, Professor of Natural History. The Duke of Gordon, Chancellor of the University, was there, an old man of eighty, not at all superannuated, and well worth seeing. We attended the same day the assizes, and heard the Chief Justice Clerk condemn a thief for burglary. The next morning we breakfasted with Dr. Glennie, Professor of Moral Philosophy, a clever man, married to the niece and representative of Dr. Beattie the poet. There is in their room a most beautiful portrait of Dr. Beattie by Sir Joshua Reynolds, which has kept its colour. After seeing everything worth examining in geology at Aberdeen, we left it in company with a young advocate, son of Sir J. Hall, an acquaintance of Buckland's, who left us the day after. He was an agreeable addition to our party, as far as he went, viz. to Peterhead, from whence to Cowl I have little to speak of, as we passed it rapidly, but Portsoy, Elgin, and Inverness presented us with some things worthy of notice.

I have no more time save to hope that, when I return to Kinnordy a week hence, I shall find papa and Tom arrived. In the meantime, believe me, with my love to all at Bartley, your affectionate brother,   CHARLES LYELL.

To His Mother, at Bartley Lodge.

Edinburgh: October 18, 1824.

My dear Mother,—In my way through Perth I learnt that my father was expected at Kinnordy on the eleventh. I conclude, therefore, that he is now there, and take an opportunity of giving you some account of my proceedings since I left Cowl. Mr. Buckland went from Ross-shire to Brora in Sutherland, and in examining that district we got within a moderate day's journey of John O'Groat's House. We then returned to Inverness, and travelled thence, in a gig, along the Caledonian Canal, by the Fall of Foyers and Fort Augustus, then visited the parallel roads of Glen Roy beyond Letter Finlay, one of the grandest natural phenomena in Great Britain. We next went by Glen Spean, Dalwhinnie, and Dalnacardoch to Blair Athole, with which and Glen Tilt we were much pleased. We then came by Killiecrankie and Perth to Edinburgh. Here we have worked very hard for a week in the geology of the neighbourhood, and in cabinets, museums, &c., and have had an excellent opportunity of seeing all the leading characters in the University. We have been at breakfasts and dinners without end, at Professor Jamieson's twice, at Professor Wallace's, Dr. Hibbert's, Mr. Allen's, four times, Dr. Greville's, &c. &c.

From Edinburgh we made a geological excursion with Dr. Hibbert to Linlithgow, Falkirk, and Stirling, which proved very successful. We then went to Dunglass, Sir James Hall's,[7] a very elegant and stylish place, about eight miles from Dunbar. The old gentleman is far past his prime, but luckily Captain Basil Hall, the author of the 'Voyage to South America,' was there, whom I had often met in town. He is one of the most gentlemanlike and clever men I have ever met with. We made some great expeditions to St. Abb's Head and other parts of the coast with Sir James and his son, and a Mr. Allison, advocate, on a visit there. Lady Helen Hall is daughter of the late Lord Selkirk; the two unmarried daughters are very pleasant, one of them very pretty. We came home yesterday morning in order to spend

Natural Philosopher, b. 1760, d. 1832.

the forenoon and dine at Craig Crook Castle, the country
house of the far-famed Francis Jeffrey.[8] This was a great
treat. He is a little man, of very gentlemanlike appearance
and manner. Shines in conversation, whether on trifling or
important topics. After his showing us round the grounds
and neighbourhood, we met at dinner, Sir H. Parnell, M.P.,
and Mr. Murray, and others. The dinner and wine in great
style. Among others at the dinner was Mr. Maculloch, who
gave the celebrated lectures on Political Economy in town
last summer, which I attended. He was an acquaintance of
mine, and pressed me to dine with him to-day, which I am to
do. I expect much amusement from the party. Mr. Buckland left this to-day for Alnwick Castle. I return to Kinnordy to-morrow.

  With my love to all, believe me, my dear mother,

Your affectionate son,

CHARLES LYELL.

[8] The eminent critic, editor of the 'Edinburgh Review,' and, later, one of
he judges in the Court of Session, Scotland.

CHAPTER VII.

JULY 1825—DECEMBER 1827.

CLATHRARIUM LYELLI—ENTOMOLOGY—LOCKHART BECOMES EDITOR OF THE
'QUARTERLY REVIEW'—BREAKFAST AT LOCKHART'S WITH SIR WALTER
SCOTT—VISIT TO CAMBRIDGE—PLEASURE IN READING LAMARCK—PLAN-
NING HIS FUTURE WORK—ON INTOLERANCE IN SCIENCE.

[Mr. Lyell's eyes having become stronger, he resumed the study
of the law at his father's request, and was called to the bar in 1825,
and went the Western Circuit for two years. He published a geological paper in Brewster's ' Edinburgh Journal of Science ' on a dike
of Serpentine cutting through Old Red Sandstone in the county of
Forfar, which is perhaps one of the best examples of Serpentine in
this country, with all its characteristic mineral peculiarities intruding
itself in the manner of trap in the sedimentary strata. A paper on
shell marl, and fossil fruit of Charæ, was printed in the Transactions
of the Geological Society this same year.

In 1826 he became a Fellow of the Royal Society, and he wrote
papers on the Plastic Clay near Christchurch in Hampshire, and on
the freshwater strata of Hordwell Cliff, Hants.

His family left Bartley Lodge in the New Forest, Hants, to reside
at the family seat, Kinnordy, Forfarshire.

In 1827 Mr. Lyell wrote in the ' Quarterly Review ' an article
on Scrope's 'Geology of Central France,' in which he showed how
entirely he had imbibed the opinions of Playfair and Hutton, and
considered that all geological monuments were to be interpreted by
reference to aqueous and igneous causes in action in the ordinary
course of nature.]

CORRESPONDENCE.

To GIDEON MANTELL, ESQ.

Bartley Lodge, Southampton : July 20, 1825.

My dear Sir,—I went to Fitton's when he came to town
for a few days in the beginning of this week, and was not a

little gratified at being there when the Clathrarium, my
namesake, was found.[1]

You might have felt secure that so beautiful a specimen
as it really is would never have been thrown away even by
mistake at the house of a geologist. Though I did not see
Chantrey[2] about the Iguanodon's teeth, yet I have no fear
of their safety, and that they will at length in his hands, like
the teeth of Cadmus's dragon, be productive of a fruitful
crop. Fitton has promised to inquire about them. He has
just returned from examining the ' Valley of Elevation,' as
Buckland calls it, at Highclere, which I consider as caused by
a continuation of that elevating force which acted in a line
along the central axis of your great saddle of Surrey, Kent,
and Sussex, and which, if prolonged, would have elevated
the firestone beds (for such they prove to be) at Highclere.

All this took place after the deposition of the London
Clay, and before it happened, the Plastic and London Clays
of the Hants and London basin were horizontally connected.
The chalk which now separates them was pushed up through
them, for, as Buckland observes, the highest chalk summit at
Inkpen Hill is still covered by decided Plastic Clay. Your
great valley is not a valley of denudation. I do not agree
with Buckland that much chalk has been carried away
between the north and south downs, for as two sides of a
triangle must be longer than the base, so when the horizontal
chalk was inclined from below London to the north downs,
and again from Lewes to Heaven knows how many miles
under the sea, how could there be other than an opening of
some miles ?

I remain here for a month or more, and perhaps may go
to Dresden to learn German this summer, or rather autumn,
but I am not sure. Buckland, you know, is made by Lord
Liverpool a canon of Christ's Church, a good house, 1,000*l.*
per annum, and no residence or duty required. Surely such
places ought to be made also for lay geologists.

Believe me, yours very truly,

CHARLES LYELL.

[1] Clathraria Lyelli, a fossil plant from the Wealden in Sussex. See Mantell's
Medals of Creation.

[2] Sir Francis Chantrey, sculptor.

To His Sister.

November 20, 1825.

My dear Marianne,—As chief keeper of the Bartleyan or
Lyellian Entomological Museum (for when we talk of small
things we should help them up with high-sounding names)
you may fairly expect me to report progress. The treasure
I brought to town remained peaceably in my cupboard till
the first evening meeting of the Linnean Society, when I
took in my pocket the small mahogany box, and you would
have been amused to see the greediness with which the
collectors examined two or three good things, and Curtis [3]
was congratulated on having met with one so easily pigeoned,
for I told him to help his collection if he could. I afterwards took him the other moths, and you will be glad to
hear that one small buff *Noctua* filled a chasm in his collection (*N. Flavilinia*) not in Donovan. He told me once
that nothing was wanting in his that was not worth a
guinea. But the grand treasure is a new moth. Which
think you? That ugly black *Noctua* of the size of *Gamma*
with white petticoats (*alias* under-wings), which 'you and
the other slaves of the lamp' caught one night. I left at
least one more of them at Bartley. This I conceive was a
guinea one. I have given it to Curtis, and in return he
presents us with *Sphinx Tilea* and many moths, and promises much more in payment besides naming them all.
But he says there are so few collectors in Scotland, that if
you choose to collect, without ever going out of the garden
at Kinnordy, you may purchase with Forfarshire *Phalœnæ*
an English collection worth many hundred pounds. Curtis
found so many things in Scotland last year, that he talks of
publishing an account of his cruise. The number of collectors of British Lepidoptera in town is very great, and makes
an extensive barter of Scotch insects the easiest thing possible. Of the latter, 30 new Lepidoptera were found by
Curtis last summer, but in England, though they have 1,500
species, they seldom get a new one, even a minikin.

With my love, believe me your affectionate brother,

CHARLES LYELL.

[3] John Curtis, Esq., F.R.S., author of *British Entomology.*

To His Sister.

London : December 4, 1825.

My dear Eleanor,—I really wish that some of you, or all of you, would write to me now and then. To confess to you the truth, I have been a little homesick since in town, and after spending six very agreeable weeks without a grievance at Bartley, I have missed the sight of a friend's cheerful countenance till last week, when the Fittons, Buckland, and Dr. Daubeny and Scrope,[4] came to town. However, I have given in my resignation as Secretary for the end of the year, and will revenge myself on Greenough & Co. by an additional week if possible at Bartley at Christmas. Coleridge informed me yesterday that he has given up the Quarterly, and that Lockhart comes from Edinburgh in a week to take it. He could not get my paper into his last number, but would have put it into the next, with certain parts abridged. He showed me how Gifford formerly cut to pieces his papers, and this encouraged me much. Indeed, had I been aware how difficult a task even for such a man as Coleridge it had been to fit an article for the Review, I should never have presumed to write. I have at all events made Coleridge's acquaintance, and though I expect to be wrecked in the attempt to pilot myself into the graces of a new editor, I shall not be ashamed with such encouragement to run the gauntlet. Scrope wants me to pay a visit to Waverley Abbey, but I shall not go, at least at present. He is a clever, gentlemanlike man of fortune, of my own age, and has just published a very creditable work on Volcanoes.—Ever your affectionate brother,

CHARLES LYELL.

[4] George Poulett Scrope, son of G. P. Thomson of Waverley Abbey, b. 1797, d. 1875. He showed an early taste for geology, and adopted Hutton's doctrines. Author of *The Geology and Extinct Volcanoes of Central France,* and other works.

To G. MANTELL, ESQ.

Temple : January 3, 1826.

My dear Sir,—I have received your book and paid for it, which I hope your subscribers will do more punctually than Macculloch swears his do.

Your eulogium on my 'profound legal knowledge,' though a severe quiz upon me, has not been quite so much a subject of amusement to my friends as I anticipated. Buckland, however, was not a little merry yesterday at my expense. I told Murchison not to laugh too freely, for I should get you by way of a set-off to omit it in the second edition, and to substitute for it 'but more particularly to Mr. Murchison, Sec. G. S., whose scientific acquirements, no less than his splendid military achievements in the Peninsula under the Duke of Wellington during the late war, are so well known and appreciated.'

Buckland has got a letter from India about modern hyænas, whose manners, habitations, diet, &c., are everything he could wish, and as much as could be expected had they attended regularly three courses of his lectures.

Yours very truly,

CHARLES LYELL.

To G. MANTELL, ESQ.

9 Crown Office Row, Temple : June 22, 1826.

My dear Sir,—I much regret missing you, particularly as I had just got a copy of my paper ready for you. I live near my old rooms. Do not talk of the G. S. As long as Warburton allows the whole to rest on his sole shoulders, and has a large mercantile business, and a hundred other hobbies besides the principal one, the London University (to which is now added M.P. for Bridport), so long publication or real utility is out of the question ; and a secretary might as well try to bring about reform, as Warburton in his new capacity to throw open all the rotten boroughs. But enough of this. I must not sport radical, as I am become a Quarterly Reviewer. You will see my article just out on

' Scientific Institutions,' by which some of my friends here think I have carried the strong works of the enemy by storm. I am now far on with a second, and hope to get it out in less than three months. I mean to help myself out of Cuvier largely, for I must write what *will be read.* The Plesiosaurus is delightful, so is Stonefield and Cuckfield. If you can send me comments on Buckland, I will use them delicately : also say how much of the great skeleton you have now got together. I would give eight or ten lines to your Museum—more I hardly can—if you would put down on this osteological topic any fact that is marvellous, also anything about the vegetation that you have gathered from Ad. Brongniart. This would come in in another place.

I am over full of work, so believe me yours faithfully,

CHARLES LYELL.

To HIS FATHER.

Temple, London : November 16, 1826.

My dear Father,—My breakfast with Lockhart this morning was exceedingly pleasant. Sir Walter Scott is in the first place a far more genteel-looking man than Phillips has represented him in his portrait, which, as I had supposed that to be flattering, surprised me considerably. His *hobbling* was much greater than I had fancied. There was only his daughter, Miss Scott, Mrs. Lockhart, a Dr. Gooch, King's Librarian, and a Dr. Macculloch, a medical man. Sir Walter was very cheerful, told a number of good stories on subjects always started by others. They were more remarkable for the rich fund he had of them, than for anything else. None of them were brilliant, but all pleasant. For instance, a Mr. Simpson was talked of. ' Ah, how is Jemmy ? I shall not forget his coming into the theatre just as Mrs. Siddons had entered her box, and been received by a round of applause from the audience. Jemmy, seeing no one on the stage, inquired the cause. " It is a mark of esteem and consideration the public have for you, Mr. S.," said a friend. Jemmy went to the front of his box and made three low bows, to the infinite amusement of all who knew him.—Now

that was a cruel joke.' His language is remarkably far
from being refined, there is positively a blunt simplicity in
it. 'The French,' he said, 'are certainly in better humour
with the English than formerly—we are *up* there.' When
the increasing Catholicism was talked of, he said, 'I think
it will not go beyond a party, but I wish the royal family
may not run their heads against that thing.' He said, 'Yesterday I was at Chantrey's, and recommend you all to go
and see his statue of Washington.' Lockhart said, 'As
portrait-painting is so profitable, I wonder there is not one
good painter now but Sir Thomas Lawrence.' I observed
'Phillips has surely painted some excellent portraits.' Sir
Walter replied, 'Yes—but Phillips is quite a hit and miss
man.' Sir Walter declared himself no judge of the art, but
went on making some curious remarks on the peculiar difficulties painters now labour under, compared to the great
masters of old, and added, 'I have seen a great advance in
the English school since I remember.' Mrs. Lockhart is a
very agreeable person. Lockhart sent me a paper in slips
the other day on our 'Public Schools,' to ask me my opinion.
I rewrote some passages relating to the introduction of
elementary information in Natural History, &c , which he
has adopted, as well as other suggestions. My plan is now
to write an article next number but one, on 'Scotch Universities,' or our whole system in Great Britain and Ireland,
which I begin to hope I shall master. In the number after
a Geological article, already on the stocks ; and then, if I can
manage it, a shorter paper on our Scotch and English
scientific journals, and Scotch and Irish scientific institutions.—Believe me your affectionate son,

CHARLES LYELL.

To HIS SISTER.

London : January 5, 1827.

My dear Marianne,—I am glad to hear that, notwithstanding a tiger broke loose,[5] so little damage was done in
the menagerie. I hope that the first week the house is free

---

[5] In the insect box.

from company you will proceed not only to arrange, but to
reset everything that stands at all in need of it, for I can
promise you that unless you get on you will have more on
your hands than you can manage when Curtis's next box,
with a certain arrival from another quarter, which I have
just heard is on the road, reaches you. I staid with Spedding[6] a week at Cambridge, at Trinity Hall, learnt much
from Professor Henslow[7] and his friends, to whom he introduced me, about the Cambridge system, for my paper ; made
an agreeable acquaintance in Marmaduke Ramsay, brother
of Sir Alexander Ramsay, a gentlemanlike Fellow of Jesus.
Grand festivities at Trinity Hall and St. John's, great
dinners, suppers, and male routs. I paid my expenses by
what I won at whist, though I revoked one night ! Copley,[8]
Master of the Rolls, was at Trinity Hall, most excellent
company. He conversed easily with every one, and upon
my mere general introduction there, he talked with me when
we met at other colleges. Lord Palmerston made me a stiff
bow, the contrast of his manner to Copley was entertaining,
for his was a perpetual canvas and acting a part. He said
that Dr. Spürtzheim (who has lately lectured at Cambridge)
was ' an ingenious humbug,' which is just what most men,
even his supporters, think of him at the University. One
night, when I went to hear the Hymn at Trinity Chapel, I
fell in with William Ramsay of Bamff, in his surplice. The
new buildings are beautiful, and will immortalise Wilkinson
the architect. I begin to think I must divide my article into
two, and write the first chiefly on the English universities.
If Lockhart cannot stand it, I will try Jeffrey. I am going
to Devizes Sessions on Monday next, three days' work. I
think you will get more chilblains than colymbetes in the
loch in winter. When you fish there take shells also, as I
should publish a list in some future marl paper. When I
called yesterday on Mrs. Smythe,[9] I caught her making an
abstract from La Place's ' Astronomy ' of those facts and
speculations which could be made intelligible to persons not
mathematicians. Her selection is so beautiful and striking,

---

[6] T. S. Spedding, of Mise House, Cumberland.

[7] Professor of Botany. [8] Afterwards Lord Lyndhurst.

[9] Wife of Admiral Smythe.

that when it is finished I shall get a copy. Professor Henslow says that the Cambridge and Huntingdonshire marshes swarm in some seasons with *Machaons* and *Lycæna dispar,* and he will get me some. He showed me a good beginning of a collection of British insects at their Philosophical Society.

Your affectionate brother,

CHARLES LYELL.

To G. MANTELL, ESQ.

London : March 2, 1827.

My dear Sir,—On my return from the circuit yesterday I found your second letter, having received the first with Lamarck at Dorchester. You know that half my time is now spent at Sessions, Circuits, &c., and must not therefore be surprised when you receive no immediate answers to your correspondence, which I always receive with great pleasure. I devoured Lamarck *en voyage,* as you did Sismondi, and with equal pleasure. His theories delighted me more than any novel I ever read, and much in the same way, for they address themselves to the imagination, at least of geologists who know the mighty inferences which would be deducible were they established by observations. But though I admire even his flights, and feel none of the *odium theologicum* which some modern writers in this country have visited him with, I confess I read him rather as I hear an advocate on the wrong side, to know what can be made of the case in good hands. I am glad he has been courageous enough and logical enough to admit that his argument, if pushed as far as it must go, if worth anything, would prove that men may have come from the Ourang-Outang. But after all, what changes species may really undergo ! How impossible will it be to distinguish and lay down a line, beyond which some of the so-called extinct species have never passed into recent ones. That the earth is quite as old as he supposes, has long been my creed, and I will try before six months are over to convert the readers of the Quarterly to that heterodox opinion. I should like to discuss these matters with you at Lewes, but between law excursions and town studies, I have

never a moment to spare. If ever I can I will, and give you
notice, and can assure you, that I know I shall receive a welcome, and you need not, therefore, repeat your kind invitation.

I wish among your new Groombridge fossils there had
been a good cetacean, for theoretically it would be of more
importance than the iguanodon. Not that I doubt some of
the oolitic cetacea. I am going to write in confirmation of
ancient causes having been the same as modern, and to show
that those plants and animals which we know are becoming
preserved now, are the same as were formerly. *E.g.*, scarcely
any insects now, no lichens, no mosses, &c., ever get to places
where they can become imbedded in strata. But quadrupeds
do in lakes, reptiles in estuaries, corals in reefs, fish in sea,
plants wherever there is water, salt or fresh, &c. &c. Now
have you ever in Lewes levels found a bird's skeleton or any
cetacea ? if not, why in Tilgate and the Weald beds ? In our
Scotch marl, though water birds abound in those lakes, we
meet with no birds in the marl ; and they must be at least as
rare as in old freshwater formations, for they are much
worked and examined. You see the drift of my argument—
ergo, mammalia existed when the oolite and coal, &c., were
formed. Broderip [1] says, that in spite of all the dogs and
cats which float down the Thames, none of their remains
have been found in recent excavations in the Thames deposits. Send me your thoughts on the subject. If I am asked
why in coal there are no quadrupeds ? I answer, why are there
none, nor any cetacea, nor any birds, nor any reptiles in the
plastic clay, or lignite formation, a very analogous deposit,
and as universal in Europe ? Think of these matters, and
believe me yours most truly,

CHARLES LYELL.

To His Father.

April 10, 1827.

My dear Father,—I was glad to hear this morning from
Kinnordy, as I had had no news since you were snowed up.
Leonard Horner is in town from Edinburgh, very *à propos* to

---

[1] William John Broderip, zoologist, especially devoted to molluscous
animals, died 1859.

keeping me right in my article, as he is a great education man, as well as geologist. His gratitude to me for having got into the ' Quarterly Review ' an article on the liberal side of geology is very agreeable. He is eager to serve me, and wanted me to let him go carefully over the article, with his friend Brougham, which I begged him not to do, as Mr. Brougham might make a good joke out of revisals of ' Quarterly Review ' articles. Horner himself is a safe man. Brougham had second business, next to Scarlett, this last circuit, a fine thing. He has published two pamphlets which I have not read, one ' On the Pleasures of Education to the Poor,' and the other on Hydrostatics, which are having a prodigious sale : they talk of ten thousand copies going off in a short time. Scrope has just published a volume on Auvergne, with beautiful panoramic and geological views of the country. As I am, with many others, indignant at an atrocious article which Macculloch [2] wrote on his late work on volcanoes, in the Westminster, I am determined to give him a moderately long article in the ' Quarterly Review,' a sort of abstract which I conceive will take one-fourth the time of an original article, and the latter, as far as science is concerned, should not be, I am clear, given to a periodical. All that Fitton means, can be done easily in an elementary book, like Mrs. Marcet's ' Conversations on Chemistry,' for which one might get well paid, without much credit, and without putting one's name, and which is greatly wanted. But if I wrote anything, I would wait longer (unless I was in want of cash), and certainly a work even elementary, which would gain one reputation, would neither be done in a few months nor easily. I have my doubts as to leaving sessions on the score of economy. I found the additional time of which I am master, in consequence of ostensibly following a profession, is very great, and perhaps the 30*l.* that sessions cost me, might be annually returned by an additional article, which I might be thereby enabled to write. It is wonderful how little mercy one's friends have on one's time, if one has no excuse deemed valid for declining unprofitable parties, or refereeships of papers, or secretaryships, &c. The circuit

[2] Author of *Description of the Western Islands of Scotland,* and other works.

costs under 50*l.*, everything included. My purse would not
have required replenishing for some time, but I am much
obliged to you for anticipating my wants as usual. I find
them diminish monthly, in proportion as I am more agreeably employed, and if with the willingness to work and
industry which I now have, I had any chance of earning what
I require by my own exertions, I should be without a care,
as far as I am myself concerned. But to be willing without
avail to work hard, and almost for nothing, is now the fate
of many hundreds of barristers, and many millions of our
labouring classes, and we must congratulate ourselves at not
being among the latter. I am quite clear, from all that I
have yet seen of the world, that there is most real independence in that class of society who, possessing moderate means,
are engaged in literary and scientific hobbies ; and that in
ascending from them upwards, the feeling of independence
decreases pretty nearly in the same ratio as the fortunes
increase. My eyes go on tolerably, and I feel my facility of
composition increases, and hope to make friends among
those that a literary reputation will procure me who may
assist me.

Believe me, your affectionate son,

CHARLES LYELL.

To HIS SISTER.

1827.

Dear Marianne,—Curtis sends me a note to say that there
are good things among the Spring insects, and says the Miss
Lyells will do wonders in Scotland. He hopes you will get
some general knowledge of botany, as a little knowledge
even of the wild Scotch plants would, he says, double the
value of your entomological information. It is a singular
proof of how much more is to be done in entomology in
North Britain than in botany, that Curtis found in his excursion last summer in Lepidoptera alone thirty insects new
to Britain, and some quite unknown and new ; while Dr.
Hooker has never found more than, how many plants ? not
ten, I think I recollect. You will be glad to hear that Mr.
Vigors has sent five specimens of the clear-winged Sphinx

' tulipiformis ' and a great many *Nocturna* and *Coleoptera*, &c., all named, as a present to our collection.   I told him that Curtis was to have the first picking of whatever you got in Forfarshire, to which he said, ' I was once in Edinburgh, and collected only for two days in June, and I would give all my duplicates for half the things I got in those two days.'   I met with the Rev. D. Cooke, nephew of the geologist.   He says when he next sends to Curtis he will send for us that beautiful insect *Carabus nitens*, and wants from us *Chrysomela Banksii* in return, of which you should send up four the first opportunity,

Your affectionate brother,

CHARLES LYELL.

To HIS SISTER.

Temple : December 9, 1827.

My dear Caroline,—Captain Smythe [3] called here, in town for a few days.   Mrs. S. and family quite well, and like Bedford much.   He is building a very fine observatory, the Astronomical Society having done him the honour of voting him the use of some valuable instruments bequeathed to them.   On Thursday I breakfasted with Dr. Richardson, and gathered much useful information in geology from a *tête-à-tête* with him.   On Wednesday I dined at Briscoe's : he had been in the tunnel that morning, which is now proceeding vigorously.   Gurney the engineer seems determined to run a steam coach from this to Southampton—the other day it went round the Regent's Park without a horse, and ran against a gate.   But he says he shall put one horse in front, who will guide it, and have no weight to pull, but merely to canter ten or twelve miles an hour.   The machine is built.   I have sent off a heavy box of fossils for the tower [at Kinnordy], a white hare stuffed would be a desirable addition.   I miss several dogs' jaws and the cat's skull in the bones which I brought from the marl. [4]   When you have time, inquire about them.   Clift pronounces the bones to be the same animals I had supposed, including the wolf, but

[3] Admiral Symthe, F.R.A.S., F.G.S.   [4] From the Loch of Kinnordy.

a few are still to be made out. Of the insects in the marl, Curtis says that except two or three he cannot pronounce even on the family, they are such mere fragments. If one leg or antenna had been found, he says he would have hit off the family. Two seemed to be *Bembidium,* and another, *Helops Lim.* His curiosity, however, is much excited by some of the elytra, and if there are many fragments of the marl still in the garden, send them, for every morsel has been pulled to pieces; and as for the rock marl with *gyragonites,* it is all gone, and I could dispose of another great box. Curtis showed me with much pleasure a *noctua* sent in your spring box. It is new to him, he thinks to all. It comes nearest to *Hadena plebeia* and *Leucostegina,* but is of a more lively grey colour than *Acronicta rumicis.* As to the small *phalœna* you are quite right. It is the one which he hopes may be new; but he thinks a new *noctua* of marvellously more importance than a *phalœna.* The Miss Walkers caught on Ben Nevis and Ben Voirlich, at about 3,000 feet high, some new and many rare Lepidoptera, which they have sent Curtis. We must positively mount next year.

Your affectionate brother,

CHARLES LYELL.

To GIDEON MANTELL, ESQ.

Southampton : December 29, 1827.

My dear Sir,—I am on a visit here to an uncle, where some of my sisters are staying, and for a short season am to be in a continued round of dinners and balls, but shall at least secure a half-hour, in writing to you, from these amusements. I hope Dr. Fitton and Mr. Murchison have said something to you about filling up the map of Sussex, and certain corrections of green sand, &c., near Portsdown Hill.

I marvel less at Dr. ——'s anticipations (as I supposed them) in geological speculation, now that I observe he followed Hutton, and cites him. I think he ran unnecessarily counter to the feelings and prejudices of the age. This is not courage or manliness in the cause of Truth, nor does it promote its progress. It is an unfeeling disregard for the weakness of human nature, for as it is our nature (for what reason Heaven knows), but as *it is* constitutional in our minds, to feel a morbid sensibility on matters of religious faith, I conceive that the same right feeling which guards us from outraging too violently the sentiments of our neighbours in the ordinary concerns of the world and its customs should direct us still more so in this. If I had been Sir A. Campbell, I would have punished those Christian soldiers who dug up the idols of the Burmese temples in the late campaign, and sent them home as trophies. To insult their idols was an act of Christian intolerance, and, until we can convert them, should be penal. If a philosopher commits a similar act of intolerance by insulting the idols of an European mob (the popular prejudices of the day), the vengeance of the more intolerant herd of the ignorant will overtake him, and he may have less reason to complain of his punishment than of its undue severity.

Believe me, most truly yours,

CHARLES LYELL.

CHAPTER VIII.

JANUARY–SEPTEMBER 1828.

ARRIVAL OF AVA FOSSILS—BOTANICAL CHAIR, LONDON UNIVERSITY—DR. FLEMING'S HISTORY OF BRITISH ANIMALS—PARTY AT SIR GEORGE PHILLIPS'S—TOUR ABROAD WITH MURCHISON—PARIS—JOURNEY TO AUVERGNE—SECONDARY AND FRESHWATER FORMATIONS OF FRANCE— VOLCANIC PHENOMENA NEAR CLERMONT—COUNT DE MONTLOSIER— BEAUTIFUL SCENERY ABOUT MONT DOR—CANTAL FOSSILS—ENTOMOLOGY —NICE—MURCHISON OVERTAXING HIS STRENGTH—PAPER ON EXCAVA- TION OF VALLEYS—DETERMINES TO GO TO NAPLES AND SICILY—LETTER TO HERSCHEL—PARTS FROM THE MURCHISONS AT PADUA—GIOTTO'S PAINTINGS.

[In 1828, when he had already planned and made notes for the 'Principles of Geology,' Mr. Lyell set out on a tour to Auvergne and Northern Italy with Mr. Murchison, and after parting from him in Lombardy, he continued his journey alone to Rome, Naples, and Sicily. The results of this tour appear partly in scientific memoirs jointly by his fellow-traveller and himself, and partly in the 'Principles of Geology.' Among the former, is a paper on the Excavation of Valleys, as illustrated by the volcanic rocks of Central France, read before the Geological Society in 1828, and published the year after in the Edinburgh 'New Philosophical Journal;' also a paper on the Tertiary Strata of the Cantal, published in France, in the 'Annales des Sciences,' 1829; and another on the Tertiary Freshwater Strata of Aix, in Provence, in the Edinburgh 'New Philosophical Journal,' 1829.]

CORRESPONDENCE.

To GIDEON MANTELL, ESQ.

Temple: January 17, 1828.

My dear Sir,—The best news I can send you is the safe arrival in Bedford Street [1] of the Ava fossils, of which you

[1] Where the Geological Society then was.

saw a brief account in Jamieson's last Journal from the
'Calcutta Gazette.' To say that they surpass in value any
collection ever brought to Europe, from any other quarter of
the globe, is to say little. In a few days the Embassy have
done us a service which a man's lifetime might have been
deemed well bestowed in rendering.

Crawford,[2] late Governor of Singapore, F.G.S., and author
in our Transactions, was sent ambassador to the Burman
King. On his return from Ava, down the Irawaddi, which
they descended in a steamboat, they were detained many
days in Lat. 21 N., the river having been half dried up, and
the boat stranded. There was a line of hills on each side,
between the river and the hills, an irregular sandy lowish
region (about 300 feet above the sea). All over the surface,
which was nearly bare, silicified fossils were sticking up,
whose weight perhaps had resisted, when loose sand and
gravel were washed away. Crawford employed his servants,
and bribed the natives, to collect, and they filled twelve
chests! Almost entire jaws of a new species of mastodon
as big as an elephant—different from the five species described by Cuvier—smaller teeth of mastodon (Clift says may
be young ones of same); silicified ivory tusk, teeth of
rhinoceros, fragment of a bone of hippopotamus; jaws and
skull of an enormous Gavial; jaws and teeth of an alligator;
large scales, &c., of tortoises. Shells apparently freshwater
—only one species—Sowerby says a Cyrene. Wood trunks
of trees—monocotyledons, and perhaps dicotyledons —no
botanist has seen them, structure beautifully preserved, as
are all the teeth, &c.

They appear to me to have been all converted into a
ferruginous chert while most of them were quite perfect,
both bones and plants. Jamieson is wrong in saying they
are not at all rolled. They are in some instances slightly
rounded by attrition, done of course when in a soft state.
Inside the hollow trunks, and adhering to the bones, is a
ferruginous gravel exactly like Tilgate aggregate bed. The
whole room and yard in Bedford Street looks as if it
were filled with magnificent Tilgate fossils. They are of a

---

[2] Mr. John Crawford, author of *The Embassy to Ava.*

yellow ferruginous colour. Saurians lying in all directions—
here an immense femur, there a long stem-like *Clathraria
Lyelli*; here the scale of a large tortoise, there a shell and
teeth of alligator, &c. &c. They will be exhibited to-morrow,
and Buckland is expected to lecture on them in the evening,
after the anniversary dinner.

Very truly yours,
Charles Lyell.

To Gideon Mantell, Esq.

Temple : February 5, 1828.

My dear Sir,—I have been out of town for ten days in
Hants, with some of my family who are there on a visit from
Scotland. We also went to stay at Dr. and Mrs. Buckland's,
at Oxford. The Canon has a glorious house, and is admirably set down for himself and geology.

I at first intended to write ' Conversations on Geology :' it
is what no doubt the booksellers, and therefore the greatest
number of readers, are desirous of. My reason for abandoning this form was simply this; that I found I should
not do it at all, without taking more pains than such a form
would do justice to. Besides, I felt that in a subject where
so much is to be reformed and struck out anew, and where
one obtains new ideas and theories in the progress of one's
task, where you have to controvert, and to invent an argu-
mentation—work is required, and one like the ' Conversations
on Chemistry ' and others would not do. It should hardly
be between the teacher and scholar perhaps, but a dialogue
like Berkeley's Alciphron, between equals. But finally, I
thought, that when I had made up my own mind and opinions
in producing another kind of book, I might then construct
conversations from it. In the meantime there is a cry among
the publishers for ar elementary work, and I much wish you
would supply it. Anything from you would be useful, for what
they have now is positively bad, for such is Jamieson's Cuvier.

Buckland has been very quiet as yet as to the manifest
difficulties of the Ava fossils, but will no doubt hold forth at
the general meeting.

Yours most truly,
Charles Lyell.

To Dr. Fleming.[3]

Temple: February 6, 1828.

My dear Sir,—Lockhart is exceedingly pleased with the salmon fishery article, and tells me it had, among other merits, that which, unfortunately for him, few papers have, and which is no small charm to an editor, the requiring no sentence mending, a drudgery which he abhors, and which he says cuts up his whole time. Murray also, when I last saw him, thanked me for having been the channel of procuring your article. He has a good publisher's tact for knowing beforehand what people will read. Dr. Hooker [4] has at last made up his mind, and rejected the botanical chair [5] here in form. Young Lindley, the Horticultural Garden Secretary, told me yesterday that even he is now no longer a candidate; thinking, that as there will be no compulsory attendance for medical students, such a class will be precarious and a failure, and that the University do not guarantee such a minimum as can warrant a man, who has anything certain to give up, in venturing. I fear they have made this mistake, that they have not determined to bribe able men for the chairs that can scarcely be profitable from the classes; for Babbage viewed it in this light when offered the mathematical chair. 'What they will secure to me,' he said, 'is no more than I could make in the same number of hours by authorship, and get more fame. They have no dignity to confer as yet, they have their reputation to make. I have not. If, as they admit, they wish to get some from me, why they ought to buy it, and pay for it.' Curtis has found many new and rare insects among our last batch. I look forward to our having a grand bout of entomology and geology in the hills next summer, and bringing back a great stock of health and knowledge.

Ever most truly yours,

Charles Lyell.

[3] The Rev. Dr. Fleming, naturalist.   [4] Afterwards Sir William Hooker.
[5] At the London University.

To His Sister.

My dear Caroline,—Curtis has positively made out the
very species of two insects preserved in the marl,[6] and what
is curious, though now known as northern insects, they are
neither of them very common, nor I believe ever caught by
us, nor are they *water* beetles. I cannot give you the names,
but one is of the *Elateridæ.* That dark clay seam between
the marls was by a summer's flood, as these insects prove.
You really must have worked hard to name so many. On
the 20th you shall have the last batch named by Curtis. I
have long looked out for a *P. cratægi* in vain, but will get
one for our *Artaxerxes.* I remember one season at Bartley
when it swarmed, and they are so stupid, you may take
them up with your hand.

Your affectionate brother,

Charles Lyell.

To Dr. Fleming.

My dear Sir,—I have delayed an unreasonable time in
thanking you for a copy of your 'British Animals,'[7] for which
I am truly obliged to you, and the Preface of which I read
immediately with much interest, and hope to make myself
gradually master of the rest. I think that by putting forth
your Geological Epochs you have done yourself but justice,
and whatever modifications may be made hereafter on such
a system as you proposed, it is one which I firmly expect
will ultimately prevail in geology ; and your priority in deve-
loping the idea, even so far in this country, will be referred
back to with honour. I had fully intended to commence forthwith an article for the ' Quarterly Review,' but was prevented
by many circumstances. One was, that I agreed to go with
Mr. Murchison at the end of the present month to France,
to be in Auvergne by the first week in May, to study the

---

[6] At Kinnordy.
[7] A ' History of British Animals,' by Dr. John Fleming.

volcanic district for some months, and connect Scrope's country with that round Marseilles, and perhaps the Vicentin. My chief defence is, that I did hardly dare review your work until I had obtained more knowledge of zoology, and of systematic arrangements, for your system has been the subject of so much criticism and cavil here, that a man need know what he is about, if he really intends to do you any good by a review.

They say that in putting mollusca before insects, you have committed the great blunder in which Lamarck had the honour of setting Cuvier right. I do not pretend to have an opinion on these subjects yet, but certainly I have more respect for a hornet than a snail, even now that they have found out in New South Wales how the former cannot *help* making true hexagons. As for omission of animals, recent and fossil, I have no doubt you are aware of the bats and other deficiencies, which will come out in second edition as *Vespertilio pygmœus.*

Believe me, very truly yours,
CHARLES LYELL.

To HIS FATHER.

London : April, 1828.

My dear Father,—I chaperoned Mrs. Somerville to Sir George Phillips's on Sunday evening, after a dinner at Dr. Somerville's. Sir G., who is one of the new Baronets, is an M.P., as is his son. A room full of Sir J. Reynolds's and other good pictures, and a famous living gallery of portraits. The party was—Sir Walter Scott, Cooper (the American novelist), Mrs. Marcet and daughter, Sir J. Mackintosh, Rogers the poet, Dumont the Genevese jurisconsult, ' Conversation' Sharp, Lady Davy, Spring Rice, M.P.,[8] Dr. Wollaston,[9] Newton the American artist, Mr. and Mrs. Lockhart, Scott's son and unmarried daughter, &c. Lady Davy, to whom Mrs. Somerville introduced me, was talking well of Manzoni's new celebrated novel, ' I promessi sposi,' so I got her afterwards upon Dante; and she said, ' I bought Rossetti,[1]

---

[8] Afterwards Lord Monteagle. [9] Natural philosopher.
[1] A commentator on Dante.

and read some, but left off for fear I should feel obliged to give way to his theories. I am a devoted admirer of Dante, and should never forgive the man who lessened him in my estimation. There was too much politics perhaps before, but an allegory in every word is horrid. I admire Johnson for saying after all his labours the honest truth, that a person had best read Shakespeare after all, quite through, before he looked at a single note, and I advise all to read Dante as I have done, three times, and as I mean to do a fourth, before they read Rossetti. There is much that I cannot understand, it is true, but there is much that delights me.' Lockhart asked me to breakfast with him next morning, where Sir Walter was, who was good-humoured enough to remember me. He is much aged since last year, but pleasant company. Mrs. Lockhart says she cannot take so many bairns north of the Tweed this year, but she was much pleased with the invitation to Kinnordy, and says she had such a fancy to see Strathmore, that nothing but positive illness prevented her.

Ever yours affectionately,
CHARLES LYELL.

To MR. MURCHISON.

Temple : April 29, 1828.

My dear Murchison,—I hope you have not been expecting a letter sooner, but I was desirous of seeing my way as much as possible before I wrote.[2] Herschel called the other day with the *camera lucida*, and told me some experiments which they have had in melting granite into a slag, like lava, in glass furnaces (he and Faraday), that are *àpropos* to us ; but more of that anon, as also an evening I spent with Scrope, ' Auvergnising.'

Herschel rather alarmed me by prophesying the plague I should have with my barometers, but I shall take them at all events. Dr. Hooker begs me to attend to the plants, as to the soils they live on, so I called on Bicheno,[3] who is now the authority in these matters. He says all we can do is to

---

[2] Projected tour with Mr. Murchison.
[3] An English botanist, secretary to the Linnæan Society.

attend to the plants which abound, as, for instance, he has
done to heath in England, which is restricted in large
quantities to two or three formations. When we come to a
recent crater, or half-covered stream of lava, it will be really
a very curious point to ascertain what is growing on it. If
there was a large tree growing on the most recent stream,
it would be almost worth having it cut down, for if it had,
as some are said to have in Norway, 1,500 rings, we should
carry the eruption back to Noah, allowing many centuries
before any tree grew. But, joking apart, the trees and prevailing plants on the recent cones, &c., will be important.
I have set up the proper boards for drying. My father
writes to persuade me to take my clerk. His chief object is
that he fears my eyes, which got weak some years ago under
the sun of Italy, will require the assistance of my well-
trained amanuensis, and that in this way some plants will
really be obtained for Dr. Hooker and himself, and some
insects for my sisters. If I take him, he will of course find his
way per coach to Auvergne, and find his way when there on
his legs principally, and I should not give him the barometers,
as this disables the bearer from work, and I should possibly
send him back when I quitted the mountains of France. He
is clever, and I doubt not I should do more with him, but I
am not quite decided if it will answer the additional expense,
and am sorry I cannot talk over with you any possible objection that you might have to another servant. However, I
will take care, if I bring him, that he keeps clear of incom-
moding the party.

Believe me, ever most truly yours,

Charles Lyell.

To His Father.

Hôtel Mont Blanc, Rue de la Paix, Paris : May 9, 1828.

My dear Father,—I left London May 4, slept at Dover,
and on Monday evening got into the mail at Calais with
Hall.[4] Next day stopped only three hours at Amiens, and
arrived at Paris four o'clock on Wednesday morning. After

---

[4] His clerk.

four hours' sleep, breakfasted with Murchison, and had a
grand day seeing people, and in the evening cut in for a
famous lecture of Prévost's on Geology—cozed till midnight
with him here, and next morning again with him at breakfast. He is very well, has two children, Madame in the
country, at Montmorency. He is as keen as ever about
Geology. Murchison has planned everything capitally, and
unless my confounded passport, not sent in time from Calais,
does not detain us, we start for Clermont to-morrow, geologising a little *en route.*

Murchison has ordered his letters to Clermont Ferrand.
Mrs. M. very well and in good spirits. Has been dining with
Lady Granville, Pozzo di Borgo, Cuvier, and other great
personages, fashionable and scientific, and no end to offers
of boxes to opera, &c.; but science and real work is made in
spite of this the main object, and other advantages all made
to turn to account in this way. They are both very industrious. I have seen Férussac, Duvan, Dufresnoy, Elie de
Beaumont, and am to meet the three former again to-day.

With my love to all, believe me, your affectionate son,

CHARLES LYELL.

To His Father.

Clermont Ferrand, Auvergne : May 16, 1828.

My dear Father,—We left Paris on Sunday, May 10,
Murchison and I on the box of his light open carriage, and
Mrs. M. and a Swiss maid in the inside. Started at half-
past six o'clock, and carriage broke down in boulevard, opposite the coachmaker's, who had made a precious job of
remedying some slight damage.

Most of the road from Fontainebleau to Moulins was
dreadfully out of repair, but the vehicle was got safe over,
and in proportion as we approached the mountains, and re-
ceded from Paris, the roads and rate of posting improved,
and at last averaged nine miles an hour, and the change
of horses almost as quick as in England. The politeness also
of the people has much delighted us, and they are so intelligent, that we get much geology from them.

The processions of 'Rogation,' and yesterday of the *fête*

of Ascension Day, have shown a strange contrast between
the religious faith of the Auvergnois, and the provinces
around, and to the north of the Loire. In the latter, there
was literally none but priests singing, monks and choristers,
hundreds of fine young men, and in most cases only four or
five women and children following, sometimes not a single
man. We heard that in some processions to distant crosses
the priest had found himself at the cross alone. Here the
church and procession is as numerously attended as any in
Germany, yet the Jesuits have no influence, as they universally return Liberal deputies. Murchison has letters from
the Deputy La Fayette's son, which ensured us attention
from a useful medical *savant* on our arrival. I never did in
my life so much real geology in as many days. A good paper
might be written if we stopped here. We have learned many
new things in regard to secondary formations in France,
north of the Loire, and since that we have positively found
that nothing was correctly known of the freshwater formation, either as to organic contents, inclination and disturbance of strata, mineral structure, &c. &c. So we think
of writing ' On the Freshwater Formation of Central France,
and its relations to the Primary, Secondary, and Volcanic
Rocks,' &c. This may be considered quite a new and unexplored field, and when investigated, it must throw quite
a new light on the age of the volcanic rocks, and of the
elevation of the country, &c. We have generally begun
work at six o'clock, and neither heat nor fatigue have stopped
us an hour. Mrs. Murchison is very diligent, sketching,
labelling specimens, and making out shells, in which last
she is an invaluable assistant. She is so much interested in
the affair, as to be always desirous of keeping out of the
way when she would interfere with the work, and as far as
I yet see, it would be impossible to form a better party. I
sent Hall by diligence here. He had a day before our
arrival, and collected insects, among which a specimen of
Papilio Antiope is the most showy. I will see what can be
done in plants, but the French say everything is well done
in the botany of Auvergne, no part of Europe better. So say
the *savans* of Paris ; but so some say of the geology, of which
we find that nothing is done. Murchison, I am happy to

find, is quite of my mind to direct all our force to where there is real work, to concentrate, and not to run, as Buckland has done, so that now he cannot literally publish on a single spot on the Continent.

Auvergne is beautiful—rich wooded plains, picturesque towns, and the outline of the volcanic chain unlike any other I ever saw. The range of Mont Dore, seen over the volcanic range from fifty miles to north of this, was covered with snow, and looked like the Alps, but they say it will soon melt.

We have been much surprised that ever since we left the chalk and oolite, and entered the freshwater formations along the Allier, we have never seen, till we came close to Clermont, *any* volcanic pebbles in the gravel, which was chiefly granitic. If most of the volcanoes were antediluvian, and Buckland's deluge caused the diluvium, how could this have happened? I verily believe that we have collected already a larger number of organic remains, from quadrupeds down to fossil seeds and shells, than have yet been published as belonging to Auvergne. Murchison is a famous hand at a bargain, and as he takes that on himself, I hope to get through very economically. He always makes a bargain before going into an inn, by which much time and expense are saved in the end. We get off within a third of what the natives pay, and for half what John Bull does *en route*; but, thanks to his campaigning of old, he loses no time in transacting this business.

Believe me, your affectionate son,

CHARLES LYELL.

To HIS FATHER.

Clermont Ferrand, Auvergne: May 26, 1828.

My dear Father,—I have just returned again to Clermont, from an expedition of five days, and we have discovered that there is no end to the work to be done in this country, and that it is of the most interesting description. The first day was spent in ascending some of the lofty volcanic Puys near here. Mrs. Murchison accompanied us, and then returned to Clermont, where she employed herself during our absence in making panoramic sketches, receiving several of the gentry and professors, to whom we had letters, in the neighbourhood, and collecting plants and shells, &c., while Murchison and I, with my man, went on in a *patache,* a one-horse machine on springs. We first visited Pontgibaud and the Sioule, to see the excavations made by that river, in the grand lava-current of Come, which descended from the central range, and dispossessed the river of its bed. The scenery was beautiful. Just as we were leaving the place, the peasants offered to take us to a volcano farther down the river. As no Puy was mentioned in Desmarest's accurate map, nor by Scrope, we thought their account a mere fable ; but their description of the cinders, &c., was so curious, that we had the courage to relinquish our day's scheme, and proceed again down the river.

You may imagine our surprise when we found, within a ride of Clermont, a set of volcanic phenomena entirely unknown to Buckland, Scrope, or the natives here. A volcanic cone, with a stream of basaltic lava issuing out on both sides, and flowing down to the gorge of the Sioule. This defile was flanked on both sides by precipitous cliffs of gneiss, and the river's passage must have been entirely choked up for a long time. A lake was formed, and the river wore a passage between the lava and the granitic schist, but the former was so excessively compact, that the schist evidently suffered most. In the progress of ages, the igneous rock, 150 feet deep, was cut through, and the river went on and ate its way, 35, 45, and in one place 85 feet into the subjacent granitic beds, leaving on one bank a perpendicular wall of basaltic lava towering over the gneiss. In the Vivarrais, where similar phenomena had been observed, Herschel had remarked a bed of pebbles between the lava and the gneiss, marking the ancient river-bed, but Buckland endeavoured to get over this difficulty by saying that these pebbles might have covered a sloping bank when the river filled the valley, and that this bank may have always been high above the river bed ; for if the sloping sides of a valley, said the Professor, be covered with pebbles, as they often are, and the valley is filled with lava, and then the lava cut through and partially removed, there will of course be a line of pebbles at the junction of the lava and the rock beneath, but these pebbles will not mark an ancient river bed. Now, unluckily for the Doctor in this case, he has no loophole. An old lead mine, said to have been worked by the Romans, happens to have exactly laid open the line of contact, and the pebble bed of the old river is seen going in under the lava, horizontally, for nearly 50 feet. This is an astonishing proof of what a river can do in some thousand or hundred thousand years by its continual wearing. No deluge could have descended the valley without carrying away the crater and ashes above. 600 or 700 feet higher is an old plateau of basalt, and if this flowed at the bottom of the then valley, the last work of the Sioule is but a unit in proportion to the other. There are several of the Clermont *savans* who, since they discovered how much we were interested with this, have given us to understand they intended to publish on it, but no doubt they will take a year before they launch out in the expense of a *patache* to Pontgibaud. Murchison certainly keeps it up with more energy than anyone I ever travelled with, for Buckland, though he worked as hard, always flew about too fast to make sure of anything. Mons. Le Coq, the botanist, a clever young man, assures me that the geology of the soils does not affect the botany of Auvergne. I shall get some specimens from him for Dr. Hooker, I expect. None to be bought, at least this year, for it seems there may be hereafter. It is a wonderful fact that *Glaux maritima* grows round some saline springs here. Busset, an engineer, who is mapping Auvergne, has forced us to dine with him to-morrow. As we know his object to be to get geology out of us, of which he knows nothing, M. fears it will be a bore, but the man is evidently clever. We shall get barometric heights from him, and a map of our little volcanic district, and if he pumps unreasonably, I shall find a difficulty in expressing myself in French. We are to meet Count Le Serres there, a gentlemanlike and well-informed naturalist, who has a property on Mont Dore, and knows more geology than any one we have met here, professors not excepted. He organised a geological society here, and they chose Count Montlosier [5]

[5] Naturalist, author of an essay on the Extinct Volcanoes of Auvergne, b. 1755, d. 1838.

as president; but the Jesuits took alarm, and, declaring that Montlosier had written a book against Genesis, got the Prefect and Mayor and Government to oppose, and at last put the thing down; at least it merged in the regular scientific Etablissement de la Ville, and Montlosier is just coming out with a book against the Jesuits, a more popular subject in France at present than geology. We are to visit him at his château near Mont Dore. We like the people and the country.

Believe me, your affectionate son,

CHARLES LYELL.

To HIS FATHER.

Bains de Mont Dore, Auvergne : June 6, 1828.

My dear Father,—I am this moment arrived here, after passing three delightful days at Count de Montlosier's, an old man of seventy-four, in full possession of faculties of no mean order, and of an imagination as lively as a poet's of twenty-five. I stayed a day longer than the Murchisons, as I was determined to have one more trial to find a junction between the granite of the Puy chain and the freshwater formations of the Limagne, and I actually found it; and my day's work alone will throw a new light on the history of this remarkable country. I believe most of the granite to have made its appearance at the surface at a later period than even the freshwater tertiary beds have, though *they* contain the remains of quadrupeds.

The scenery of Mont Dore is that of an Alpine valley, deep, with tall fir woods, high aiguilles above, half covered with snow, and cataracts and waterfalls. A watering-place with good inns at the bottom of the valley. I shall send Hall back from here, as, although he has been useful, I do not think the advantage will overbalance the additional expense. Le Coq has promised some plants for certain, and Hall has done pretty well in insects.

Believe me, your affectionate son,

CHARLES LYELL.

To His Mother.

Bains de Mont Dore, Auvergne : June 11, 1828.

My dear Mother,—We have been so actively employed, I may really say so laboriously, that I assure you I can with great difficulty find a moment to write a letter. This morning we got off, after breakfast at five o'clock, on horseback, to return from St. Amand to this : arrived at seven o'clock. But one day we rode fifty-five miles, which I shall take care shall be the last experiment of that kind, as even the old Leicestershire fox-hunter was nearly done up with it. But I have really gained strength so much, that I believe that I and my eyes were never in such condition before ; and I am sure that six hours in bed, which is all we allow, and exercise all day long for the body, and geology for the mind, with plenty of the *vin du pays*, which is good here, is the best thing that can be invented in this world for my health and happiness. Murchison must have been intended for a very strong man, if the sellers of drugs had not enlisted him into their service, so that he depends on them for his existence to a frightful extent, yet withal he can get through what would knock up most men who never need the doctor. He has only given in one day and a half yet. On one occasion we were on an expedition together, and as a stronger dose was necessary than he had with him, I was not a little alarmed at finding there was no pharmacy in the place, but at last went to a nunnery, where Mdlle. La Supérieure sold all medicines without profit—positively a young, clever, and rather good-looking lady, who hoped my friend would think better of it, as the quantity would kill six Frenchmen. M. was cured, and off the next morning, as usual. The mischief is, that he has naturally a weak though a sound stomach, and if he possessed a more than ordinary share of self-denial, and was very prudent, and after much exercise did not eat a good dinner when set before him—if, in short, he would take the advice which many find it easy to give him, he would be well.

He has much talent for original observation in geology, and is indefatigable, so that we make much way, and are thrown so much in the way of the people, high and low, by means of our letters of introduction and our pursuits, that I am getting large materials, which I hope I shall find means of applying. Indeed, I really think I am most profitably employed on this tour, and as long as things go on as well as they do now, I should be very sorry to leave off; particularly as, from our plan of operation, which is that of comparison of the structure of different parts of the country, we work on with a continually increasing power, and in the last week have with the same exertion done at least twice as much in the way of discovery, and in enlarging our knowledge of what others had done, as in any preceding. I expect it will be at least three weeks before we can have done with Central France, and then we hope to work south towards Nice, down the Rhone, keeping always in analogous formations, and then to the Vicentin if possible, though this is very uncertain, as we can never see far before us, either as to time or place, directing our course according to the new lights we are gaining.

We shall leave this place in a day or two. I like it well enough, but it is certainly too early in the season to enjoy it, and Mrs. Murchison suffers from the cold and damp, though she has not often complained in this tone.

Mont Dore is partially covered with snow, and almost always with clouds, and the transition in coming up here from the low country is violent. Yesterday we rode up from the climate of Italy to that of Scotland. It is the most varied and picturesque country imaginable. There are innumerable old ruins for sketches, with lakes, cascades, and different kinds of wood, so that we wonder more and more that the English have not found it out. The peasantry are very obliging, industrious, well-fed, and clothed, and to all appearance are the happiest I ever saw. We have crossed the chain of Puys, the Limagne, and the valleys leading from Mont Dore, in all directions. The people in the higher regions begin to talk French—at least, there are generally some who have served in the armies, and their children catch some from them. Their own language has a good deal of the old Provençal in it, and a great many of the terminations are Italian. In short, we often find a demand in Italian succeed when French misses fire; but all our am-

munition often fails to produce any impression. The population is dense, and bears no resemblance to other parts
of France that ever I saw. In the mountains a large portion
do not believe that Napoleon is dead, especially the old
soldiers. There is an almost entire want of gentry here, but
as it does not arise from absenteeism, but from the great
subdivision of property, it evidently produces no ill effects on
the character and well-being of the people.

Give my love to all at Kinnordy, and believe me,

Your affectionate son,

CHARLES LYELL.

To HIS SISTER.

Puy en Velay : July 3, 1828.

My dear Caroline,—I was glad to find your letter awaiting
me here yesterday, though now of a somewhat ancient date,
and hope that another will arrive before we quit this, in
about a week, from some one. We have now seen the Cantal
district, of which Aurillac is the principal town, and I was
there particularly interested with the freshwater or rather
lacustrine deposits ; as although of an older date than all
the volcanoes of that country, and constituting chains of
hills, their correspondence, when considered foot by foot with
the beds which are at our door in the marl loch, is as complete as you can imagine. The same genera of shells and
compressed vegetables, and seeds of chara, &c., of which we
hope to add many new species to the fossils hitherto known.
I hope, by the way, you will not forget, when the marl is dug,
to keep up inquiries for jaws and teeth of animals, and if you
can get some good specimens of the shells, it will help me ;
for now that I see the perfect analogy of these older formations, I am resolved to give a notice on the minor details of
our loch. I am in great expectation that some of my Cantal
fossils will prove the elytra of beetles. They were in marl
near Murat in Cantal, upon which rested an enormous load
of different volcanic materials, such as currents of columnar
basalts and other lavas and breccia, no less than 800 feet in
a very precipitous height. Yet the limnei and planorbis and
other lake remains in the foundation beds were not injured,

and looked just as marls might do in Angus. The most
characteristic people we have seen are perhaps those of Mont
Dore, but they are an oddish race throughout these regions of
Central France. The language varies as we move about,
and is always equally unintelligible ; it is more Spanish here
than farther north. The communication between the capital
and the interior is wonderfully small, but to our surprise the
roads have been far better than near Paris, and the carriage
has never been broken since we crossed the Loire. They
macadamise their roads in good style. This is a most extraordinary and picturesque valley, whereas the Cantal, though
interesting to a geologist, has not much to boast in that way.
Mont Dore is fine and Alpine, but then it is not equal to the
Swiss Alps and in the same style. But the chain of Puys of
Clermont is so perfectly unlike any other European chain,
and its Prince, the Puy de Dôme, is so handsome, that this
is worth a visit by all who travel in France.

We flatter ourselves we have got more ample materials
for the geology of these parts of France, at least on the
modern rocks, than have as yet been published on any districts of equal extent and equal distance from Paris ; and as
this has all been done in seven weeks, we should have something to show at the end of six or eight weeks more, if all
could go on at the same pace. But I do not think the Murchisons will stand fire. Symptoms of flinching from the
heat, which makes scarcely any impression on me, begin to
betray themselves. Thoughts of a retreat to the Alps, con-
sultations with me whether I think it practicable to proceed
farther south in the dog-days, have been mooted, and I suppose the whole scheme will hardly be persevered in just as
if Sedgwick and I were here together, determined to hunt
down thoroughly one subject at a time. But as far as we
have gone, I was never with a better man for doing work well
than Murchison, and as we have two weeks before us here,
that will be a good deal to have secured. The French, since I
was last on this side the water, have come round marvellously
to our costume, and they are on the high road to become
just as great politicians, and in many respects to be more
like us. If our people would have the same good sense to
take from them what they surpass us in, it would be a good

compromise. Their system of early rising, and spending a moderate time at dinner, agrees famously with those who wish to make use of their evenings and still see society. If Prévost had known these parts of his own country, he would not have regarded the scene at Penzance on Guy Fawkes's day as particularly John Bullish. For on St. Peter's day we saw at St. Germain in Auvergne, and other towns, bonfires and fireworks, and the peasants pushing each other into the fire and jumping through it. We are to visit the country house here of Bertrand Roux, author of a geological work on these parts. He is a merchant, and an intelligent man.

I sent back Hall from Mont Dore. The last two days were the first in which he signalised himself by doing anything whatever in entomology, for he then took about two dozen of *Blandina*. We saw numbers afterwards.

With love to all, believe me, your affectionate brother,

CHARLES LYELL.

To HIS SISTER.

Aubenas, Ardêche : July 15, 1828.

My dear Marianne,—We found Puy en Velay a very extraordinary and beautiful town, with some remarkable rocks standing up in it like Edinburgh Castle, and as well covered. Our introduction to M. Bertrand helped us not only to the best geological cicerone we have yet met with in France, but also an agreeable opportunity of seeing two large French establishments, his own and his mother-in-law's, Madame de Laintenot, who, in being able annually to visit Paris, is a great personage among the provincials. We spent a day at her place, and two at Bertrand's, who inhabits a fine old monastery called Doue, which his father, a rich merchant, purchased in 1791 out of the Revolutionary plunder. He is himself a merchant. The ladies, young and old, displayed a curiosity about Mrs. Murchison's dress and English wardrobe that would have diverted you. Considering the natural gaiety of the various persons we met there, we were somewhat amused at one favourite theme of regret, expressed by many of the old people, that the French were becoming too *serious.* They attribute this to the English style of education, which they say is now being introduced, which, although right in many respects, is carried to an extreme. *C'est plus solide,* they say, than theirs was, and there was once some reason to accuse them of *légèreté,* but there is a medium. Even our friend Bertrand seems decidedly to take the same view of the matter, though he accounts for it in another way, for he says every young man now looks early to a profession. None go out of good society, as formerly, into the Church and monasteries, and there being no elder sons, none can afford to be gay and thoughtless. But even he appears to see too much puritanical severity in the rising generation, which I have certainly not seen, and as far as a traveller can judge, they are still an infinitely more lively people, high and low, than we are, and look at least happier, especially the peasantry of Central France. From Le Puy, where we stayed eight days, we went to the Haut Vivarrais. The descent from the granite mountain of Velay to Vivarrais (now Ardêche) is exceedingly fine, the outline of the hills very Alpine, and the deep valley clothed with rich chestnut trees, and vines dressed in the Italian style. We stayed three days at Thueyts, a small town in the Ardêche, whence we were able to examine carefully what Scrope in a letter calls ' the pet volcanoes of the Vivarrais ; ' and such they really are, very far surpassing all the two or three hundred we had seen in Auvergne, Cantal, and Puy. All lovers of the picturesque, and who have got no farther than ' a taste for geology,' should come to these, for they will find it all so clear, as well as beautiful. The craters are so perfect, yet the cones covered with such fine zones half-way down of chestnuts, and the lava currents eaten into by the rivers, present such splendid colonnades of basaltic pillars, that it is a country to make everyone desire to know something ; and as the granitic schists in which the valleys lie are all of one kind, there is no danger of the volcanic matter being confounded by anyone with the older rock through which it has burst, or over which it has flowed. The proofs of their immense antiquity are quite enough to bear out Scrope. But Murchison and I must take care how we recommend any English to come here, as there is no post over much of the road, and the *voituriers* have no harness, and cannot

drive two horses abreast, and scarcely in a single instance in
Central France can they be trusted.   For ten miles there is
a broad road worthy of the magnificence of the *Grande
Nation*—bridges quite splendid, rock blown through, &c.
Then all at once stones and ruts and a narrow lane, with a
frightful precipice at the side, and no parapet.   As far as
steady geological work goes, I have not had a single day to
regret since I entered Auvergne, two calendar months
yesterday.   Amidst solemn declarations on the part of Murchison that he cannot go south in the dog-days, he has
ordered letters to be addressed to Nice, where I hope to
hear from you.   You will not suppose that any attention
has been paid to entomology, except what has been unavoidably done in the *Aptera* and *Diptera*, the former of
which have been sacrificed in large numbers by the Murchisons, having robbed them of many a night's rest.   They
have spared me, but in the day-time the *Diptera* have
attacked all indiscriminately, both man and beast, ever since
we passed the first Spanish chestnuts, on our descent from
the high central plâteau towards the Rhone.   This country
breeds many silkworms.   They tell me that two caterpillars,
and frequently three, spin the same cocoon, which then
contains three pupæ.   The greater part of the *Lepidoptera*
are English acquaintances, and were remarkably so till we
descended from Central France, as were the plants.

Believe me, your affectionate brother,

CHARLES LYELL.

To HIS SISTER.

Marseilles : August 3, 1828.

My dear Caroline,—At Aix there has been discovered,
within a year, a thin band of fissile, compact, calcareous marl
full of insects.   It divides into thin laminæ, as does our
Kinnordy insect bed, but is a solid stone.   The insects, even
the gnats, are beautifully perfect, wings, antennæ, and all.
Until we came all had gone to M. de Serres of Montpellier.
He has made out fifty genera.   The beetles are chiefly
Curculionidæ; most of the rest are *Diptera*, *Musca*, and
Culex, &c.   A perfect *Gryllus* and a *Forficula* are amongst them.

There are so many, that I am in hopes they will somewhat aid the question as to the climate of these parts when
the tertiary strata were formed.   The series of strata in
which they occur is 400 feet thick, and the fish and shells
show them to be freshwater, probably found in a lake.   The
country has been much thrown about since, and the insects
must be of great antiquity, though not old, geologically
speaking, for hills and valleys have since been formed.
Murchison kept saying, when we were hunting about the
quarries and getting them from the man, 'How I wish your
sisters were here, they would enjoy the chase, and then we
might be at work at the geology.'   I think we have got at
least fifteen genera, and more will be sent us.   We find
provincial geologists almost everywhere who know enough
to be useful.   Nismes is a glorious remnant of Roman grandeur, with an amphitheatre and temple nearly entire.   The
old aqueduct at the Pont du Gard, north of Nismes, is as
fine as almost anything in Italy.   There is a triumphal arch
and tower, like the lantern of Demosthenes at St. Remy,
between Nismes and Aix, of great beauty, and of which, if at
Rome, antiquaries would talk through folios.   Montpellier
is the most luxurious place we have seen, and this the
cleanest seaport.   The troops are passing in daily, returning from Spain, much quizzed by their countrymen and
women, for being burnt as black as Spaniards.   The generality here are fair, spite of the sun.   Other troops are going
out to the Morea, such a stunted race!   By accurate calculation of the height of men of the levy since the peace, it is
found that the mean height of Frenchmen has been
diminished several inches by the Revolution and Napoleon's
wars.   These are now the sons of those who were not
thought by Napoleon strong and tall enough to fight and
look well.   But they will rout the Egyptians for all that,
and I suppose be a match for the Russians.

With love to all, your affectionate brother,

CHARLES LYELL.

Nice : August 20, 1828.

My dear Eleanor,—There has not literally been one drop
of rain here for eight months ! So all the rivers are quite
dry, and the stagnant waters are not a little odoriferous,
and were it not for some protecting saint, the malaria
would have its head-quarters here, according to the modern
doctrines on that subject : and why it never visits Nice, is
certainly a puzzling problem. The only French journals
admitted here are the ultras, who are defending the cause
of the persecuted Jesuits, and are just now very angry with
the liberal papers for some profane quizzing of the Archbishop of Paris, who has been praying for rain. While he is
about the matter, he may as well have it sent here, for they
hardly hope for a drop for a month to come, and do not
think it at all unprecedented. Murchison has not regained
his strength from a severe attack at Fréjus, so as to be able
to take the field again. But we have been hard at work in
writing, from our materials, a paper on the excavation of
valleys, which is at last finished, and after two evenings'
infliction, is intended to reform the Geological Society, and
afterwards the world, on this hitherto-not-in-the-least-
degree-understood subject. Besides this mighty operation,
we have performed two jaunts with Mrs. Murchison, each at
half-past four o'clock in the morning, to see certain deposits
of fossil shells, and collecting these, with which she has been
much pleased ; and this and the cessation of eternal bustle,
while the campaign was at its height, while we were as yet
only crossing the Balkan, and before our descent upon these
hot latitudes, has restored her health and spirits, which
had failed sadly. Her lord has a little too much of what
Mathews used to ridicule in his slang as 'the keep-moving,
go-it-if-it-kills-you' system, and I had to fight sometimes
for the sake of geology, as his wife had for her strength, to
make him proceed with somewhat less precipitation. You
may suppose it was not over prudent to attempt hard work,
and only to sleep, or rather to be in bed, five hours at most.
I expected a break-down before. I trust still to start in a

few days to the Vicentin, either by Genoa or the Col de Tenda, for that country must afford some curious analogies and lights to what we have seen in Auvergne. The people are polite and agreeable here, the Government unpopular, though most absolute. Numbers of soldiers and priests everywhere. The principal librarian has no books but what he sells to 500 English who reside here in winter. He says the inhabitants of the great city never read scarcely even novels. What, then, do they do? ' Can you not see? They sit in cafés, play billiards, hear the bands play, go to mass, and give and attend fêtes.' All this makes the place very gay. Everything is quite gone back to the old *régime* in good earnest. And when the Jesuits are exiled from France, if they dare execute that order, they will find the receptacle prepared by the King for them here, a congenial home. They are positively hiring apartments now for French pensioners, youths who are to follow them into exile for education. Our kind old host Montlosier's works are so attacked by the ultra journals here, that it is clear they produced no small effect. I caught a curious *Mantis* this morning, the peasants call it ' prega-dio,' and when it erects its neck, and holds up its chelæ and clasps them, it certainly seems to be saying its prayers. I shall try and catch some gnats for him in my room, for, in spite of a mosquito net, they have bitten me all over. One day of winter in Italy would do more for geology than a week of this everlasting blue sky : even the Italians are sick of seeing it. No wonder we are so much in the dark in geology as to India and the tropics ! If I ever attempt to understand the south of Italy as well as I hope I now do Central France, it shall be in winter, or before May. All our best men made the blunder we have. We went out in the bay on Sunday with a famous diver, who brought up sponges and other creatures that I had never seen alive before. He took down a hammer, and, when out of sight, knocked off the stone-perforating shells.

With love to all, believe me, your affectionate brother,

CHARLES LYELL.

To His Father.

Nice : August 24, 1828.

My dear Father,—Before we cross the Alps, and enter as it were upon a new expedition, I will send a few words on the fruits and adventures of our past tour, and the plans I wish still to realise before I return. The paper which we have written here, and which for the first moment almost has had time given it to think over any of our data, has encouraged us to be sanguine as to what has been obtained. The whole tour has been rich, as I had anticipated (and in a manner which Murchison had not), in those analogies between existing Nature and the effects of causes in remote eras which it will be the great object of my work to point out. I scarcely despair now, so much do these evidences of modern action increase upon us as we go south (towards the more recent volcanic seat of action), of *proving* the positive identity of the causes now operating with those of former times.

At this very place, which Brongniart and Buckland have been at, without seeing, or choosing to see, so unwelcome a fact, we have discovered a formation which would furnish an answer to the very difficulty which Sedgwick when at Kinnordy put to me. He said, ' You who wish to make out that all is now going on as formerly, help me to conceive a sea deep enough and disturbed enough to receive, in any length of time, such a series of strata of conglomerate and sandstone as you have shown me in Angus.' Now here we have just such a series as that in Forfarshire, only very much thicker, and in the intervening laminated sands are numerous perfect shells, more than 200 species in Risso's cabinet, eighteen in a hundred of which are *living Mediter-ranean* species, whose habits are known. By this grouping of these shells and their state, the sea is proved to have been in a perfect state of tranquillity, except at those periods when the pebbles were washed down. This conglomerate is intersected by a valley and a gorge, displaying it for fifteen miles ! The hills of conglomerate are some 800 feet high, and at this height are fossil species, which may be the identical ancestors of others that now live opposite here, where the sea within gun-shot of the shore is 3,000 feet deep, measured by Saussure, and since verified. Into this the Magnacon, after not flowing a drop for eight months, brings yearly multitudes of pebbles. But I feel, as Murchison says, that if I wish to get out, by November year, such a book as will decidedly do me credit, and probably be a source of profit, it is to the south of the Po I should hasten, and before Christmas everything might be done. He said yesterday, ' At Milan or Verona, in three or four weeks or thereabouts, the operations connected with our paper will be over, but Sicily is for your views the great end : there are the most modern analogies, volcanic, marine, elevatory, subsiding, &c. I know the island as a soldier, and if you make straight for Etna, will just time it right for work, for the season will be exactly suitable.' . . . I feel so decidedly that three months' more steady work will carry me through all, and tell greatly both to the despatch of finishing my work, as well as to the power I shall have in writing it, that I should greatly repent if I did not do all in my power to accomplish it now.

With my love to all, believe me, your affectionate son,

CHARLES LYELL.

To SIR JOHN HERSCHEL.

Milan : September 9, 1828.

Dear Herschel,—I arrived here with the Murchisons yesterday, and to-morrow we start for the Vicentin, where we hope to finish successfully the remainder of the plan which was talked over with you, and which has been thus far executed much to our satisfaction. We profited much by your hints and memoranda, and Murchison begs me to say that he intends still further to benefit by your advice in the Val de Fassa, though his time will be shorter than he had intended. We part at Verona, and I then hope to make out a tour in S. Italy and Sicily. I now regret much that I had not formed this scheme before I left England, for although I believe I remember pretty accurately the principal geological facts in regard to Sicily which you and others gave me, yet for want of noting the localities I shall be little able

to avail myself of them. You would render me a great
service if you would send me, in a letter to Florence, any
hints and advice on points of geology which you think
should be investigated, and which you either saw, or think
you omitted. As we have not been idle as far as our time
permitted in Auvergne, Velay, and Vivarrais, in working at
the relations of the volcanic to the tertiary and other
associated rocks, I am prepared with a moderate stock of
queries, and of difficulties, to be answered I hope by studying a more modern volcanic district. The effects of earthquakes on the regular strata, and the light thrown on the
excavation of valleys by lavas, are subjects to which I have
directed a large share of attention. I should therefore be
anxious to examine such parts of the coast of Sicily or
Calabria as afforded evidence of elevation or subsidence,
either by the aid of buildings, &c., raised or sunk, as at Baiæ,
Temple of Serapis (if the latter be not otherwise explicable),
or by help of modern species of shells lifted up, or sea
beaches. Wherever you observed signs of such effects of
earthquakes in Sicily, I should be glad to have the localities,
and where you saw the greatest fissures, if any, still open.
The disturbance in the freshwater strata of Auvergne and
Cantal, due to volcanic action, is so much greater than I had
been led to expect, and that of the sub-Apennine beds from
Montpellier to Savona, containing as they do nearly twenty
per cent. of decided living species of shells, that I cannot but
think that Calabria and Sicily must afford proofs of strata
containing still more modern organic remains, raised above
the level of the sea. If you will inform me how you moved
about in Sicily, I shall be much obliged to you, and any
desirable head-quarters, from which much may be learnt,
besides Catania and Palermo and Messina.

As I know how well and how fully your time is occupied,
I am aware that this request is somewhat unconscionable,
but however few the memoranda you can favour me with, I
shall prize them much. The names of any localities, or
persons or things to be seen, or to be avoided, in a tour of a
few months, is what I am most in need of, or any books
which I should buy or read. We asked here for the Council
of Mines. ' It is done away with.' For Brocchi's successor.

'The chair exists no longer.' For his collection. ' It has been pillaged by the Germans, and they have left nothing worth seeing.' Is there no scientific establishment left? 'There is only the library.' So much for the paternal government of Austria.—Believe me, ever most truly yours,

CHARLES LYELL.

To HIS FATHER.

Padua : September 26, 1828.

My dear Father,—The Murchisons left this for the Tyrol this morning, and I start for Verona this evening. I came here that I might examine the Euganean Hills, the geology of which is interesting.

The new road by the Maritime Alps is quite a fine addition to the English tourists who wish to vary their way over the Alps. The scenery is remarkably identical in character with that which we saw along by the Gulf of Spezzia, supposing greater height of the mountains ; and it is throughout the same. The road is always in sight of the sea, often overhanging it on a precipitous steep of some thousand feet. Hills covered with old olive trees, and now and then large orchards of lemon trees. These here and there exchanged for vines. The palms in many gardens of considerable height, and great prickly pears hanging over walls. Sea, deep up to the rocks, which rise boldly out of it: scarcely ever a sail.

Between Savona and Modena we did an agreeable piece of geology, the Cadibona coal basin, evidently a lacustrine deposit. As it was unoccupied ground, and of interest from producing a curious fossil quadruped, called by Cuvier ' Anthracotherium,' we expect it will make a good subject of a notice for Jamieson's Journal as soon as I return. The neighbourhood of Turin may perhaps lead us to another notice, for our passage over the Apennines from Savona to Alexandria certainly gave us a key to the environs of Turin, on which such blunders are current, that we may be excused if we set some right, even though we cannot pretend to have done the district. But this will not be for the Geological Society, to which we give the paper I mentioned, and one on the freshwater formations of Auvergne, if our specimens all reach London in safety.

From Borelli, the conchologist at Turin, and others at Vicenza and this place, we learnt much. But the chief advantage from a few days' hard work at Monte Bolca and other parts of the Vicentin will be, that it renders us quite competent to judge the merits of and to understand the various works both of the ancients and moderns on this country. The volcanic phenomena are just Auvergne over again, and we read them off, as things written in a familiar language, though they would have been Hebrew to us both six months before. The trachyte of the Euganean Hills is the porphyry of Lintrathen and the Clune,[6] so is that of Mont Dore—all identical—and the felspathic lavas of the volcanoes of S. America come quite close to them. It is here pinkish, as in Forfarshire, and sometimes regular as there : so it is in the lava streams from Mont Dore.

I asked Giapelli, an eminent engineer to whom Count Marzari Pencati of Vicenza gave us letters, to take me to see Giotto's pictures, on which Phillips gave a good lecture in Somerset House last spring. I suppose we saw them in 1818, and that you remember them. To me they were quite new. The Chiesetta dei Cavalieri Godenti has all the walls painted by Giotto. It is in the amphitheatre (Roman), or rather joins and looks into it. These knights were a powerful and religious order, contemporary with the Templars. They called themselves ' Rejoicers in the mysterious joy of the blessed Virgin.' Fresco I believe was then unknown, and the plaster was coloured in a manner which you know, yet the colours remain. On one side over the door is a view of Hell. The devil is a faithful portrait, full length, of the Burmese idol lately set up in the British Museum. (As there are many Eastern figures in other parts, Giotto may through Venice have got some imagery from Eastern superstition.) He is eating a naked figure, and griping another in his claws. On his right is a Pope, with the white tiara, holding a purse full of money with a fast grasp in one hand, and stretching out the other to *bless* a starving woman, who is begging of him. Not far off is a naked cardinal, with a

[6] Forfarshire rocks.

red cap. Two demons near the entrance gate are pulling in head-foremost a priest in his black gown. On the devil's left hand are numerous demons pushing various figures into a gulf of mud. Giapelli tells me that Dante and Giotto were friends, and that it is not known whether the painter borrowed from the poet, or the poet from him. But that a religious order of knights should have had this satire blazoned on their walls, is a trait of the times worth knowing! Of course the vulgar were not admitted. The form of the room is that of many meeting halls of Freemasons. When it became a church, they could not destroy it, without its making the satire more public, as a chancery injunction, or K. B. prosecution for blasphemy, circulates the evil tenfold. But it is singular that they chose to make it a church, and shows that it must have been received doctrine that the wicked among the popes and cardinals might expect equal justice. All this is of course in print somewhere. I am going to Parma with a letter from Buckland to Professor Guidotti, from whom he says he learnt most of any north of Apennines, in Italy. I hope I may hear from Kinnordy at Florence. All tell me that if I would do Etna well, I must lose no time, because of snow.

With my love to all, believe me, your affectionate son,

CHARLES LYELL.

CHAPTER IX.

OCTOBER 1828–JANUARY 1829.

PARMA—PROFESSOR GUIDOTTI—FINE COLLECTION OF FOSSIL SHELLS— ADVENTURE AT THE DOUANE AT PARMA—GEOLOGY OF THE VALLEY OF THE ELSA RIVER NEAR SIENA—ANCIENT VOLCANOES NEAR VITERBO— GEOLOGY OF ROME—NAPLES—EMBARKS FOR ISCHIA—PÆSTUM—TEMPLE OF NEPTUNE—CATANIA—ASCENT OF ETNA—THE CASA INGLESE—THE GEMELLAROS — HARDSHIPS AND ADVENTURES — SYRACUSE — AGRI- GENTUM — A FESTA — PALERMO—BANDITTI—DIFFICULTY OF LEAVING SICILY—RETURN TO NAPLES.

CORRESPONDENCE.

To His Sister.

Florence : October 10, 1828.

My dear Caroline,— I went from Padua to Verona, and made a good geological expedition to the celebrated Ponte di Veja, a natural bridge over a valley, which is a very striking object. As I found no letters at Verona, I went to Parma with a letter to Professor Guidotti, a gentlemanlike and agreeable man, who welcomed me in a most delightful manner. He has the finest collection of fossil shells in Italy, and as they chiefly belong to the most modern formation, it was of first-rate interest to me to get from him a multitude of facts. You may suppose I was not a little happy to find that he was anxious to have as much geological instruction from me which my tour had enabled me to give him. So we spent three days, from six o'clock in the morning till night, exchanging our respective commodities, and I believe those which he gave me were at all events the most substantial, for he liberally presented me with a set of most of those shells which occur in the sub-Apennines, *identical* with species now living in the Mediterranean, to me a useful gift. He was so excited by our geological researches and conclusions, that the very day I left Parma he started on an expedition planned by me. Imagine what fossil conchology has become in this country, when I tell you that he has 1,100 extinct species as perfect as when they lived, which is infinitely beyond the number now living in the Mediterranean and Adriatic, and he finds at the rate of 50 new species a year, which fishing in those seas would not do. At Bologna I introduced myself to Professor Ranzani, who showed me the Museum, &c., and gave me a recommendation to Professor Targioni here. In the two days which it required for a vetturino to crawl over the Apennines, starting at 5.30 and arriving after dark, I made out the series and succession of rocks very well, for I always walked when the horses did, being unable to sit upright in the cabriolet. But after all, one gains in information, as well as in pocket, by travelling thus : you hear Italian talked, a great point, and see many curious characters. Our party consisted of a clever and pleasant French lawyer, a young Greek gentleman educated in Italy, a Sardinian priest, and a French merchant, all men of my own age. Targioni introduced me here to Professor Nesti and Baron Bardi, who showed me the Museum. With their instruction, I rode to the upper Val d'Arno, a famous day for me—an old lacustrine deposit corresponding delightfully with our Angus lakes in all but age and *species* of animals; same genera of shells. They have just extracted the fortieth skeleton of hippopotamus ! have got above twenty elephants, one or two mastodons, a rhinoceros and stags, and oxen out of number. I forgot an adventure at the Douane at Parma. Dr. Lorenzone of Vicenza had given me some Monte Bolca fish as a present. I told the officer they were not minerals, and after a debate he said, as there was no metal in them, perhaps they were not, but I must declare what they were. ' Pesci petrificate.' ' Dunque sono cose vegetanti,' and showing me that the law enumerated all vegetating things, he insisted on taxing me. As I knew the joke would be relished at Parma, I told him to make out a bill, for I meant to appeal. But when it was drawn up, the soldiers and clerks laughed and quizzed the Douanier, so that he threw the bill into the ditch, and made out another, calling them *minerals*, to my great disappointment. He only made me pay threepence English! I am rather stiff to-day, for I had to ride to Val d'Arno and back, 37 miles, and walk ten to fourteen hours,—hard work. I got a letter from Murchison here; he has worked hard in Tyrol. I got a famous letter also from Scrope, full of geological and practical hints on Sicily. I must make up my mind to rough it in earnest, it seems. He recommends boating a good deal off the coast, because the sailors are more honest than the landsmen there, and because my boat, drawn up on the sand, will be a far better bed than any I can get there. How comfortable! He also declared that as they gave him dogs to eat, and stones, miscalled bread, that portable soup would be as indispensable as to Captain Parry. He speaks of course not of the Grande Route, but of a geologist's. I feel more and more every day, that unless I see some district with my own eyes which has suffered from recent earthquakes and recent volcanoes, the first start of anything I write would be at a halting pace, and built on the facts of others, which might perhaps, as applied by me, afterwards give way, and down would come the theory, true or false.

To-morrow I start for Siena. I shall be very glad to hear from Kinnordy at Naples.

Your affectionate brother,

CHARLES LYELL.

To HIS SISTER.

Rome : October 20, 1828.

My dear Marianne,—I believe I gave Caroline a history of my progress to Florence, in the two days' journey from thence to Siena. I contrived, by dint of a small gig and boy, to do a great deal of geology, going round by Colle and the valley of the Elsa River, which was a favourable tour for me, as bringing down the chain of geological events to a later period than any of equal extent which I had seen, except, perhaps, in the Upper Val d'Arno. But the lacustrine formation of Elsa was new to me, not only as containing in it

much travertin, just like the Bakie limestone,[1] but because
there are hot springs still producing the same rock. I
believe the shells (freshwater) of Elsa will prove all recent
ones ; the *Neritina* have their colour well preserved. Yet
there are valleys 100 feet deep, and towns in them in this series
of strata, which overlie others, analogous to our most modern
in England, excepting of course our Angus Lake deposits.
That part of Tuscany is very picturesque. At Siena I had
not the same good fortune I had had at Florence to find the
geologists in town. One to whom I had a letter, Mazza, was
just getting on his horse for an absence of several days, but
stopped an hour, and showed me a collection of fossil
shells, which, if I had not seen Guidotti's at Parma, I should
have thought much of. After one day's exploration of the
geology of Siena, I took a gig for four days to Viterbo, in no
small fear that I should repent having defied the rain ; but
as Murchison and I were never stopped once in three months
by rain, so I had three splendid days for my work, and,
indeed, half the fourth, when I got inside a *voiturier's*
machine for the rest of the way to Rome, and then it rained
plentifully. Almost the whole way from Siena to Viterbo is
a theatre of extinct volcanic action. Such repeated layers
of ashes, pumice, tufa, and dust, cemented with stones in
very thin seams, that the eruptions must have been numerous
beyond measure, as high hills are formed entirely in this
manner. These and the lavas were just like Auvergne, but
there are more fine lakes in ancient craters than one can see
in France. Bolsena is the largest. All these spent
volcanoes threw their ashes and lavas over that formation
which contains the 1,100 species of marine shells which I mentioned I had seen in Guidotti's collection. Some geologists
pretend that there are proofs that the volcanoes began before
these strata had done forming, which may be true of some; but
I have seen none yet, although I have been from morning
till eve immersed in matters relating to times beyond the
flood. I must of course have had many adventures of post-
diluvian date, and among those I may enumerate the vile
Douane, where all my shells and specimens were opened and
disturbed. In the coach at Viterbo I found a Roman priest,

---

[1] In Forfarshire.

head of a college, who had been thrown by accident with two
Scotch clerical students (going to the Irish Catholic College
for their education), who could not talk Italian or French, and
were getting on in moderate Latin with the old gentleman,
in which language for a whole day, for the first time in my
life, I had to talk all the way to Rome. At Rome I found the
geology of the city itself exceedingly interesting. The celebrated seven hills of which you have read, and which in fact
are nine, are caused by the Tiber and some tributaries, which
have cut open valleys almost entirely through volcanic
ejected matter, covered by travertin containing lacustrine
shells. The volcanic strata repose in the Monte Mario on
the sub-Apennine formation, or strata containing shells
analogous to our *most recent* formation in England.
The shells in the travertin are all real species living in Italy,
so you perceive that the volcanoes had thrown out their
ashes and pumice, &c., and these had become covered with
lakes, and then the valleys had been hollowed out all before
Rome was built 2,500 years and more ago. Thus giving us
a grand succession of events later than the formation of the
Monte Mario beds, in which *few living* species of shells are
found.

Naples, October 29, 1828.—After two rather fagging days,
I arrived here safe. I persuaded the whole party (four
Frenchmen who were with me posting to Naples) to stop
and see Lake Albano, the crater of an old volcano, with
which they were fortunately delighted, as very picturesque.
The Pontine Marshes are nothing but the swampy deltas of
three rivers, like the lower part of the plains of the Po.
Arrived here, I went straight to Colonel Visconti, the geo-
grapher, Capt. Smythe's friend, who immediately put me in
the way of everything. As all the geologists were gone to
the island of Ischia, where I always intended to go, I
embarked immediately, and just arrived in time to *miss*
them ; fortunately perhaps, for in three days' active examination of that beautiful isle, I flatter myself I made some
discoveries, which if I had been lionised I should certainly
have missed. I found shells, among other things (marine),
2,000 feet high on the old volcano, which had not been
dreamt of here. But the people !—nothing you ever heard of

them is I believe exaggerated. I do not mean as to the graver charges, for of course those only who live much among them can judge of that, but as to what one sees characteristic in passing through them. It seems as if one half of an immense population, noisy and cheerful, had nothing to do, and the habits of course of those to whom time is of no value are most inconvenient to anyone who wishes to make good use of his. On an expedition, the hours spent in bargaining is a nuisance to which even the north of Italy is a joke. In the shops there are no fixed prices. Whatever contract you make, your driver or boatman finds he has done something more than was specified, into which he has designedly led you. The Ischians are just the same, and famous gasconades. The following was my *rencontre* yesterday with a chasseur, on the hills, about my age. 'Where does this aqueduct go to?' 'To Ischia, sir; five miles long, and cost 25,000 piastres.' 'How old?' 'Eighty years; built by the town.' 'Was it injured by the earthquake of February?' 'Not a stone; had it been thrown down, Ischia would be dead; they must drink salt water.' 'Were any houses in Ischia hurt?' 'Not one, by St. Nicholas! I wonder at it.' 'So do I, it was very close to Casamia.' 'No, no, no; I don't mean that; but in Casamia they are honest folk, but in Ischia all rascals, robbers, assassins, 5,000 of them, and all thieves, by St. Nicholas, I wonder they were spared.' The Government have sent off the steamer to Marseilles, and I must work here for a week. No other way of going, for the pirates of Tripoli have taken so many Neapolitan vessels, that no one who has not a fancy to see Africa, will venture. The rain has begun to-day.

Your affectionate brother,

CHARLES LYELL.

To RODERICK MURCHISON, ESQ.

Naples : November 6, 1828.

My dear Murchison,—I am just returned from a safe expedition to Pæstum, and as far as shells in the travertin are concerned, I hope to do a little, but in general the geology between this and the plain south of Salerno is very flat, the Apeninnes, as usual, presenting that dearth of fossils which renders them everywhere so monotonous. They are very siliceous in these parts, cut into deep valleys, and near Vietri is a fine example of an old lacustrine formation, which was gradually deposited when the valleys between that and Salerno were two or three hundred feet less deep than now. It was in the style of Cadibona, and though I found nothing but small pieces of carbonised wood in some of the red clays (red just like the cement of Mediterranean breccia), I have no doubt that remains which might rival the *Anthracotherium* would reward a good search, of which in this country there is no chance.

As I passed rapidly from Rome to Naples, I shall reserve my observations in that country till my homeward journey enables me to fill up certain hiatuses. Here I found the steamboat, by which I was to have sailed forthwith to Palermo, detained by the Government on special service, and as I was condemned to the Phlegrean Fields for twelve days, I sailed immediately to Ischia, after passing the Solfatara, Monte Nuovo, Lake Averno, &c., on my way to Puzzuoli, and seeing them all. I found Ischia just what I had hoped, a most admirable illustration of Mont Dore, with the difference which we sometimes wished for, as the substitution of marine for freshwater, and the consequent abundance of organic remains. The older trachytes are positively many of them undistinguishable from Auvergne, as are the white tufas, sometimes pumiceous. But the analogy which struck me most forcibly was a quantity of streaked and ribboned altered marks, like those which we found so numerous on Rigolet, and of which we took some specimens as curious instances of ferruginous yellow and brown stains. I was not so lucky as to get a shell in any fragment of this, but I found, on ascending Epomeo, the same occurrence of clays just like the conchiferous beds which form the fundamental strata of the island, as we found high up in Mont Dore marks like the queer Cypriferous marls of the Limagne. At last, at an enormous height, corresponding in Ischia to what the great cascade is in Mont Dore, I found, in a mass of this clay, marine shells unaltered, and belonging to the same class as those in the lower regions of Ischia. I feel no doubt now that we were right in our speculations on the *preten-due breccia*, &c.; besides which, I have now the satisfaction of having an example of marine remains at a greater height than any of Brocchi's sub-Apennines, and belonging to a formation decidedly more recent. It is annoying to find that Monticelli and Corelli have been poking about, and buying shells in Ischia for years, and that they have not got half as many shells as I got during my visit, which amount to thirty species, without counting microscopics. The proportion of recent species is exceedingly remarkable. They alternate with the volcanic products of the old trachytic volcano Epomeo, which is surrounded by many minor volcanoes of a date comparatively as inferior as is that of Tartaret or Chaluzet to Mont Dore. At least Epomeo is to me as unintelligible as Mont Dore, and the others are as perfect as the craters of Vivarrais. Unluckily their lavas go straight to the sea in ridges, and then form promontories, and are nowhere cut by torrents; otherwise, as the date of some can be nearly determined, they would have been invaluable documents of comparison.

When I have seen the parasitic volcanoes of Etna, I will resume the conclusion as to Auvergne which Ischia suggested. I was much gratified by finding that the dikes of basalt and compact lava, which are so numerous in the great escarpment presented by Somma towards the Atreo del Cavallo, agree with the dikes of the Hebrides and the lavas of Vivarrais in presenting us with coal-black pitchstone at their contact with the intersected beds of scoriæ.

Necker [2] must of course have observed this, though I had forgotten it, but we must take care to insert his observation in our paper, or he will have a right to complain. It occurs in four or five dikes; an inch or two of pitchstone (obsidian) separates the vertical dike from the horizontal tufas and scoriæ through which it passes.

Vesuvius after the eruption of 1822 exposed seven dikes in the interior of the crater, quite analogous to those of Somma, but I am very sorry to say that the small sputtering cone which is now constantly filling up the interior of the

---

[2] A Swiss geologist, grandson of Saussure, who passed the last years of his life in the Isle of Skye.

great crater of 1822 prevents my descending to examine this splendid modern analogy.

Corelli has been writing on the present state of Vesuvius for two years, and never went down to see the dikes in the crater. Nay, he has not been up Vesuvius at all in that time, and Salvedra swears he was never up Somma in his life! Remember me kindly to Mrs. Murchison.

Believe me, ever truly yours,

Charles Lyell.

To His Sister.

Naples : November 9, 1828.

My dear Eleanor,—My expedition to Ischia was of use to me here, for as I procured three times as many fossils in three days as they had ever got from the Isle, they regard me as a good workman. Costa, an ex-professor of Otranto, has named them all, so let the box go to the bottom, or the shells be annihilated by Douaniers, I will let the world know that the whole Isle of Isk, as the natives call it, has risen from the sea 2,600 feet since the Mediterranean was peopled with the very species of shell-fish which have now the honour of living with, or being eaten by us—our common oyster and cockle amongst the rest. Costa is to get my boxes of shells from Messina, &c., and examine many before I get back to Naples, which is getting work done rapidly, and will encourage me much. The grand difficulty in England is that our conchologists know nothing of fossils, but in a country where part or all of a formation contains recent species, this impediment is removed. My wish was to find this peninsula get younger and younger as I travelled towards the active volcanoes, and it has hitherto been all I could wish, and I have little fear of bringing a great part of Trinacria into our own times as it were, in regard to origin. Poor Costa, with several other naturalists of decidedly superior knowledge to any of the present professors in the university here, have been deprived of their chairs and persecuted to beggary for their 'constitutional' opinions. The police discovered lately that they were earning some bread by teaching languages, natural history, &c., and the

public were prohibited from employing them even as private
teachers, so that now their ruin is complete. In all the
great states of Italy except Tuscany, the inquisitorial suppression of all cultivation of science, moral or physical, is
enforced with unrelenting rigour, and considerable success.
Among eighteen persons shot for political offences, Carbon-
arism, &c., the day before I passed through Salerno, several
were priests. On the road to Pesto I met three detachments
of similar prisoners with guards. They were strung together
by ropes, like those met by Don Quixote. The driver of my
gig to Pæstum, an arrant coward, kept comforting himself
on the number of troops sent to quell political disorders,
and then came to me in the middle of the night at Eboli, to
try to make me go in the dark, to take advantage of the
French Ambassador's escort, who was going to visit the
temples. As I wanted to geologise the recent travertins of
the intervening plain by *daylight*, I did not stir till 6 A.M.
The Temple of Neptune is the finest relic of antiquity I ever
saw. The travertin of which it is built was in such enormous blocks that they cannot be used for other buildings,
unless carved and cut, and this is far more expensive, as they
are very hard, than quarrying fresh travertin. This I believe
has saved them partly. The same size of blocks would enable
them to resist earthquakes, which have troubled this coast;
for when Messina was shaken down, the jail alone stood,
built of large stones. I have not seen the spectacle inside
of Vesuvius yet, for in high winds, which have prevailed, so
many loose stones are detached, that the cone cannot be
scaled, but I have done the geology pretty well, and boated
it along the section of the cliffs of lava facing the Bay of
Naples. Pompeii afforded me some good geological hints,
besides, of course, much amusement and instruction. The
paintings frescoed on the walls of some newly disinterred
houses are in such Chinese perspective, and so many ages
behind the bronzes and statues and alto-relievos in the same
houses, that my faith in the excellence of ancient painting is
somewhat shaken. The head man who was my cicerone
there was ignorant to a diverting degree, not often the fault
of Italian lionisers. 'This, sir, is the wife of Pompey the
Great, named after Pompeii; she is weeping her husband's

death, who was killed at the siege of Troy!' I have, of
course, seen the museum here of things found in Herculaneum
and Pompeii. It is of great interest, but of less than it ought
to be made, if the Government was spirited. Scarce a fourth
of Pompeii, or a hundredth part of Herculaneum, is explored.
There has not been enough done in excavating small private
houses, and preserving the furniture. The innumerable
bronzes and busts, &c., one sees equal and superior elsewhere.

Dr. Daubeny's [3] letter came yesterday, with many good
hints on Sicily, and a joint letter from Dr. and Mrs. Buckland this morning is full of good practical hints, as well as
scientific. It is a most kind service to have done me, for as
they are persons who make no difficulties, I am sure that
whatever they recommend is indispensable. So I have bought
tea, sugar, cheese, and four bottles of brandy, which Mrs.
B. says will keep off malaria, and their weak wine will not.
It seems that even in winter this evil attacks those who
live poor, and where inns are few and bad, you cannot live
well unless you provision your mule. Between Messina and
Syracuse lies all that is of the greatest importance to me,
and I have still a fair chance of doing it before the December
rains set in seriously. Naples is not so dear a place as
Florence or Rome, for those who are a match for the natives.
If you are inclined to be absent, as I am, it is a good place
to cure you, for you must have your wits about you every
moment. I asked yesterday if it was too late to frank a
letter. ' No, sir; mind if it is to England, you only pay fifteen
grains (sous).' I thought the hint a trait of character, as
they are all suspicious of one another. The clerk demanded
twenty-five. I remonstrated, but he insisted, and as he was
dressed and had the manners of a gentleman, I paid. When
I found on my return that I had been cozened, I asked the
head waiter, with some indignation, ' Is it possible that the
government officers are all knaves?' ' Sono Napolitani,
Signor; la sua eccellenza mi scusera, ma io sono Romano.'
When the Austrian troops retired, the king gave the officers
a choice of any new uniform they pleased, and they actually
chose the Austrian! After their dastardly stand, which

[3] Professor of Chemistry at Oxford.

followed all their vapouring about a constitution, I think this is as memorable a proof of want of national pride as was ever known.   They are a very fine-looking people, and become the lion's skin far better than the troops from whom they borrowed it.   There is a Virgin, with candles always burning, in almost every café and wine-house.   The noise in the streets is dreadful.   As in ordinary squabbles, men and boys bawl louder than our criers, all those who cry various goods for sale, quite deafen you, and as many are hoarse habitually, it is painful to hear them.   They roar as if they would bully you into buying, and numerous idle boys spend hours in mimicking them.

I shall perhaps get to Rome still in December.   With my love, believe me,

Your affectionate brother,

CHARLES LYELL.

To HIS SISTER.

Catania : November 29, 1828.

My dear Marianne,—After my successful expedition to Val del Bue and return here, I went this morning to see the Islands of Cyclops, eight miles from hence, and made a good cruise.   The chief of these small isles opposite Trezza is to an ancient lava of ancient Etna what the Needles in the Isle of Wight are to the chalk of that Isle—detached portions in the sea.   This, Virgil said, the giant Polyphemus threw at the Trojan fleet.   It still remains, but is wasting so rapidly, that in some 10,000 years it will be annihilated.   It consists of basaltic columns, like the Giant's Causeway, covered by strata of clay.   I saw that same clay was on the mainland at the foot of Etna, and some peasants assured me, that it contained ' roba di diluvio,' so I hastened thither, and found 700 feet and more up Etna, in beds alternately with old lavas, sea-shells, fossils, but many I know of modern Mediterranean species.   This is just what everyone in England, and at Naples and Catania, told me I should *not* find, but which I came to Sicily to look for—the same which I discovered in Ischia, and what, if my geological views be just, will be found near all recent volcanoes, and wherever earthquakes have prevailed for some thousand years past. I have set a man and boy to work, at so much per day, and if they do their duty, I shall find, when I return from Etna, something that will fix the zoological date of the oldest part of Etna. I go up again to old Gemellaro's to-morrow at mid-day, if weather fine, to ascend Etna next day.

Catania, Nov. 30.—Dr. Gemellaro [4] called this morning, who has published on Etna. Confesses that he never hunted up the ' roba di diluvio,' because the shells they brought him were Mediterranean ones, and he supposed they were taken from the sea to strengthen walls!! A good joke, as I found them for miles, and in strata 100 feet thick! He is not a little annoyed at seeing of how much importance I think it. I shall start for his brother's this afternoon.

Nicolosi, Dec. 2.—I was most fortunate in my expedition to the highest cone. Old G. started me in the middle of a coolish night, by moonlight, so that we reached by daybreak the highest part of the woody region. There a fire was made of boughs from the old chestnuts and oaks, which, though one or two thousand years old, stand on *modern* lavas of the volcano. I had taken my tea-kettle, and, as a preservative against cold, made a breakfast of hot tea, following Captain Parry's advice. We did not cross much snow before Casa Inglese (9,000 feet high) : then we ascended the cone. I had determined to think with Ferrara, who has published the best guide-book of Sicily, that the sickness felt from the rarity of air on the summit of Etna was all nonsense, but I had no sooner ascended 200 feet from the hut called ' Casa Inglese,' than my head ached, and I felt squeamish. My guide said, ' We must go slower; I have known many vomit violently during this part of the ascent.' The fact is, you go up gradually to the foot of the cone, your lungs get pretty well weaned of breathing more condensed air ; but when you rise almost perpendicularly from 9,000 to 10,000 feet, the air within you is so condensed, compared to the atmosphere you enter, that you are ready to burst. Such

---

[4] Dr. Giuseppe Gemellaro, who lived on Etna, and published a map of the early eruptions. His two brothers also devoted themselves to the study of the mountain, and Professor Carlo Gemellaro contributed important treatises, one of which, published in 1858, was dedicated to Sir Charles Lyell.

vapours were issuing from the crater, that I saw nothing distinct in it; but from it I saw into the Val del Bue, a view which Buckland had begged me to make an outline sketch of, and which in about half an hour I effected, I believe decently. Inside the crater, near the lip, were huge masses of ice, between which and the scoriæ and lava of the crater issued hot sulphurous vapours, which I breathed in copiously, and for six hours after I could not, even after eating and drinking, get the horrid taste out of my mouth, for my lungs had got full of it. The wind was so high, that the guide held my hat while I drew ; but though the head was cold, my feet got so hot in the cinders, that I was often alarmed that my boots would be burnt. I descended to examine the ' Cisterna,' a curious case of recent subsidence at the foot of the crater ; then to ' Torre del Philosophia,' which overhangs the head of Val del Bue; then back. Old G. said, 'I congratulate you on having done this on December 1, for the wind has changed, and you are probably the last who will mount this year.' Accordingly this morning Rosario, my muleteer, called me, and said, ' Signore, Mongibello e bianco come una colomba ' (they call Etna ' Gibello ' in Sicily), and sure enough I found that down to the bottom of the woody region, for 4,000 feet, the mountain was one mass of snow, such a metamorphosis in eight hours ! and sleet falling fast here. I am a willing prisoner, needing a recruit after such a fag, with a good library of books on Etna, and a companion who has made the volcano the study of his life, and who has published the only scientific accounts of the late eruption.

Nicolosi, Dec. 3.—Rain and snow. Old Gemellaro and his youngest brother (not Dr. G.) delighted to hear old stale news, which I read a month ago in French papers, of the Russian campaign, French politics, &c., for the Government lets them know nothing in Sicily. The young G. who has come here, finding I had only a brazier of charcoal in my room, has ordered a fire, God bless him, and I am at last warm. The elder G. is a kind of laird and chief magistrate. Cannot talk French or English, so we talk half Latin, half Italian, and so always get on with one or other. He is really a good Latin classical scholar.

Catania, Dec. 7.—Left Nicolosi December 5. Visited only
a fine instance of a lava of Etna, cut through by a river
which it had crossed. Took a peasant for a guide from
Aderno, a stupid officious bumpkin, whom I could not
persuade to stay with my horse, while I went down to the
waterfall. The lad, I believe, wanted to see what the devil I
wanted to look for, so having, as he assured me, secured the
beast, and tied him to a great stone, he came with me.
After exploring, I became anxious, and told him to look if
the steed was safe. We climbed a rock, and saw he was
just as we left him, but the nag, who was weary of standing
still on a lava current, no sooner caught sight of us, than he
plunged and pulled till he broke loose. He fancied we
were on the opposite bank, as we appeared to be, from the
winding of the river, so he plunged into the stream to join
us. It was the Simeto—the old Simethus—the only one in
all Sicily which deserves to be called a river, in a narrow
gorge, a violent current, and just above a cataract. I thought
it was all over with him. The saddle-girth gave way, and
down went the saddle over the waterfall; but the horse was
a powerful creature, and landed safe, neighing all the time
with his head turned to me, but he could not climb up the
steep rocks on the other side. When I got down opposite
him, he wanted to cross again to me, but as it was too
dangerous, I prevented him with stones. I found the lad
frightened out of his wits, and got at last in such a rage
with him, that I threatened to throw him into the river,
upon which he went off for assistance. I waited for an hour,
when I saw two woodmen coming on the opposite cliff, who
descended by my directions to the horse (the gorge was
somewhat like the Isla at Airlie Castle, but less high). They
actually had to construct a path, and got him up with great
difficulty. In a word, he was not wounded, and we found
the saddle, stirrups and all, floated three-quarters of a mile
down in the middle of the river on a rock. The steed was
brought over a ford, and I only lost three hours and a half,
and 3s. for the men. I did some geology the same day, but
could not reach Catania. Night comes on here like an
eclipse of the sun, all in an instant, without twilight. The
roads are getting horribly muddy. Rosario advised sleep-

ing in a low inn, or rather stable, with the horses, twenty-
five mules in the same, half of them loose. I dreaded a kick.
'No fear,' said he ; 'you shall have clean straw, and I will
sleep between you and the horses!' I insisted on proceeding to a village two miles on, which we reached at last. I
went (*à la Murchison*) to the Pharmaçia—the apothecary
always sets up for a *savant*, and is much flattered at being
applied to for scientific information. After some talk, he
sent all over the village, and at last procured me a bed,
better than Aderno, for the inn there was vile enough. In a
country where they have nine months' heat, there are no
protections against cold. When I ordered a fire at Aderno,
they brought a sort of garden flower-pot, broken, with some
charcoal, for there are no chimneys. 'Have you not another?
This will not warm the room?' 'Yes, sir ; but you had better
not, for you will be stifled.' 'Impossible! One might as well
talk of being stifled in the open air as here, with four doors,
not one of which shuts close.' 'Four doors, sir! Bless me!
Maria Santissima! Why this is only one opening into another
room, and this into a passage ; and this, sir, is the door, and
this a window.' Now the last you must know was an opening with two shutters, for except in the capital towns of
provinces the inns have no glass; so if you want light you
must open the shutter, and let in rain and wind too.
Rosario understands the commissariat department, and
though the room is bad, I am never without prog. He
believes in all the saints and miracles. I forgot to say that
when he heard of the horse's escape, he crossed himself and
said a prayer. When he learnt how he was got up the rocks
on the opposite bank, and when long afterwards I related
how the saddle was recovered, he drew his bridle, crossed
himself, and said another prayer. On my return here, I
found I was elected honorary member of the Geronian Society
of Catania, the only body which publishes any memoirs on
natural history and philosophy, and theirs are very respectable. If detained by rain, I shall make use of their library,
ours I should say, and reading-room.

Ever your affectionate brother,

CHARLES LYELL.

To His Sister.

My dear Caroline,—I passed over the plain of Catania, a delta of the Simeto, early, following the sea-beach, near which were hedgerows of Indian figs and aloes. Before the sun rose from the sea, all the snow of Etna was glowing with red, like another dawn in the west. I have seen this sight three times, a proof of fine weather and early rising. Slept at Lentini; inn good, for Sicily; glass window, and really not many panes broken. Next morning the churches began as usual to fire their batteries. They cannot pray in this isle without gunpowder. In Catania I was opposite the Duomo, and the smoke and din were most warlike. About fifty 'maschi' are set in a row—small short iron tubes, with touch-holes—and a man or two stand with long canes lighted at one end, to fire them one after the other, like a *feu de joie*, the instant the priest has said the benediction. A greater gun is often put in the middle or end, according to the fancy of different churches. One night at Catania I heard a drum and two clarionets, followed by a rabble, shouting and throwing up squibs and fireworks. 'What is this?' 'Per la gloria della Vergine.' 'Della Vergine?' 'Signor, si, della immaculata.' When I reached Syracuse, it was the first of thirteen days' festa in honour of Santa Lucia. 600 'maschi' were let off in the piazza. As the Sicilian gentry are great sportsmen, they send their steeds to face the battery, for if they stand this they will not easily flinch from a gun. So the groom rides up, and the town boys stand behind with whips and switches, to prevent a retreat. As they stand a good chance of a kick, and hope to see the grooms thrown, they are never tired of this sport. The horse rears and plunges his nose in the smoke, with the noisy mob behind, and I could not help thinking what religious associations the lads must acquire in regard to 'maschi.' From Lentini to Melilli. I had a letter to the vicar, a *brother*-member of the Geronian Society. His invitation to sleep there was such as I did not choose to accept. The room I got (for inn there was none) was a queer place, and everything to be begged and borrowed but the bed. Next day I found that a crater discovered by the priest was a circular valley cut out of *limestone* strata, with three veins of volcanic rock through them. I must write to Dr. Gemellaro, for if he publishes this in the transactions of a Society at the *foot* of Etna, it will be a joke against them they will never get over. Went to Sortino, to see grottoes of Pentalica, where a vast population once lived underground. Made a man dig, by Buckland's direction, into stalactite, but the instrument was so bad, I was obliged to go away after four feet were dug into, and no signs of a bottom, or bones. Got too late to enter the gates at Syracuse; such bogs as I have had to get through! No funeral ever went slower than we for two days, and some fords not quite so shallow as could be wished. After seeing the paths called roads, near the great towns in Sicily, I shall respect the worst in the Grampians, and honour the muddiest road in Angus. It requires force to extract the hoof, and a leap to jump down the stones. In summer they go by other tracts, now bogs which would swallow us up. At Floridia, a small town near this, they showed me a room to which the caves I had seen at Pentalica would have been luxurious, so cold and stinking; so off I went, determined to beg at the best house. There was only one with two storeys, a Pharmacy, and the owner, the Mayor or Syndico, a Maltese, who gave me an excellent apartment, was glad in return to hear stale news which I had read a fortnight before at Naples, the taking of Varna, &c., and he told me of some new antiquarian discoveries at this city, and then politely left me alone. Rosario brought everything for dinner, and I gave the servant what a bed would have cost—a shilling—and got here to breakfast in light rain, yet have geologised and antiquarianised all day with much success. An old monk, who collected shells and fossils, died lately, and I purchased the latter for a trifle; not many, but good, and more than any other geologist has got to throw light on the old volcanoes of Val di Noto.

Licata, December 16.— From Syracuse by Noto to Cape Passaro, and then, to avoid some fords which the rains had swollen, by *Spaccaforno* to *Scicli, S. Croce, Terranuova* to

Licata; to-morrow to *Girgenti.* Beautiful weather for seven
days, and I have made such progress in the geology of the
isle, and have such a near prospect of completely mastering
the difficult points, which no one ever did before, that I
mean after the late fine weather to penetrate to *Castrogiovanni,* which I had entirely given up. The inns are execrable beyond description, and bread often requiring all the
digestive powers which I have gained by being on horseback for ten successive days for eleven hours each.

 Girgenti, December 18.—An uncomfortable inn at *Noto,*
but next day threatened with so much worse at *Pachino,*
that in spite of Rosario's remonstrances I pushed on to
Capo Passaro, but the old priest who had received the Buck-
lands was dead. His young successor however, a simple
frank young man, welcomed me in a dwelling like a humble
Scotch Highland manse, and cold enough. Saw the fine
rocks of the Cape that evening, and again before breakfast,
and then on to *Spaccaforno.* My man had built on my
staying, because the *wind* was very high, and turned out
sulky. No inn at *Spaccaforno,* a great place bigger than
Salisbury. Went to the Jesuits' convent. 'Would have
been happy, &c., but the Bishop was arrived on a triennial
visitation, and they and the Carmelitans crowded with his
suite.' Went to beg of the Mendicant Capuchins. 'They
had a room vacant, but the beds were all ordered off to the
other convents for the Bishop.' At last Rosario made them
understand we should pay as at an inn, and a monk gave me
his cell. Such a villanous place, hardly room for a small bed
and table; unswept, &c. The bread here and next day black,
and half baked, and scarce digestible, and no meat, not even
a fowl to be bought. Population 9500 ! Most of the brethren
rude and uncouth. My host quite otherwise. On my com-
plaining of the wine we had procured, he went out and came
back with a bottle under his cowl, and a cheese. I thought
of the friar in 'Ivanhoe,' but unluckily for my dinner, and
the joke, both kinds of prog were very bad. He made a
decent show next day of not having expected payment, but
I said it was to sing masses for my soul. Next night on to
S. Croce. Road horrid ; no inn, but a camera attached to a
'fundaco' or mule stall; and as you may be obliged to sleep

in the latter, this chamber ranks here in a muleteer's estimation as respectable. Each of the dirty family sleeps in it
by turns, and turns out for the stranger. The cobwebs were
thick and pendent, and the hostess said, ' Cercate, Signor,
ma non c'è altra,' with such an air of presuming on her
monopoly, that I applied to the Syndico, who ordered me an
apartment belonging to the commune. It was decent but
gloomy. Next night, at *Terranuova,* I petitioned for more
bedclothes at the inn. ' Non c'è altra coperta, Signor.'
' Can't you borrow one ? ' ' No.' In such weather in Scotland
you might do without fires, but here the houses are planned
for being cold. The sun is very warm from 11 to 2 o'clock,
even a few lizards out, and *Papilio cardui, edusa, phlœas, bras-
sica,* and *rapœ,* and the moth *popularis* (all British), without
any foreigners mixed. Many flowers in full bloom, a blue
iris, daisies, marigold, rosemary, and numerous splendid
shrubs by the sea-side, some of which I remembered in the
garden at Bartley. The olives are old and picturesque, and
their light green, and the rich green of the Carubiere,[5] with
very rarely a deciduous tree, make it quite unlike winter
in appearance. As the inn at Licata yesterday was the best
since Syracuse, I did not sleep, as I might have done, at a
rich merchant's, Mastroeni, to whom I had a letter, but went
to a soirée there, very gay. Saw the fine old Doric temples
of Agrigentum on my way to this,—fine geological sections near them. I found the people here, as indeed I had
found them everywhere in Sicily, celebrating a festa, but
the cause was novel. Eleven years ago the citizens and
populace threw down the King's statue, and broke every
limb to atoms. His Majesty, of his royal clemency, bethought
him, eleven years after, of punishing them by transferring
the tribunals to Caltanisetta, a smaller city, where a powerful garrison kept all loyal, as they did in *two* other places
only, in *all* Sicily. Hearing of this threat, the magistrates
ordered a new marble statue from Palermo, which was
being set up when I entered. But the fatal decree arrived
first, so the senate, to propitiate the King, ordered a three
days' festa, and all the usual ceremonies are to be performed before the statue, while they are to obtain inter-

---

[5] *Ceratonia Siliqua.*

cession of a patron saint when they carry out or go in procession to the image of the Virgin, or their protector St. John. The *Bishop* and clergy attended, and sung solemn mass ; maschi were fired off, and in the evening above a thousand were to be exploded. Service also in the cathedral. I looked in, and while the people were on their knees, the organ playing and the bell sounding, in came a military band playing a *march*, while a file of soldiers fired off a volley just within the porch, and made a terrific noise in the Gothic aisles. There was to be two more days of this, and they still hoped that when the King heard of such devotion he would relent and reverse his decree. The Sicilians have the character of being very sincere among Italians, and compared to Romans or Neapolitans I believe they are so. The King and Government are unpopular, and the former known to be a mere gourmand and sportsman, yet what did the Roman senate of old ever do more servile when they decreed divine honours to Augustus and other living emperors. I forgot one droll sight. Fifteen orphan boys were paraded before the statue stark naked on a windy day, and then clothed by the Bishop in the name of the King.

Ragusa, near Modica, December 25.—From Girgenti I rode through the heart of the island by Cannigatti, Caltanicetta, Piazza, and Caltagirone, to Vizzini, within a day of Catania again, to get new sections of the old limestone and volcanic district of the Val di Noto, and decide the difficult problem whether the rocks of the east or west of this isle were first formed, for certainly they were not contemporaneous. I do not wonder so many have doubted, for the convulsions have been tremendous, though the strata, geologically, is very modern. I have made up my mind on the subject, and am glad I returned, though I half repented when I found in so many large fine cities such wretched accommodation, especially on two of the days when I did less geology than on the coast, but the Val di Noto redeemed all. The centre of the isle is without wood, downs, with grass and wheat on clay soil, the latter a few inches high. It ought to be $1\frac{1}{2}$ feet, and the farmers are alarmed at the extraordinary want of rain, which has been so favourable to me, as the clay districts have become hard again. Short as are the days, ten hours or less,

Rosario says that few English travellers can ever work more hours daylight, for they are obliged to be indoors in the heat of mid-day, as in other seasons. Near Piazza, where there was no geology, I went after an *Hipparchia*, the first I had seen. It was *pamphilus*, and I also saw *Magæra*. Rosario must needs imitate me, and has really acquired so good an eye for a fossil shell as to be of great use. He dismounted, caught a *pamphilus*, and holding it battered 'twixt finger and thumb, pronounced after a stare this notable opinion : ' Questo e un grillo alato, ma ce ne sono in Sicilia che volano senza ale.' Shortly after this he was pitched headlong over the mule's head : his hand swelled with the fall. I conclude he was snoozing, though he gives another explanation. He tells me I am taken for a merchant travelling for *trovatura*, or treasure-trove, and the peasants offer themselves for guides to the grottoes where the Saracens are said to have left enormous treasures, but where no one has yet had the art to extract anything. At Palagonia the best room was occupied by a Sicilian baron. Innkeeping must be a bad trade, as I had never found the best room taken before, so I was given the barn. Had it been a windy night I should have suffered much. Here the loft of the inn was so small, and had been uncleaned for so many years, that I went out and applied to the first peasant I met. He showed me his cottage, very clean and comfortable, gave me the key, and left me and Rosario in full possession. Most of the inns in Sicily must have been designed for travellers below the rank of ordinary peasants, judging by the comparison. I got a good many shells at Caltagirone, most of any place since Syracuse. Not a drop of rain now for fifteen days !

Palermo, December 31.—Arrived to-day, somewhat fagged, but the beautiful appearance of this city and the fair promise of the inn have quite set me up ; and as I shall be busy every moment till I sail for Naples, I will employ a few minutes, if but to tell you that in the honest peasant's bed above mentioned I slept not, though tired, ten minutes in the whole night. They had added, at my asking for covering, an old blanket, in which were more fleas than all the insects in our collection, and I suppose they had starved since last winter, for next morning I was marked all over from head to foot with red spots like one of the Ancient Britons with his body painted. I never passed such a night since I had the measles at Sarum, yet the host and Rosario thought it a capital joke, and I heard them laughing about the 'pulci' having found a piece of 'carne fresca,' and they gave me pretty clearly to understand that they thought one who could not stand a fleabite was unfit to travel. Rosario must needs pull out the blanket, to show where they were sleeping next day, and they were in thousands, yet the owner and his wife use this blanket without inconvenience when it is cold.

I shall send Fanny a letter from Naples or Rome, in which I shall write some adventures of the last part of my tour, that you may have a complete idea of the luxury in Sicily; which, however, as far as my objects are concerned, has answered remarkably well, and for the time has yielded more than I could have expected.

January 1.—A happy New Year to you all at Kinnordy. With love to all, believe me, your affectionate brother,

CHARLES LYELL.

To His Sister.

Palermo : January 1, 1829.

My dear Fanny,—I find that I ante-dated by two days my last letter to Caroline, and wished her a happy New Year before the old one was over. I was glad to find, when the cannon of this port fired a grand round at daybreak, that I had two days more to get back to town, which I hope to do before February is over. As for the two seas and the Alps which lie between England and me, and the vetturinos, and the winter, they all seem trifling inconveniences to me now, after five weeks' roughing it in Sicily, for I am sure it will all be luxurious comparatively. As they say that those parts of a tour read best which were least agreeable to the traveller, I shall go back to *Ragusa*, though you must not flatter yourself that you will hear of banditti, for every part of the island I went through, and that was about two-thirds of the whole, was, I am sure, as safe as the best parts of England. However, I believe I may absolve Scrope and others who wrote me discouraging instructions on this head about the prudence of taking pistols, a fierce dog, &c.; for when they were here, the late revolution, by suspending all law for a time, had rendered highwaymen numerous. Rosario saw twenty-six shot in one day, and being courier then, was allowed additional time for the letter-carrying, as he often went round, to avoid dangerous places. There is a strong military police now, for the poverty and misery of the population are at their height, and without a strong force to suppress them, they would be more honest than any in Europe if they did not turn robbers. I found that a chief of banditti from the side of Marsala (where I was not) has just been taken here, and shot, they tell me, without trial, a Turkish proceeding, but I hope this is not correct. These men were convicts in the islands off here, who escaped. Guidotti of Parma, a somewhat timid gentleman, never expects me to return, as a Prussian naturalist of his acquaintance was murdered three years ago. I learn that it was his own guide who was the assassin, and he the only traveller who for many many years has suffered. The population of Sicily is not one-fourth of what it was once. Syracuse alone had a number equal to the present total, yet the Greeks were always fighting each other, town against town. But they were active, industrious in commerce, and every acre was cultivated. Now one-third of all the land belonging to the convents is ill-cultivated, for the friars never reside long enough in one convent to acquire a local interest. The bishop changes them from one to another, as often as they quarrel with him, or one another. Another third of land belonging to the nobles is said to have been also much neglected. But a strange revolution is going on. The old law of primogeniture is abolished, and the French system established of equal division, males and females. As this is only eleven years old, it has only begun to take effect, but the rich Prince of Paterno died lately, and I found the inn full of land surveyors, estimating his lands, which had been divided into eight portions, and many were selling their shares to Catanian merchants. The nobles were against the king in the revolution, and it is now the policy of the court (a short-sighted one, no doubt) to degrade them. The taxes are complained of as very exorbitant, raised to pay the debt

incurred by the Austrian occupation. After my sleepless
night at *Ragusa*, mentioned in my last, I went to *Castrogiovanni*, and on my way was revived by finding in some
marls some fine impressions of fish, the first fossils hitherto
found in the formation which produces sulphur and salt in
Sicily. This will, I am sure, be of no small interest to Buckland and Daubeny. I found the same again in four other
localities on my first two days' journey from *Castrogiovanni*
to *Palermo*, when at length I rejoined the carriage road, and
continued on it for three days, making, in my whole tour,
exactly five days, that I was on a road on which any machine
with wheels, however rude, could move. Not that Sicily is
at all unfitted to have beautiful roads at the same cost as
any other country. As I had good reason to suspect that
Rosario was planning being a day longer than I intended, I
rode on before one day, starting early, and the night and the
rain came on, and he could not quite come up with me. So
besides the ordinary luxury of a Sicilian inn at *Rocca
Palumba*, I had the benefit of a night without my baggage
and portable provisions. But next day reached this, and no
sooner had we arrived, than heavy rain, snow, hail, and
thunder came on incessantly. Fords that we crossed daily
during the last fortnight are now impassable, and journeys
which we performed in a day, with as much ease as it is done
in summer, are now two days' hard work through the clay
and bogs. Yet this rain *ought* to have come three weeks
earlier. The American Consul, Gardner, has given me some
beautiful fossil shells. I drank tea there, and was glad to see
again a regular English fireside and tea-table.

January 3.—Made acquaintance with Professor Ferrara;
learnt much about earthquakes; bought his books; dined at
Lord Northampton's. The Marchioness finds good marks
for her wit in the numerous and illiterate princes and dukes
who pay great court to them. I learnt much of the state
and feeling of the Sicilian *noblesse* by her criticisms. I have
been to see the convent of the Capucins, where there are
great vaults full of dried bodies. What an Egyptian piece of
superstition! Dukes and bishops, &c., dried, and in coffins;
monks and paupers innumerable, dried and set up in their
clothes; some 150 years old, ghastly spectacles; long files of

these mummies, ten times more frightful than skeletons. These Capucins give soup to a hundred or more beggars daily. The mendicity here is distressing, not a noisy set of idlers making a trade of it, as at Naples, but the misery of a starving population; children exposed in frightful nakedness positively howling at night in my street, and now and then well-dressed persons begging piteously. This faithless packet will never sail, they always promise solemnly to-morrow. A bad cold and slight sore throat, and my impatience at being detained, makes me low-spirited.

January 5.—Bought some good fossil shells, saw the museum, cathedral, &c. Cold worse, captain will not sail, and threaten him to wait for steamboat. No keeping warm without stifling oneself with charcoal. The merchants here are gloomy about the end of the war, for they say the Black Sea will be open by demand of all the Powers, and their corn will sell no longer. The retreat of the Russians from Silistria is the theme of politicians.

January 6.—The wind fair, and the scoundrel will not budge, for he is afraid as long as there is a cloud. Steamboat expected. Dined with the Northamptons. Met some Sicilians. The old lady made me go to the play with her and her unmarried daughter: it was one of Goldoni's, by good luck, and we got back to tea, and I had a talk till 12 o'clock with Lord N., who reads up to the day, and is a most agreeable companion. I had not exhorted him in vain to collect fossils, his old hobby, and he has already got some, and taken full instructions. Among other things, he means in February to go to all my localities of fish, and will probably send for Rosario as a guide from Messina. Lady N. gave me a hint, as from experience in Sicily, that I should often have slept sounder if I had had some straw taken from the rick, and slept on it in my cloak, which I shall remember another time in southern regions.

January 7.—Is it not strange that the best houses belonging to Sicilians have no fireplaces or chimneys? yet the snow now covers the hills which overhang the rich plain in which this beautiful town stands. The Northamptons have built, since their arrival, chimneys in all the rooms, and famous fires they make, to the great satisfaction of the natives. They

have got a beautiful mansion. Almost all those of the
nobility are now let out in floors to merchants and others!
Wind fair, and cannot get away. The excuse is infamous,
after so many promises. ' Prince Butera has not yet sent his
horses to embark!' The diploma which I found for Scrope
and myself, from the Geronian Society of Natural History in
Catania, must have been devised by a ' Lepidopterist,' as
Stephens would say, for among the emblems, a great butterfly at the top cuts a greater figure than a range of basaltic
columns at the bottom.

January 10.—Gulf of Naples, on board 'Costanza' brig.
Although the steamer arrived on Thursday, my impatience
to be moving brought me on board, but instead of sailing at
midday, we were kept five hours, a fair wind blowing,
waiting for a passenger who never came. This lost us just a
day, for we kept the wind till the middle of last night, and
were then opposite the Isle of Capri. Three hours more of it
would have taken us to Naples. After eight of dead calm,
we moved on one mile per hour, the whole of to-day, then
six miles the last hour, which brought us in at 24 o'clock, *i.e.*,
6 o'clock, or sunset, and the blessed police will not allow our
landing till 8.30 o'clock to-morrow. ' Basta che siamo nel
porto, Signor,' says the captain, coolly. Vesuvius, since I
was here, put on a covering of snow down to the middle,
which seems great presumption for a monticule not much
higher, I fancy, than those of Glen Isla, and in this climate.
I am the only passenger in the best cabin, and my knowing
how crowded the cabin in the steamer is, was one of the
principal causes of preference. But this is as rough as the
interior of Sicily, and no bedclothes but a blanket, which
is a bore, now a third night is added so unnecessarily.
Talking of rough work, there were two things of which I
deemed it advisable to break my valet Rosario. First, not to
bring in the salt in his hands, often none of the cleanest;
and secondly, not to stuff the meat remaining after dinner,
and intended cold for next morning, into my slippers, without at least some paper round it. From these he desisted,
but a few hints on minor subjects were met with, ' A la guerra
come a la guerra!' in which he was right as a traveller in
his isle. We had quite a gale for the first twenty hours, nine

knots an hour, without a sail but two very small ones. I
was very ill, but recovered so as to read and write ever since.

Your affectionate brother,

CHARLES LYELL.

To RODERICK MURCHISON, ESQ.

Naples : January 12, 1829.

My dear Murchison,— Thrice did I entreat the lazy rascal
who is intendant of the post at Palermo to look back over
the letters of a month before, as I knew I had some waiting.
I now find to a certainty, from the post office here, as well
as the landlord of the hotel, that no less than three letters
were there for me. As I have never once heard from home
since we parted, I cannot tell you how great my disappointment was, after being kept nine days in Palermo. They must
now travel after me to Paris. One of them no doubt was
yours.

I have so much to tell you, that where shall I begin ?
My last letter was from Licata. Thence I went to Girgenti,
and found that, notwithstanding what the Catanese said to
the contrary, Daubeny was most indisputably right in
identifying the calcareous breccia, as he terms it, of that
place, with Syracuse, and as I afterwards found of *Castrogiovanni.* Then this formation (indubitably far, very far more
recent, than the sub-Apennine beds) attains the prodigious
height of 3,800 feet ! I got so astounded by the results I
was coming to, in my tour from Etna, round by *C. Passaro,*
that I began to doubt them ; and not without some struggle
with my desire to get out of the inns and horse-paths and
other evils, I struck back again to Val di Noto right through
the centre of the isle. I found, as I believe I told you before,
that the so-termed Blue Clay formation must be divided
into two very distinct and in some cases unconformable
deposits. The upper resembles the blue sub-Apennine marl,
covered near *Caltagirone* with its true Sabbia giallonula. The
lower is without shells, white marls, beds of gypsum just like
those we saw at *Cagliari* from *Asti,* beds of sand and sulphur,
rock salt, &c. Now in the foliated marls (white) of this
last formation, after searching in vain on the sea-coast, I

found at last organic remains. First near Radusa, a small hamlet in the interior, where fishes' scales, and some entire fish like M. Bolca, turned up, and some impressions of leaves. Whether fish be marine or freshwater, I cannot say. Thirty miles off, near *Castrogiovanni*, and in three other localities between that and Palermo, I found the same fish in the same beds. But the first marked locality, only four days' ride from Palermo, is the richest, and would probably prove a M. Bolca with some labour, *i.e.*, if the formation be not lacustrine or fluviatile, against which I know no argument yet. As a whole, it is uncommonly like our Superga *marls*, not a grain of green sand however, but positively identical with the beds I saw in my solitary expedition on the *cheval de poste* to the gypsum beds of [name omitted], where also I found laminated marls with leaves between two gypseous strata. But these are mineralogical identities, and go for nothing. What will you say if I tell you that even the blue marl with its capping of yellow sand cannot be Brocchi's sub-Appenine beds? I am come most unwillingly to this conclusion. But the numerous extinct species which characterised the sub-Appenines are wanting here, and living shells are present too plentifully to admit a doubt that it is more related to our own epoch than that remote one when the Parmegiano and Placentino beds were deposited.

I think from what I saw at Pæstum and elsewhere that O. d'Halloy strained a point to make out some travertin older, but if I can find a day for Tivoli without hard rain, I will get more data. In the meantime, Sicily will throw all such proofs in the shade. I am much disturbed in mind about Siena, and must go there from Rome, to try and satisfy myself. My notion is that the *tout ensemble* of shells there will not be that of the Parmezan. I am beginning now to be able, when I see large collections, to distinguish between any marked difference in the proportion of lost species and genera. Sicily is as unlike *Asti* as possible, though there are some lost species in the former. I should build no system on such glances, but it might check me from reasoning too fast on the data obtained from zoologists as to Ischia and Etna.

We must preach up travelling, as Demosthenes did

' delivery,' as the first, second, and third requisites for a modern geologist, in the present adolescent state of the science. De la Beche is going just when he should.

Yours ever most truly,

CHARLES LYELL.

To RODERICK MURCHISON, ESQ.

Naples : January 15, 1829.

My dear Murchison,—I leave for Rome to-morrow morning, and shall be there the day after. The rain is incessant and tropical. I have got *all* my shells named, except some which *unfortunately* I sent direct from Palermo. The results of my Sicilian expedition exceed my warmest expectations in the way of modern analogies. I will tell you fairly that it is at present of no small consequence to me to get a respectable sum for my volume,—not only to cover extra expenses for present and future projected campaigns, but because my making my hobby pay its *additional* costs, which it entails, will alone justify my pursuing it with a mind sufficiently satisfied with itself, and so as to feel independent, and free to indulge in the enthusiasm necessary for success. I shall never hope to make money by geology, but not to lose, and tax others for my amusement; and unless I can secure this, it would in my circumstances be selfish in me to devote myself as much as I hope to do to it. I have little fear of accomplishing so much, and with that view, the instant our Auvergne paper is off our hands, I shall set steadily to the task. My work is in part written, and all planned. It will not pretend to give even an abstract of all that is known in geology, but it will endeavour to establish the *principle of reasoning* in the science; and all my geology will come in as illustration of my views of those principles, and as evidence strengthening the system necessarily arising out of the admission of such principles, which, as you know, are neither more nor less than that *no causes whatever* have from the earliest time to which we can look back, to the present, ever acted, but those *now acting*; and that they never acted with different degrees of energy from that which they now exert. I must go to Germany, and learn German geology and the language, after this work is published, and before I launch out in my tables of equivalents. I shall, I expect, get out my book before the tardy G. S. gets out our Auvergne paper ; but no matter, as I shall always refer continually to the MS. as read, and 'ordered to be printed.' This year we have by our joint tour fathomed the depth and ascertained the shallowness of the geologists of France and Italy as to their original observations. We can without fear measure our strength against most of those in our own land, and the question is, whether Germany is stronger. They are a people who generally 'drink deep or taste not,' &c. Their language must be learnt ; the places to which their memoirs relate, visited ; and then you may see, as I may, to what extent we may indulge dreams of eminence, at least as original observers. If I can but earn the wherewith to carry on the war, or rather its *extraordinary* costs, depend upon it I will waste no time in bookmaking for lucre's sake. At least I will answer for myself for many a day. Try to get me off being V. P. G. S., if intended again. I have, as you have, sacrificed something to official duties, and I am prepared to do so again at some future time, but I will not *for the time coming*, except as referee and councillor. As Sedgwick is to be President, there ought to be working V. Ps, and Fitton ought to be free now, he has done so much. I have a deal to tell you of news in geology, which the relation of what concerned me more nearly has left no room for, but you shall hear from Rome.

Ever yours, most truly,
CHARLES LYELL.

CHAPTER X.

JANUARY TO DECEMBER, 1829.

CORRESPONDENCE.

To RODERICK MURCHISON, ESQ.

Rome : January 17, 1829.

My dear Murchison,—In the diligence I had time, as we travelled from Naples, to re-read your letters. I believe all you mean by modifying the caveat is to avoid our appearing to announce as *new* an old well-known fact of lava currents. This is well, but the *caveat* part remains necessary. Your term ' stone walls ' is forcible, and not too strong, and should be inserted with ' as if between two,' or something in that way. I am anxious, as I told you, to throw off our principal paper, to clear the ground for other work ; and, besides, I should like them to see that *we* do not require a year or two to hatch our materials. We will write it for reading, just as if we expected it to be put into the printer's hands the day after. That miserable old style of making a difference between a paper ' to be read ' and one ' to be published ' has, we know, been the secret of retarded volumes, overworked editions, and dull thin meetings, and it is very bad economy of time in authors.

Getting eighty species (only part of my Palermo collec-

tion) named at Naples, enabled me, by working in the
museum of their Academy, to identify a great part of the
tertiaries of Calabria with certain parts of Sicily, chiefly
with the tertiaries of Messina, which are in that island of
very limited extent. I shall write a letter from hence to
Lord Northampton, telling him that when he arrives at a
hundred species, it is only his starting point, as I was there.
He will be glad to hear the news of the scientific world, in
yours and Fitton's letters. I do not despair of stirring him
up to such a point, that he will send to town 300 species
before spring, which will ensure me against the chance of
having to renounce my present opinion on the proportions of
Sicilian tertiaries extinct and identiques.

Rome, January 19, 1829.—There is a deluge of rain here,
after the drought. All tell me that I might have gone ten
years to Sicily, and not experienced the same absence of rain
in the same season. What with the fine season in Auvergne,
and all summer and autumn, was there ever a greater proof
that Providence watches over geologists? The chief danger,
.perhaps, in philosophical subjects is, that the majority may
feel them cold, and without sufficient excitement. Mix up
with this abstract quality a little human passion, a slight
dose of personal interest, and you draw in ten times as many
who look into the thing, and some of them come over to you,
and transfer the stimulus at first supplied by extraneous
considerations to the real subject itself. Truth once per-
ceived, never I am persuaded wanted supporters in this
world, even if men were sure to be losers by their support.
The only difficulty you have to struggle with is the enormous
indolence of mankind, who will never look into evidence till
goaded on to it by some stronger motive than the mere love
of abstract truth. I speak of the οἱ πολλοί. Do not build
too much on the Col de Tenda section, for next to the green
sand I believe the pretended universality of the ' New red '
to be among the emptiest bubbles in geology. I have just
learnt from Scarpellini, the astronomer, that Brocchi's *com-
pagnon de voyage*, Ricasoli, is in Rome, and has some collections made in that tour, and I am to beat up his quarters
forthwith.   Ever most truly yours,

CHARLES LYELL.

To HIS SISTER.

Rome : January 21, 1829.

My dear Marianne,—Although I cannot write you a long
letter, I will not leave this for Siena, for which I start to-
morrow, without acknowledging yours, which I was very
glad to find here. My letters from geological friends are
very satisfactory, as to the unusual interest excited in the
Geological Society by our paper on the excavation of valleys in
Auvergne. Seventy persons present the second evening,
and a warm debate. Buckland and Greenough furious,
contra Scrope, Sedgwick, and Warburton, supporting us.
These were the first two nights in our new *magnificent*
apartments in Somerset House. Their letters are all full of
Dr. Wollaston's last moments, whose place they justly
observe there is no one to supply in town. He has left
2,000*l.* to the Royal Society, and 1,000*l.* to the Geological
Society, a bequest, as Fitton says, which although of great
consequence in itself, will do us more good as showing the
deliberate approbation of such a mind. I will send you,
when I return, Murchison's two letters, with the account of
the discussion on our memoir, which will amuse you. It is
a very small part of our work. I am pledged to set steadily
at the rest the moment I return, and expect that one month
will do it, for M. has got all the organic remains named,
and the mechanical work over. Longman has paid down
500 *guineas* to Mr. Ure of Dublin for a popular work on
Geology, just coming out. It is to prove the Hebrew cosmogony, and that we ought all to be burnt in Smithfield. So
much the better. I have got a rod for the fanatics, from a
quarter where they expect it not. The last Pope did positively dare to convoke a congregation, and *reverse* all that
his predecessors had done against Galileo, and there was
only a minority of one against; and he instituted lectures
on the Mosaic cosmogony to set free astronomy and geology.
How these things are so little known in Paris and London,
heaven knows. They are golden facts, and I find the state
of the question here to shame the Granville-Penn school of
England. I have had great light thrown on Sicily and

Auvergne by the travertins of *Lakes Tartari* and *Tivoli.* On a fine day I saw the cascade, then full of water. I find the marine formation here just the same as Sicily, a new light to the geological world, and I have now got full proof that half Sicily was formed since the Mediterranean was inhabited by present species of testacea. The present Pope is a bigot, but has quite failed in sundry attempts to undo what the last great man effected. He issued an order, which was positively published, to shut up the English chapel, but our Consul waited immediately on the Cardinal Secretary, and said he must, if the news was true, in pursuance of his duty, send off a messenger to St. James's ; but he hesitated till assured again, for he wished to remind his Holiness that there were numerous Catholic chapels and churches in London, and although he would not take upon himself to say that his Britannic Majesty would instantly shut them all up, yet he would not answer for him. The order was repealed. You may imagine how glad I shall be to get home after such a continual accumulation of materials, without time to digest and apply. I have, besides four books sent home by Murchison, nine others and a half entirely full of geological sections, notes, &c. It is very encouraging to perceive by my letters how much more every year the subject is taking hold of the public mind. My own opinion of its future capabilities has been very much raised by my tour in Sicily.

Believe me, your affectionate brother,

CHARLES LYELL.

To RODERICK J. MURCHISON, ESQ.

Rome : January 22, 1829.

My dear Murchison,—Monte Mario turned out, as I expected, my Sicilian formation, and not sub-Appenine. Tivoli and its cascades gained in beauty by the late heavy rains, but I had my usual good fortune in a bright sunny day to view them, and study the grand section of travertin, 500 feet perpendicular, there displayed. That this was formed in an ancient small lake, and not, as some have dreamt, in the course of the Anio, I have full proof. Some of the pisolitic and oolitic travertin there throws much light on some of my Sicilian beds, as does still more Omalius d'Halloy's older travertin of the plain near Tartaret. That variety of travertin which they call 'tartaroso' is precisely our great tubular travertin near La Serre, so like, that if they changed places, the migration would never be detected. This contains, with the plants (but very rarely indeed), helices. The older travertin, of which a great part of Rome is built, contains no organic remains. It affords me some delightful Sicilian analogies, but I hardly think it does any to Central France, at least I must see our specimens again before I answer for it. It is entirely different from the other travertin, or that which resembles the indusial. If I were to return to Auvergne, the indusial travertin and the other, and probably some of the rest of the compact calcareous limestones, would now appear as volcanic to my eyes as the augitic sands. It seems to me an immense oversight in all who have passed through Tuscany, the Campagna Romana, and Sicily, to have brought hot springs so little into play in their systems. There is in the Campagna di Roma a beautiful chain of alluvial phenomena, which, but for Moses and his penal deluge, would have thrown more light than any part of Europe on the modern ages of the earth's history, and they tend to confirm I think the Boulade [1] case as viewed by us. I have got news again of the Viterbo bones, and envy the first *Sedgwick* who gets at them. They are elephant and co., and are in an alluvium *overlaid* by *prismatic* lava of one of the extinct Tuscan volcanoes! The elephant and rhinoceros in alluvium about Rome, which is geologically thousands of years younger than 'crag.' This alluvium is younger than most, if not all of the extinct volcanoes of that district, as appears beyond a doubt both by its contents and its relation to the present valleys, and to the alluvial volcanic tufas. Ricasoli, who travelled much with Brocchi, is a capital worker, and if the poor devil was in a more favourable situation would do much. Babbage or Herschel would be diverted to hear that Scarpellini could not lionise the

---

[1] Boulade bone-bed. See paper by Lyell and Murchison on Excavation of Valleys, illustrated by the volcanic rocks of Central France.—*Edinburgh Philosophical Journal*, 1829.

museum to me this morning, as a Cardinal had requested
him to assist at the solemn benediction and christening of
a great new *bell*, to be put up to-morrow in St. Mary's.
But Ricasoli went with me to the museum, and the varieties of
travertin there are most instructive. Lord Northampton
had formerly presented Padre Ricca with a handsome set of
Scotch minerals, and his letter of introduction procured me
duplicates (very fine) of Sienese fossil shells from Ricca.
Then I hunted up the widow of the late porter of the
museum, and for a dollar and a half got forty species more.
Then last, not least, Ricca introduced me to an active
intelligent young Belgian, who has been sent by the King
of the Netherlands to collect in natural history, by name
Canraine. I found him very anxious to learn something of
geology. He knew where the sections were, and went with
me. I taught him the geology, and he gave me a list of the
shells found in said localities. By degrees he became so
interested with the new views, that he begged me to stay a
day more, which I will do, for I shall get more fossils,
and he has already undertaken that the University of Leyden
shall send to our Geological Society *a complete set of Sienese
duplicates*, and in return receive English fossils. We may
hope for this next summer, and expect at least 500 or 600
species. The result is that here the *Sabbia giallonale* is not
one formation with the blue marl (which from here to south
of Sicily is called by all, chalk, creta, and when they talk
French, *la craie*). The sand has an immensely nearer relation to the Mediterrranean than the inferior formation, and
comes very close, if not quite, to any Sicilian fossil beds.
Now I told Guidotti that he ought to find something of
this at Parma, where the whole thickness exceeds 3,000 feet,
which could not be the work of a day, even in a geological
calendar, but he protested it was not so. However, I have
now stirred him up with a letter, and as he told me he
should collect this winter quite with new eyes after my
cramming, I shall look with much curiosity for his answer.
Between the yellow sand here, which is 300 to 400 feet, and
the inferior blue marl, which is thousands of feet deep, is a
beautiful formation, consisting of alternations of fluviatile
and marine beds. If the tertiaries from the north of Italy

to Cape Passaro shall ultimately correspond to my present view of them, they will I think establish for ever the elevation theory; but I am the more afraid of their certainty, in proportion as they suit, so beautifully, the system I anticipated. But when all my specimens are examined, when the rest are collected by my agents and Lord Northampton at Palermo, when Costa sends me the results promised of his Calabrian expedition next summer; when I get from Ricasoli the Roman, and from Canraine the Tuscan list, and from Guidotti a fresh examination of sub-Alpine, &c., then I will make all necessary modifications; for in the main I am sure I cannot now be shaken, and Ischia and Etna form one extreme of the series, and perhaps Parma, &c., the other; but more when we meet. In the meantime do not take off the freshness of my revolutionary prospects by too early disclosures.

Ever yours,
Charles Lyell.

To Roderick J. Murchison, Esq.

Geneva: February 5, 1829.

My dear Murchison,—In my last I mentioned my Siena operations, which ended in my bringing away a respectable box of fossils, but unluckily we cannot at present get any collector there to work for money. At Genoa I found Viviani, gentlemanlike and intelligent, pretends not to geology, but would soon make any one like natural history, and we owe to him Paretto and Sasso. The former was, unluckily for me, still in France; the latter, a young physician, has written a very fair memoir on the basin of Albenga, west of Savona and Genoa, and of which our Savona blue marl was an outskirt, and he has given a list of 200 shells. Some Brongniart saw, and the rest Bronn, a German, corrected. I found that Paretto has lately published an article on the Cadibona coalfield, which, as well as Sasso's, I purchased after a toilsome hunt in the stalls. Paretto leaves little for us to do, but as you will soon read it, I say no more; but the way he has hit off the Siena basin in a note, convinces me he will make head soon of all the Italians. He is only twenty-eight, independent, and full of zeal and talent.

I found at a Pharmacien's a decayed jaw of Cadibona,
which I strongly suspect was a stag's. It was not worth accepting, nor would it have travelled, but I found a noble
piece in the Museum, which they seem to have been hardly
conscious of themselves, for Paretto positively overlooked it!
One molar and two canine teeth of an animal about the cut
of a tiger, all fixed in the jaw, decidedly a large carnivorous
beast. I got Viviani to order it from under the glass to his
own house, and there to let me draw it; and when he saw
how badly I got on, he took his *camera lucida*, and with
enviable rapidity and faithfulness finished a drawing of it
of natural size, distinguishing also between the coal and
bony parts.

Viviani has some splendid views, founded on a multitude
of facts, as to the distribution of plants in the Mediterranean
islands, for botany is his *forte*. I struck up a geological theory
to account for them, which, although it shook his nerves,
pleased him so much, that he gave me two of his works on
botany. I pushed on, per courier, through deep snow and
hard frost to Turin, and was there only part of the day,
closeted with Borelli nearly all the time. He showed me
his list of the shells which La Marmora got this year in
Sicily. Some species which I had not, but the whole not
numerous. I was delighted to find that the conclusions from
the list as named by Borelli were the same I came to from
Costa's catalogue. Over the latter, B. looked with amazing
interest, and after cutting up his names, &c., said, ' But still
you need not be afraid to reason on them as a geologist!'
He was astounded at the result, and said, 'I begin to think
the day *may* come when the retiring of the ocean will be
doubted and disputed by many.' He has at last been curious
to know what difference there would be in the inferences from
the sub-Apennine north of Tuscany, and at last proposed that
when I published my Sicilian lists I should add one from
him. I told him I should have craved this before, but did
not, because I knew his intention to publish such a list.
He looked over Sasso's Albenga list, and after swearing it
was a string of blunders, surprised me by saying, ' Yet, after
all, I have no doubt, from some Albenga shells which I have,
that your conclusions from Sasso's list were quite just, and

though he has mistaken genera, &c. &c., yet there are not more than ten blunders in 200 which could mislead a geologist, which would not be to him serious in the main.'

There was nothing but snow and biting frost between Turin and Chambery in diligence. Thence to-day I came in a char, and did some interesting geology. A beautiful tertiary basin of green sand, facsimile in mineralogical appearance to Superga, and also with some blue marl surrounded by those marble-like, amorphous, dolomitic, or calcareous rocks like Nice. The pebbles are not Serpentine in this instance, but chloritic schist and green hornblende schist. A conglomeration just like Superga tops all. I shall get up this concern here. This green sand and sandstone, as well as the superincumbent conglomerate, is provincially termed ' molasse.'

February 6.—I have just had this morning a famous geologico-botanical discussion with Professor De Candolle, and am almost certain that my spick-and-span new theory on this subject will hold water.

Always affectionately yours,

CHARLES LYELL.

To HIS FATHER.

Geneva : February 7, 1829.

My dear Father,—I arrived here on the evening of the 5th. The hills between Genoa and Alexandria were covered with deep snow and ice, and the frost intense. I got a place with the courier, so that it was soon over, but though I wrote a week before to Turin, all the places per courier for Mont Cenis were forestalled for five weeks. The diligence performed twenty-four miles on sledges, and ought to have done much more. Two days before, an avalanche stopped the diligence for twenty-four hours, just at the very coldest part: they had just cut through it when we arrived. Fifty stout workmen were employed. The manner in which the road is kept up is admirable, and when one sees the Alps in winter, it is wonderful to think that one can pass in sixteen hours. I was not a little relieved when I got to Chambery, without a frozen toe or finger, for the cold is unusual this year. On

descending to Alexandria, the chariot fell into some deep
frozen ruts, and shook me and the courier out of our sleep.
The poor old gentleman believed it was another earthquake,
for he had been bruised by the shock when *en voiture* on the
same mountains. The gloom still sensible in the minds of
the Genoese in consequence of the earthquake is very remarkable—it is now six months since. Almost every house
was rent, the inn where I was had sixteen fissures in the
walls. For eight days every vessel and boat in the port,
every square, and all the suburbs, were filled day and night
by the inhabitants. All the troops were insufficient to act
as a police, as almost every door was left open, as each slight
shock made them rush out. One more like the first would
have brought half the town down. I have profited far more
in my homeward tour, rapid as it has been, than I had the
least idea I should have done. At Siena I got a handsome
present of fossil shells from the excellent Padre Ricca, to
whom Lord Northampton had given me a letter. I also
struck up an alliance with M. Canraine, a naturalist now in
Tuscany, charged with a mission by the King of the Netherlands on scientific subjects, collecting, &c. His letters will
give me much information which I yet want on the tertiary
basin there. My facts on these recent beds in different
parts are beginning to point to curious results. At Genoa,
besides several geologists, I found Viviani, the botanical
professor, a very superior man. His generalisations on the
distribution of plants in the isles of the Mediterranean lead
to splendid theories. I fitted them to my new system of
geology for Sicily and Southern Italy, by which he was
pleased, and gave me two of his botanical works.

I was anxious to come here, instead of by Lyons, to get
some ideas from De Candolle on my new geologico-botanical
theory. He did not of course remember me, but when he
heard my name, he asked after you, and said I should take
une petite brochure, as a souvenir from him. He is in full
force, and has been most useful to me, having given me what
cannot be bought, his splendid essay on Geographical Botany,
the most beautiful generalisation of a multitude of facts
which I think was ever produced in natural history. I am
now convinced that geology is destined to throw upon this

curious branch of inquiry, and to receive from it in return, much light, and by their mutual aid we shall very soon solve the grand problem, whether the various living organic species came into being gradually and singly in insulated spots, or centres of creation, or in various places at once, and all at the same time. The latter cannot, I am already persuaded, be maintained. Viviani was puzzled to account for Sicily having so much less than its share of *peculiar* indigenous species, but this should be, for I can show that three-fourths of this isle were covered by the sea down to a period when nine-tenths of the present species of shells and corals (and by inference of plants) were already in existence. Such an isle like Monte Nuovo has been obliged to borrow clothes from its neighbour, having scarcely had time to furnish any yet for its own nakedness. It has not yet seen out a tenth, perhaps not a twentieth part of a revolution in organic life. Give it the antiquity of the high granitic mountains of Corsica, and it will also boast its indigenous unique plants, unknown elsewhere either in the Mediterranean or other part of the globe.

Necker, grandson of Saussure, is a good geologist here, young and independent, just returned from a tour in Dalmatia. With him I have already mastered much of the geology of these parts. We have defied the snow in our excursions. I start to-morrow, and shall be in Paris in four days.

With love to all, believe me, your affectionate son,

CHARLES LYELL.

To GIDEON MANTELL, ESQ.

Paris: February 19, 1829.

My dear Sir,—Deshayes, now the strongest fossil conchologist in Europe, has lost so seriously by his fine work on the shells of the Paris basin, that he is not only obliged to stop, so miserably is it encouraged, but his circumstances are injured for a time by it. I have bought a copy (seventy francs), to help him, and as I find he has amassed in his researches for new things a multitude of fine duplicates, I trust through Prévost to negotiate a purchase for you. The

grand thing will be to get the names of perhaps 100 *genera*
from such a man, who is acknowledged to be the Cuvier of
tertiary shells, no mean acquirement now that they amount
in *his museum*, including those living in the Mediterranean
and the Channel, to above 3,000 species. His drawings of
microscopic shells of the Paris basin are most curious. Such
incredible forms of multilocular shells, and some so elegant.
God knows whether the work will ever come to these. It
is one-third finished.

February 20.—Just heard a good lecture of Prévost's to
a numerous class. It was on diluvium and caves, a good
logical refutation of the diluvian humbug. The news of the
day is that a Dr. —— has just read a memoir at the Institute
on a new small tapir, and produced the head and jaws. It
inhabits the mountains of Upper India, and they swear that
the species is not distinguishable hardly from one of the
Montmartre Paleotheriums. In return I announced Mary
Anning's[2] new Pterodactyle of Lyme. How grand your
museum will look when, under every bone of which the
ornithologists could make nothing, you write Pterodactylus
Dorsetianus and Tilgetanus! Prévost has a beautiful large
mâchoire of an Anoplotherium, which he showed me imbedded in gypsum, intended as a present for you, and if we
succeed in getting the shells, it is to go with them. He
observed, ' This will be better, for it will cut a mighty figure
among the cockles ; but if it is thrown into Cuvier's large
caisse it will be drowned.' I have no fear of getting the
shells, and if it answer expectations I shall be able to enlarge the commission in future, should you incline to be
sumptuous. In the meantime prepare to disburse 100 francs,
which I shall pay Deshayes before I leave Paris, which I
suppose will be in two days, for town. Young Brongniart
protested at breakfast on Sunday at his father's, before a
large party of *savans*, that you knew more of fossil plants
than any man in England. I could have told him that a
certain Mr. Stokes[3] had recently accumulated a quantity of

[2] Miss Mary Anning, of Lyme Regis, who was zealous in excavating the
fossils of that region.
[3] Charles Stokes, b. 1783, d. 1853, a respected member of the Stock Exchange, full of vast research in the Natural History Sciences, and remarkable
for literary and antiquarian, musical and artistic, knowledge.

fossil and recent woods, plants, &c., which with no small
study, and that too aided by *Brown*,[4] had placed him in a
situation to become the critic of Adolph B. himself, which
moreover he will be, as I learn—but I kept this to myself,
as it will be time enough when he learns it.

Ever truly yours,

CHARLES LYELL.

To HIS SISTER.

Paris: February 23, 1829.

My dear Marianne,—I was glad to find your letter here
on my arrival seven days ago, after a fagging journey of four
days and four nights from Geneva, having only had six hours'
rest at Dijon. Sledges for forty miles over deep snow on the
Jura, and ice the rest of the way, so we were continually
summoned by the conductor to turn out of the warm inside,
while the horses tried to pull the diligence up. Notwithstanding this, the company of a clever and lively French
officer in the interior, and some good memoirs given me at
Geneva, rendered the journey not without some amusement.
The difficulty of keeping warm was great enough. Here I
found a better climate. The very evening of my arrival I
caught Prévost just going to give a lecture, and met there
other geologists, who took me next morning to a lecture at
the École des Mines, and thus I got thrown in a few hours
into the heat of the battle, which is carried on with vigour,
notwithstanding the complaints of the *savans* that the public
will think of nothing but politics and romances, and their
evident fear that we shall soon distance them in England in
this branch.

I have learnt very much from various old and new
acquaintances, and although they have pumped me somewhat unmercifully, I am sure I have reaped most from the
exchange. Some of them have come by most opposite routes
to exactly the same conclusions as myself, and we have

---

[4] Robert Brown, b. 1773, d. 1858, President of the Linnæan Society, accompanied Captain Flinders as naturalist in the 'Investigator' round the shores
of Australia in 1802, &c. An indefatigable and profound botanist. Named by
Humboldt *Botanicorum Princeps.* He inherited Sir Joseph Banks's herbarium,
which is now in the British Museum.

mutually felt confirmed in our views, although the new
opinions must bring about an amazing overthrow in the
systems which we were carefully taught ten years ago. I
have established a regular correspondence with Deshayes,
now by far the principal conchologist in fossils, and he is to
name all my fossil shells from Sicily. I have also been getting
the necessary works, and many are German. How glad I
am to hear you are determined to persevere in it : do not be
frightened by the barbarous aspect of it. I would give a
hundred pounds if I knew it decently, indeed it would soon
be worth that to any author who writes on natural history.
I shall work at it hard before a year passes away. Cuvier is
in great force, and gave a famous soirée the other day. He
has been chosen by the ministers to defend their municipal law
in the tribune : two months of his time will thus be lost to
science. He talked to me of the Catholic question, our cor-
poration rights, &c., and not a word could I get on natural
history. Yet this year he has come out with four volumes
on fish, and eight more will appear in two years. He has
received between 2,000*l.* and 2,200*l.* for it, whereas no other
person in Paris could persuade a librarian to publish one
volume for nothing. He is also publishing another edition
of his ' Règne Animal,' and other things. He has been very
obliging to me, for on my applying for casts of animals for
Mantell, who has been begging in vain for a long time, he
gave me an order for whatever I liked, so I have sent off from
the museum a huge box with casts of *every* thing. I got into
Cuvier's sanctum sanctorum yesterday, and it is truly characteristic of the man. In every part it displays that extraordinary power of methodising which is the grand secret of
the prodigious feats which he performs annually without
appearing to give himself the least trouble. But before I
introduce you to this study, I should tell you that there is
first the museum of natural history opposite his house, and
admirably arranged by himself, then the anatomy museum
connected with his dwelling. In the latter is a library disposed in a suite of rooms, each containing works on one
subject. There is one where there are all the works on or-
nithology, in another room all on ichthyology, in another
osteology, in another *law* books ! &c. &c. When he is

engaged in such works as require continual reference to a
variety of authors, he has a stove shifted into one of these
rooms, in which everything on that subject is systematically
arranged, so that in the same work he often takes the round
of many apartments. But the ordinary studio contains no
book-shelves. It is a longish room, comfortably furnished,
lighted from above, and furnished with eleven desks to stand
to, and two low tables, like a public office for so many clerks.
But all is for the one man, who multiplies himself as author,
and admitting no one into this room, moves as he finds
necessary, or as fancy inclines him, from one occupation to
another. Each desk is furnished with a complete establishment of inkstand, pens, &c., pins to pin MSS. together, the
works immediately in reading and the MS. in hand, and on
shelves behind all the MSS. of the same work. There is a
separate bell to several desks. The low tables are to sit to
when he is tired. The collaborateurs are not numerous,
but always chosen well. They save him every mechanical
labour, find references, &c., are rarely admitted to the study,
receive orders, and speak not.

Brongniart, who in imitation of Cuvier has many clerks
and collaborateurs, is known to lose more time in organising
this auxiliary force than he gains by their work, but this is
never the case with Cuvier. When I went to get Mantell's
casts, I found that the man who made moulds, and the
painter of them, had distinct apartments, so that there was
no confusion, and the despatch with which all was executed
was admirable. It cost Cuvier a word only.

Murchison writes me word that Mary Anning has immortalised herself anew by finding at Lyme a skeleton of one of those
flying lizards (pterodactyles), of which two only were known
of distinct species, and none before in England. Buckland has
written a paper, and there has been much merriment at the
Society about the Dragon of Wantley having devoured all
the birds, for they now declare that the Stonesfield *soi-disant*
birds' bones all belonged to this winged reptile. Murchison
and his wife have been with Mrs. Somerville, spending a
week at Christ Church, and he laments that Buckland voted
for the anti-Catholic petition, which conduct, he says, Mrs.
Buckland assured him was to be attributed to his Sicilian expe-

dition, and he trusts my journey has not made me intolerant.
Certainly, if any country could make me abhor Catholicism
in its worst form, it is that island, and you may have gathered
this from my letters ; but I confess I think it no small want
of logic to confound the emancipation question with the
state of Spain and Sicily. Besides, if fanaticism were to be
the measure of disability, I fancy that no small part of our
sectarians and even more orthodox saints would be entitled
to their share. I wish you were in the way of getting on in
conchology : the extent and beauty of these objects is truly
wonderful, and the results are becoming most important
every day. It is the ordinary, or, as Champollion says, the
demotic character in which Nature has been pleased to write
all her most curious documents. We will try to get a more
active exchange with Mrs. Buckland of insects for shells.
Murchison tells me that seventy more fossil insects have
been sent off for us from Aix.

    With love to all, your affectionate brother,
                                        CHARLES LYELL.

                    To HIS SISTER.

London : February 26, 1829.

    My dear Caroline,— I have not announced my arrival
except to Murchison, and shall appear at the Royal Society
this evening for the first time. I find the Geological Society
and Sedgwick, the new President, have persuaded Murchison
to become Home Secretary for a year, as they dubbed me
Foreign Secretary, which I trust will give me very little
trouble. Sedgwick quite astonished them, it seems, in the
chair at the general meeting, which was very full. Among
innumerable good hits, when proposing the toast of the
Astronomical Society, and Herschel, their president, he said,
alluding to H.'s intended marriage (for he is just about to
marry the daughter of a Scotch clergyman), 'May the house
of Herschel be perpetuated, and, like the Cassinis, be illustrious astronomers for three generations. May all the con-
stellations wait upon him ; may Virgo go before, and Gemini
follow after.' Poor H., notwithstanding his confusion, got
up after the roar of laughter had continued for three minutes,

and made a famous speech. You say you should like to have *chased the Edusas* in Sicily, as you have only two tattered ones. Pray explain, Mrs. Conservatrice del Museo, what has become of the splendid specimens I sent last year, and let me have your *Artaxerxes.*

Lieutenant Graves has sent a letter, saying that two boxes are in town for me from him, one all skins of birds from Straits of Magellan, and one geological. All the Auvergne and French boxes necessary for the joint paper of Murchison and me are arrived safe. I have already made progress in working up materials here, and in Paris I did much. Deshayes is positively to give me tables of more than 2,000 species of tertiary shells, from which I will build up a system on data never before obtained, by comparing the contents of the present with more ancient seas, and the latter with each other. He is to name all my Sicilian shells.

With love to all, believe me, your affectionate brother,

CHARLES LYELL.

To GIDEON MANTELL, ESQ.

April, 1829.

Dear Mantell,—A splendid meeting last night. Sedgwick in the chair. Conybeare's paper on Valley of Thames, directed against Messrs. Lyell and Murchison's former paper, was read in part. Buckland present to defend the 'Diluvialists,' as Conybeare styles his sect, and us he terms 'Fluvialists.' Greenough assisted us by making an ultra speech on the importance of modern causes. No river, he said, within times of history, has deepened its channel one foot! It was great fun, for he said, 'Our opponents say— "Give us time, and we will work wonders." So said the wolf in the fable to the lamb: "Why do you disturb the water?" "I do not; you are further up the stream than I." "But your father did." "He never was here." "Then your grandfather did, so I will murder you. Give me *time,* and I will murder you." So say the Fluvialists!' Roars of laughter, in which Greenough joined against himself. What a choice simile! Murchison and I fought stoutly, and Buckland was very piano. Conybeare's memoir is not strong by any means.

He admits three deluges before the Noachian ! and Buckland
adds God knows how many *catastrophes* besides, so we have
driven them out of the Mosaic record fairly.

Yours very truly,

CHARLES LYELL.

To GIDEON MANTELL, ESQ.

London : June 7, 1829.

My dear Mantell,—The last discharge of Conybeare's
artillery, served by the great Oxford engineer against the
Fluvialists, as they are pleased to term us, drew upon them
on Friday a sharp volley of musketry from all sides, and such
a broadside at the finale from Sedgwick, as was enough to
sink the 'Reliquiæ Diluvianæ' for ever, and make the second
volume shy of venturing out to sea. After the memoir on
the impotence of all the rivers which feed the 'main river of
an isle,' and the sluggishness of Father Thames himself,
'scarce able to move a pin's head,' a notice by Cully, land-
surveyor, was read on the prodigious force of a Cheviot
stream, 'the College,' which has swept away a bridge, and
annually buries large tracts under gravel. Buckland then
jumped up, like a counsel, said Fitton to me, *who had come
down special.*

After his reiteration of Conybeare's arguments, Fitton
made a somewhat laboured speech. I followed, and then
Sedgwick, who decided on four *or more* deluges, and said
the simultaneousness was disproved for ever, &c., and declared that on the nature of such floods we should at present
' doubt, and not dogmatise.' A good meeting.

I am to start for Scotland June 20, and shall be there two
months.

Ever truly yours,

CHARLES LYELL.

To DR. FLEMING.

Temple : June 10, 1829.

My dear Sir,—I was glad to hear from you, and can assure
you that I have been so busy since my return that I had
no correspondence with any one except on business, though

I would gladly have written to you at any time, if I had not been always hoping to have sent you a paper, *we think a floorer*, of Buckland's diluvial question. You will get a separate copy, and I wish it may be an antidote to a sharp attack which I hear Conybeare and Buckland have levelled at you, in the same number, about ' climate,' &c. Buckland was so amazingly annoyed at my having had such an anti-diluvialist paper read, that he got Conybeare to write a controversial essay on the Valley of the Thames, in which he drew a comparison between the theory of the Fluvialists, as he terms us, and the Diluvialists, as (God be praised) they call themselves.

Of course, in defining the Fluvialists, they (for Buckland wrote half the memoir) took care to build up their man of straw, and triumphantly knocked him down again. But in the animated discussion which followed the reading of the first half of the essay, at the Geological Society, we made no small impression on them. And when, last Friday, the remainder came on, we had a hot rencounter. Buckland came up on purpose again, and made a leading speech. But after we had exposed him, and even Greenough, his only staunch supporter, had given in in many points, Sedgwick, now president, closed the debate with a terribly anti-diluvialist declaration. For he has at last come round, and is as decided as you are. But you must know that Buckland now, and Conybeare, distinctly admit three universal deluges, and many catastrophes, as they call them, besides ! But more of this when we meet.

I hope to be at Kinnordy the first week in June, and to see you then soon after. Your discovery of the fish may probably confirm the opinions given by Sedgwick and Murchison on our district, as they compared the older sandstone of the Sidlaw to the Caithness sandstone containing fish. One reason why I did not write was, to tell the honest truth, that I have seen so much, and so greatly altered my views, that I know not where to begin, and nothing but a good three days' confab, *tête-à-tête*, will suffice. I shall see, in my way up, much of the geology of the east of England, and perhaps, in my way down, of the west.

I am preparing a general work on the younger epochs of the earth's history, which I hope to be out with next spring.

I begin with Sicily, which has almost entirely risen from the
sea, to the height of nearly 4,000 feet, since all the present
animals existed in the Mediterranean !

Ever most sincerely yours,

CHARLES LYELL.

To DR. FLEMING.

Kinnordy : October 11, 1829.

Dear Dr. Fleming,—When I showed your vegetable impressions to the quarrymen of Kirriemuir, many declared
that a great number exactly similar were got this year from
a quarry called Bron-head or Bren-head at Locheye, between
Dundee and Auchterhouse, which is now working on the
property, I believe, of Mr. Clayhills. I have written to
Lindsay Carnegie of Kinblethmont, who of all our lairds
promises fairest to be a keen geologist, to go to his neighbour Mr. Muir, and take instructions. In the meantime,
Carnegie has written to me to say that he has offered pre-
miums for fossils in his own pavement quarries. I shall leave
with Blackadder, for a time, the specimens of scales and
plants, to show about to the men. I am glad to find lately
that pavement stone is worked successfully at Ferne, on the
Grampian side of the trough of Strathmore, between which
and Arbroath the pavement and a powerful conglomerate
are each thrice repeated; each twice with a southerly, and
once with a northerly, dip. I have just received a famous
letter from Murchison. He and Sedgwick have worked very
hard in the basin of the Danube, Hungary, Styria, &c.
Sedgwick was astonished at finding what I had satisfied
myself of everywhere, that in the more recent tertiary groups,
great masses of rocks, like the different members of our secondaries, are to be found.

They call the grand formation in which they have been
working, sub-Apennine. Vienna falls into it. I suspect it
is a shade older, as the sub-Apennines are several shades
older than the Sicilian tertiaries. They have discovered an
immensely thick conglomerate, 500 feet compact marble-
like limestone, a great thickness of oolite, not distinguishable from Bath oolite, an upper red sand and conglomerate,

&c. &c., all members of that group zoologically sub-Apennine. This is glorious news for me. They tell me to hold my tongue till our next session. It chimes in well with making old red transition mountain limestone and coal, and as much more as we can, *one epoch*, for when Nature sets about building in one place, she makes a great batch there, and old Werner's universal envelope was, as my correspondent says of some of Brongniart's extension of the Parisian type, a humbug. All the freshwater, marine, and other groups of the Paris basin are one epoch, at the farthest not more separated than upper and lower chalk. They have also found a great gypsum and salt formation in the Alps, over lias fossils, which charms me.

Mrs. Murchison's letter contains the painful news that poor Conybeare has been thrown out of his gig, and received a dreadful concussion of the brain. When the accounts came away, he was delirious ; no hopes entertained of his life. A ride to-morrow with Blackadder, and an examination of the whole section of the Noran river, will be a finale to my outdoor geology for 1829.

Very truly yours,
Charles Lyell.

To Dr. Fleming.

9 Crown Office Row, Temple : October 31, 1829.

Dear Dr. Fleming,—Conybeare, after being fourteen days insensible, not even recognising his wife, is recovering ; and there are hopes that he will get quite over it, but will be stopped from working for a time, of course. ' The father of stercoraceous chemistry,' as Buckland called himself in a letter, has strengthened his theory, but had to retract also on one or two points.

Sedgwick and Murchison are just returned, the former full of magnificent views. Throws overboard all the diluvian hypothesis ; is vexed he ever lost time about such a complete humbug ; says he lost two years by having also started a Wernerian. He says primary rocks are not primary, but, as Hutton supposed, some igneous, some altered secondary. Mica schist in Alps lies *over* organic remains. *No rock* in

the Alps older than *lias!* Much of Buckland's dashing
paper on Alps wrong. A formation (marine) found at foot
of Alps, between Danube and Rhine, thicker than all the
English secondaries united. Munich is in it. Its age
probably between chalk and our oldest tertiaries. I have
this moment received a note from C. Prévost by Murchison.
He has heard with delight and surprise of their Alpine
novelties, and alluding to them and other recent discoveries,
he says, 'Comme nous allons rire de nos vieilles idées!
Comme nous allons nous moquer de nous-mêmes!' At the
same time he says, 'If in your book you are too hard on us
on this side the Channel, we will throw at you some of old
Brongniart's " metric and peponary blocks," which float in
that general and universal diluvium, and have been there
depuis le grand jour qui a separé, d'une manière si tranchée,
les temps ante-des-temps Post-Diluviens.'

In some sandstone sent from Glammis I find some white
stains with a bluish tinge, having clearly some analogy to
your fish scales, but much decomposed.

Remember me to Dr. Fleming, and believe me very truly
yours,

CHARLES LYELL.

To LEONARD HORNER, ESQ.

November 24, 1829.

My dear Sir,—My friend G. Eyre, a barrister who is
attending law lectures at the London University, happened
to say this morning, that when Lardner spoke of the vacant
Chair of Mineralogy and Geology in his introductory lecture,
a brother student had mentioned to him that there had been
a report of *my* being thought of for it. I ought therefore to
tell you that I have always felt it, and do now, to be out of
the question; and if the council wish to fill it up, you must
consider me quite out of the way, both in justice to the
Institution and other candidates.

I have received this morning a letter from young Allen,
full of *mineralogical* zeal for an Icelandic tour. I find that
we must go the latter end of April, in which case I should

VOL. I.   s

only be able to get out Vol. I., which I believe I shall do, for I should like to make sure of Hecla.

I have not waited till we meet, because I did not feel comfortable under the idea that I had let you go away the other day with a notion that I only wanted to be *pressed* a little. I should like for your sake, and for the science, to see a good man in, but it ought, I believe, to be a man of fortune, who did not hope to pay his travelling expenses by what he made. In the meantime I am quite sure, whoever may be the candidates, that you do good by deferring an election, because every lecturer who has not a good class rather does harm to the Institution, and one cannot expect it in geology yet.

Believe me, ever most truly yours,

CHARLES LYELL.

To GIDEON MANTELL, ESQ.

Temple : Saturday, December 5, 1829.

Dear Mantell,—We were all disappointed at your not being here yesterday, for Murchison told us you were to have been here. Sedgwick and his wind-up on the Alps went off splendidly in a full meeting. You and the iguanodon treated by Buckland with due honours, when exhibiting some great bones of a little toe from Purbeck. He greatly amazed my friend Sir T. Phillips by his humour about the size of the said giant, compared to the small *genteel* lizards of our days.

I am bound hand and foot. In the press on Monday next with my work, which Murray is going to publish—2 vols. The title—' Principles of Geology ; being an Attempt to Explain the Former Changes of the Earth's Surface by Reference to Causes now in Operation.' The first volume will be quite finished by the end of the month. The second is in a manner written, but will require great recasting. I start for *Iceland* by the end of April, so time is precious. You must let us see you here. What particular scheme have you at present in geology ? Your list of shells will be grand things.

Very truly yours,

CHARLES LYELL.

CHAPTER XI.

FEBRUARY–AUGUST, 1830.

DR. FLEMING'S REMARKS ON CONYBEARE—DETERMINATION NOT TO ENTER
INTO CONTROVERSY—OPENING OF THE ATHENÆUM CLUB—MILMAN'S
'HISTORY OF THE JEWS'—PROPOSED TRIP TO THE CRIMEA—LETTERS TO
POULETT SCROPE ON THE 'PRINCIPLES'—LEAVES ENGLAND WITH CAP-
TAIN COOKE FOR THE PYRENEES—EXCURSION INTO CATALONIA.

[The first volume of 'The Principles of Geology' was published in
1830, and after having relinquished sundry plans of going to Iceland,
the Crimea, and Black Sea, &c., he finally settled on accompanying
Captain Cooke, R.N., in the summer of this year, to explore the geology of the south of France and the Pyrenees, and from thence he
went alone to Catalonia, to examine the volcanic region round
Olot.]

CORRESPONDENCE.

To DR. FLEMING.

Temple : February 3, 1830.

Dear Dr. Fleming,—My eyes have been at times somewhat weakened by the small print of my notes, and this
prevents my being so good a correspondent as I would otherwise most willingly be; but I am in great spirits to find that
by help of a good amanuensis my eyes hold out so well, as
they have certainly gained strength for many years, and
were never so equal to the work since I first went to school,
as now.

The answer to Conybeare is severe enough, and both instructive and amusing; and to those who, like true Englishmen, love to see a good fight, it has afforded more sport
than any round fought for many a year. Mantell, whom I
visited at Christmas, was mightily delighted with it, but
thought you might have spared Cuvier, who, by-the-bye, had

lately sent him a magnificent present of fossil casts. Daubeny, who has just come up from Oxford, told me that the Oxonians made just the same remark, congratulating Buckland that Monsieur le Baron had been put on the pillory with him, and had come off with perhaps the greatest expressions of contempt. I will fairly own that this struck me on first reading, and I was the more surprised when I found that you meant to call him as your principal witness in your cause. As he and Lamarck have their worshippers, as well as Conybeare, it would be good policy to be more courteous towards them, until they provoke you as he did.

I sincerely hope you will keep in well with some party, but if you go on much longer, with an anti-Bucklandite, anti-Mackayite, anti-Lamarckian warfare, I begin to fear the odds; and as Sedgwick prophesies that the Professor's book will never lose the title of 'idola specus,' so I dread your retaining a name which I heard applied to you yesterday by a zoologist, ' the Zoological Ishmael.' The ' idola specus ' is allowed to be as clever a hit as ever was given, and the Cambridge Professor stood up manfully for you, when others were saying you were too savage, by saying that a man had a right to say hard things in a controversy. But your indiscriminate attacks are what I allude to as alone injurious. To show that you can strike home may save you from many an adversary, but I am sure many new ones will now be called forth.

As a staunch advocate for absolute uniformity in the order of Nature, I have tried in all my travels to persuade myself that the evidence was inconclusive, but in vain. I am more confirmed than ever, and shall labour to account for vicissitudes of climate, not to dispute them. I have only corrected the press to page 80, and get on slowly, but with satisfaction to myself. How much more difficult it is to write for general readers than for the scientific world, yet half our *savans* think that to write *popularly* would be a condescension to which they might bend if they would.

I daresay I shall not keep my resolution, but I will try to do it firmly, that when my book is attacked—as it will be a greater hornets' nest than a small sally of yours in Jamieson can be, however pointedly against popular doctrines—I

will not go to the expense of time in pamphleteering. I
shall work steadily on at Vol. II., and afterwards, if the work
succeeds, at edition 2, and I have sworn to myself that I will
not go to the expense of giving time to combat in controversy. It is interminable work. However, in your case,
Conybeare, although now recovered, is not preparing, and
Buckland, who likes to fight under the shield of Ajax, will
not, I suspect, step forth himself. If any smaller people
make a noise, you can leave them to do their worst. Buckland is working so hard at organic remains—iguanodons,
pterodactyles, and fifty other things—that he may build up a
second report in the time he might have spent in fighting
you about the old question. To our amusement, at our club
last week, when a stranger whom I took there, asked Buckland who Dr. Fleming was (some one had mentioned you),
Buckland said, ' Oh ! he's a man who ought to have been in
Jamieson's chair, or some other in Scotland, &c. &c. ; and
among other things, you will see in Jamieson's last that he
has thrown me into the gutter, as he says !' I cannot help
boring you with all this, because I dread half next year
being spent in answering the wasps who are preparing
their stings, and you can well afford to let them all alone.
Sedgwick has his paper on magnesian limestone ready for
you. If not sent, I will get it from him. I will look out
some organic remains of freshwater, when I have a minute
to spare. You talk of not being prolific, but here they
think you quite the contrary. If you would not divert your
forces from the main point, you will carry it with profit and
reputation with one half the labour which these infernal
squabbles will involve you in. So ends my epistle. *Qualis
ab,* &c.

Believe me, ever most truly yours,

CHARLES LYELL.

To GIDEON MANTELL, ESQ.

Temple : February 15, 1830.

Dear Mantell.—I have scarcely used a sheet of my prepared materials since I saw you, and am now engaged in
finishing my grand new theory of climate, for which I had

to consult northern travellers, Richardson, Sabine, and others, to read up Humboldt's and Daniel's last theories of meteorology, &c. &c. I will not tell you how, till the book is out—but without help from a comet, or any astronomical change, or any cooling down of the original red-hot nucleus, or any change of inclination of axis or central heat, or volcanic hot vapours and waters and other nostrums, but all easily and naturally. I will give you a receipt for growing tree ferns at the pole, or if it suits me, pines at the equator; walruses under the line, and crocodiles in the arctic circle. And now, as I shall say no more, I am sure you will keep the secret. All these changes are to happen in future again, and iguanodons and their congeners must as assuredly live again in the latitude of Cuckfield as they have done so.

You have always some new discovery to announce. I believe with you it would be difficult to exterminate pentacrinites, yet don't make too sure ; for if my new heating and refrigerating theory holds water, I may perhaps give you a cold at the line which may freeze up the pentacrinite. I have been reading at the Admiralty the last despatches of Captain Forster from S. Shetland. They mention fathoming the sea at the borders of the antarctic circle to the depth of 900 fathoms, with a self-registering thermometer. What a glorious depth. Ben Nevis would not make an islet if sunk there, and they found the warmth increase from the surface downwards, as you know in the Spitzbergen seas. Our new hydrographer, Beaufort, is very liberal to all geologists, and you may get what unpublished information you like from the Admiralty, and there is an immense deal there.   Yours very truly,

CHARLES LYELL.

To HIS SISTER.

Temple : February 26, 1830.

My dear Eleanor,—I expect p. 128 home from the printer to-night. Every word of advance since you were here has been written, and the matter in great measure collected ; so I hope next month, when I fall upon old stores written at Kinnordy, &c., that I shall make great way. I enjoy the work much, as the excitement is great, and I can get such assistance as

perhaps no place but London could afford. I wish you had
been in town at the opening of our new club, the Athenæum,
which is reckoned the most elegant turn-out of all, and for
a fortnight gay soirées were given to the ladies between
nine and twelve o'clock, and it still continues to do so every
Wednesday, which I hear is to go on for months. It is really
worth seeing, and fitted up in a style which I must say would
be ridiculous, except for receiving ladies. There has been a
great deal of fun about it, verses innumerable. Some of our
members grumble at the invasion, and retreated into the
library, which was respected at first, but now the women fill
it every Wednesday evening, as well as the newspaper room,
and seem to me to examine every corner with something of
the curiosity with which we should like to pry into a harem.
They all say it is too good for bachelors, and makes married
men keep away from home, and talk of a ladies' club, &c.
As the house was much admired, the number of candidates
increased prodigiously. The ballot, which in a smaller house
was a nuisance, is now an agreeable muster.

Milman has got into a worse scrape than I suspected by
the German latitudinarianism (by way of a laconic term)
which he has indulged in in his 'History of the Jews.'
Murchison says it is a diversion in my favour, but I trust
mine will be thought quite orthodox. At all events the
bishops cannot stop my promotion, as they certainly will
Milman's. His own parishioners have taken it up, and the
once popular preacher is in danger of having his gown torn
off. He has challenged the bishops to find *one* opinion in
his work not in the notes to *Mant's* Bible, and Lockhart
tells me this is true; but the bishops, never having read their
own edition of the Bible, are still more annoyed at this discovery. The crime is to have put it forth in a *popular* book.

Yours ever affectionately,

CHARLES LYELL.

To DR. FLEMING.

London : April 18, 1830.

Dear Dr. Fleming,—Sedgwick, when I asked him if he
should send his lashing of Ure in his anniversary speech, to

you—said of course there is one for him, for Dr. F. is one of those men who feel not only a love for truth, but an honest indignation against all those who conceal it or tamper with it, or are guilty of any dissimulation. I will add to these, that Mantell is a great admirer on the whole of your last paper, and I give you this to let you see both sides, for I did not in the least exaggerate the storm on the other side. You are one of six or seven writers in this country who combine profound scientific knowledge with a power of presenting it in a popular, powerful, and eloquent style to the public; and if by any circumstances your facility of doing good to the cause is injured by a series of personal altercations, with which I certainly think you have been in a great measure drawn, it is a national loss. Conybeare is giving divinity lectures at Bristol. Buckland, I trust, is going to repose for years under your last fire, as he did under a former one.

The 'dirt bed' which separates Portland from Purbeck rock is now proved to be a peat moss and forest, which grew on an exposed surface of Portland marine oolite. The cycades are all upright, with their roots *in situ*, the tall stems broken and horizontal. The case is more clearly demonstrated than in any submarine forest. What a subsidence *afterwards* to let the green sand and chalk accumulate above !

Believe me, ever most truly yours,

Charles Lyell.

To Gideon Mantell, Esq.

London : April 23, 1830.

Dear Mantell,—It is nearly the same to pursue a hobby as I am doing, and to be professionally engaged, and so you must excuse bad correspondence ; but I assure you I am a mile or two further on by aid of the encouragement held out in your letters, that the production will excite a sensation, which Murray says is the great virtue of a superior author. So much for the market value of commodities. By the way, Sir Philip Egerton said at R. S. last night, that you were the author of the grand catalogue of fossils in the 'Transactions,' which, knowing your other works, he selected as an extraordinary proof of your being *bien fort.* He brought parts of thirty-one individuals of *Ursus speleus* from German caves, and a great jaw of *Felis spelea,* pronounced by Clift & Co. to be equal in size to a horse ! Buckland first pronounced your bone to be that of a deer, and stood out for a long time against its being that of *Goliah.* He declared the bone to be *post-diluvian* and not mineralised, and made light of it, but did not *scratch* it. He is gone down to Lyme, so there is something in the wind—a paper on the new beast perhaps, that fish-like concern which Mary Anning wants to make a grand wonder of, and the Dr. a memoir, as I suppose. His and De la Beche's on Weymouth read last time—good, but some diluvial heresy tacked on, at which I fired a shot. The iguanodon's bones brought by them from Isle of Wight are rolled, ugly, unmeaning pebbles, save one ' sub-quadrangular vertebra,' as Dr. Buckland says, which he declares proves it to be an iguanodon. Even that is imperfect.

Most truly yours,
CHARLES LYELL.

To SIR PHILIP DE MALPAS GREY EGERTON.

9 Crown Office Row, Temple, London : April 26, 1830.

Dear Sir Philip Egerton,—In conversations with Dr. Lee, whom I introduced to you at the Royal Society, I have learnt the extreme practicability of a geological expedition which I feel extremely tempted to undertake this summer ; and although my short acquaintance with you would not authorise me to propose that you should take part in it, yet I am sure you will excuse my doing so on the score of being equally attached to geological pursuits.

The tour which Lee made when physician to Count Woronzoff, was from the Sea of Azof, through the Crimea, along the great steppe to the mouths of the Danube, a district full of grand physical features never yet made out, nor visited by a geologist since Pallas. The country is perfectly safe, and with such introductions as we could immediately get, no impediment would be encountered.

The time required not long, and the summer the proper season. The route would be by steam to Hamburg, and then

from Lübeck by steamboat which communicates to Petersburg, where we arrive on the eighth day from London. A
day or two would be ample at P., and the diligence is only
four days from Petersburg to Moscow, where a splendid city,
and the examination of the *tertiary* strata there, would refresh
the traveller, and prepare him in a carriage, bought at
Moscow, to make direct to Azof, or some place at the north
of the Don, where we should be without difficulty on the field
of action in a month from our starting from London. Give
seven or eight weeks to the Crimea, and the *great steppe*,
which latter might be done quick, boxes of shells being
shipped easily at Odessa. From Odessa to Vienna, where
the carriage is sold at no loss, and from Vienna in eleven
days to London, having been absent about four months if
you were pressed for time. I will not enter into the numerous
geological reasons for selecting that country before any
other—the light which the steppe when understood would
throw on all the steppes of the Caspian, &c.; the study of
the effects of recent earthquakes in Bessarabia, the alluvions
of the Don, Dnieper, and Danube, and the Crimea, which is
full of various interest. Murchison, who (as I know by experience) is a good campaigner, declares the scheme excellent, but fears your engagements will prevent you from
stirring. I should probably start about June 10, but if you
are disposed to join, I would wait, if by exertion you could
get over your business and go somewhat later. I need not
say that to participate in the first making out of an unknown
district is the only way to learn to unravel countries for
oneself, and that the new facilities now opened have rendered
a district accessible in a few weeks which seemed but lately
as far removed as the interior of Asia. A courier who can
be trusted is easily got at Petersburg. I should make the
expedition as strictly geological as our friend Murchison
does his, and it is from my confidence that your enthusiasm
would equally dispose you to do so, and would carry you
through the slight inconveniences, which would after all be
of short duration on the tour, which induces me to make
this proposition. I shall probably go alone, if you should
decline, and I do not wish it to be known, for when I
intended to go to Iceland this spring (before I engaged to

supply Murray with a volume on geology), I had two offers
of companions, excellent men, but one a mere mineralogist,
and the other a sportsman, who would have greatly interfered,
and whom I should still, as old friends, have offended by
declining. I would enter into many more particulars about this
scheme, if I did not feel that it would be going too far before
I am aware whether there was any insuperable obstacle to
stimulating you to such an enterprise. In regard to health,
I have made due inquiries, and you may trust me, an old
campaigner, that I should not think of it rashly. It is one
of those things that is not open to me every day, and would
probably lead to some new and grand views.

Believe me, most truly yours,

CHARLES LYELL.

To HIS SISTER.

Temple : May 11, 1830.

My dear Eleanor,—I have been intending for a month to
write and say you were a good-for-nothing gipsy for not
sending me a letter till I wrote one myself, as it takes no time
to read a letter, though I am really almost too busy to write
one. Have you heard from Tom ? His things arrived here
to-day, by sea from Devonport.

Murray of Simprim asked me to take him to the Geological Club. As luck would have it, he sat next Buckland,
with whom he was much delighted, as with a most entertaining lecture delivered by him afterwards, at a full meeting of
the Geological Society, on Bones of the Mastodon brought by
Basil Hall from N. America, and by Beechey & Co. of Extinct
Elephants from Behring's Straits. Also on my Angus Kel-
pie's feet, which were so much admired, that when I offered
the Professor one, he would not accept, saying he should rob
the Geological Society, for each differed from the other in
something.

Sir George Rose is in great force, and longs to see my
book. Pray heaven he may not think me, as he tells me nine-
tenths of the German clergy are, ' sold to the devil.' But I
assure you I have been so cautious, that two friends tell me
I shall *only* offend the ultras. After divers schemes, Iceland,
Pyrenees, West Indies, the Crimea, Black Sea, &c., I have

definitely settled, positively for the last time, for the Pyrenees. I find I shall be just relieved at the nick of time for them, can get there in eight days, and Captain Cooke, R.N., who is just returned, volunteers to go with me, and be a Spanish guide. Country cheap, no danger, bad fare, bad beds, &c., but that I can stand ; but don't like West Indian fevers, Crimean agues, nor Portuguese robbers, and worse adventures. Murchison gave me a seat in Lord Darlington's box to see Miss Kemble's Isabella. Did not think it half equal to her Juliet. I hope to have done with my volume by middle of June, and whether it is out then or in October, I begin to feel somewhat indifferent now.

I hope to get my tour over so as to be with you for a time at Kinnordy, but I cannot answer quite for so much.

Believe me, your affectionate brother,

CHARLES LYELL.

To POULETT SCROPE, ESQ.[1]

9 Crown Office Row, Temple : June 14, 1830.

My dear Scrope,—You ought to have received more sheets long ere this : we hope to be out in ten days. As I had nothing to do with your being my reviewer, I am certainly right glad of it; and not the less so, that having withdrawn from the fight some time, you may be a more disinterested umpire than many others who, like you, have committed themselves in print to many opinions. I am sure you may get into Q. R. what will free the science from Moses, for if treated seriously, the party are quite prepared for it. A bishop, Buckland ascertained (we suppose Sumner), gave Ure a dressing in the ' British Critic and Theological Review.' They see at last the mischief and scandal brought on them by Mosaic systems. Férussac has done nothing but believe in the universal ocean up to the chalk period till lately. Prévost has done a little, but is a diluvialist, a rare thing in France. If any one has done much in that way, I have not been able to procure their books. Von Hoff has assisted me most, and you should compliment him for the German plodding perseverance with which he filled two volumes with facts like

[1] Who was about to review the *Principles of Geology.*

tables of statistics; but he helped me not to my scientific views of causes, nor to my arrangement. The division into aqueous and igneous cause is mine, no great matter, and obvious enough. Von Hoff goes on always geographically. For example, he will take as a chapter ' changes in the boundaries of sea and land,' and under this may come alterations by earthquakes as well as by currents, &c. Von Hoff took his ' waste of sea-cliffs,' as far as Britain is concerned, from Stevenson—very meagre. I have done mine from actual observation, principally in coast surveys. My division into destroying and reproductive effects of rivers, tides, currents, &c., is, as far as I know, new—my theory of estuaries being formed is contrary to Bakewell and many others, who think England is growing bigger. In regard to Deltas, many facts are from Von Hoff, but the greater part, not. All the theory of the arrangement of strata in Deltas and stratification, &c., is new, as far as I know, and the importance of spring deposits. Von Hoff thinks all that is now going on, a mere trifle comparatively, though he has done more than any other to disprove it. My views regarding gneiss, mica schist, &c., could not come in Vol. I. Sedgwick found in the centre of Eastern Alps an encrinital limestone alternating with genuine mica schist and the *white stone* of Werner. This made a sensation here at G. S. this session. It was before known to E. de Beaumont. Think of this fact. Whether so made originally or not, it is clear that mica slate owes its stratification to deposition, because the limestone did. *Ergo* after a *cool* sea existed, with zoophytes enjoying the light of heaven, and feeding on some animalcules which lived as now—these rocks were formed no matter how. Graywacke in its most ancient *mineralogical* form is proved to be posterior to *vertebrated* animals at Glaris, and to fuci and some coal plants lately in Ireland. In controverting, just allude to ' having heard something of the Alpine discovery,' because Sedgwick was most unwilling last year to admit such a thing ; also hint that my reasons are yet to come, as I say in several passages.

Probably there was a beginning—it is a metaphysical question, worthy a theologian—probably there will be an end. Species, as you say, have begun and ended—but the analogy is faint and distant. Perhaps it is an analogy, but all I say is, there are, as Hutton said, 'no signs of a beginning, no prospect of an end.' Herschel thought the nebulæ became worlds. Davy said in his last book, 'It is always more probable that the new stars become visible and then invisible, and pre-existed, than that they are created and extinguished. So I think. All I ask is, that at any given period of the past, don't stop inquiry when puzzled by refuge to a ' beginning,' which is all one with 'another state of nature,' as it appears to me. But there is no harm in your attacking me, provided you point out that it is the proof I deny, not the probability of a beginning. Mark, too, my argument, that we are called upon to say in each case, ' Which is now most probable, my ignorance of all possible effects of existing causes,' or that 'the beginning' is the cause of this puzzling phenomenon?

It is not the beginning I look for, but proofs of a *progressive* state of existence in the globe, the probability of which is *proved* by the analogy of changes in organic life. 'Tis an easy come-off to refer gneiss to 'the beginning, chaos,' &c., and put back the finding an encrinite for half a century. That all my theory of temperature will hold, I am not so sanguine as to dream. It is *new*, bran *new*. Give Humboldt due credit for his beautiful essay on isothermal lines : the geological application of it is mine, and the coincidence of time 'twixt geographical and zoological changes is mine, right or wrong. Sedgwick and Murchison have found an intermedial formation in Eastern Alps 'twixt chalk and oldest tertiary, helping to break down that barrier, to fill that lacune. Until Rennel's posthumous work on currents is out, I could not have a good copper-plate of their course. Thanks for the hint, which shall not be lost, if your review helps me in spite of the saints to a second edition, and in spite of 1,500 copies, a number I regret but could not avoid. My labour has been greater than you would suppose, as I have really had so little guidance. *Your little valley paper was one of my best helps.* I mean as guide in classification of facts. I was afraid to point the moral, as much as you can do in Q. R., about Moses. Perhaps I should have been tenderer about the Koran. Don't meddle much with that, if at all.

If we don't irritate, which I fear that we may (though mere history), we shall carry all with us. If you don't triumph over them, but compliment the liberality and candour of the present age, the bishops and enlightened saints will join us in despising both the ancient and modern physico-theologians. It is just the time to strike, so rejoice that, sinner as you are, the Q. R. is open to you. If I have said more than some will like, yet I give you my word that full *half* of my history and comments was cut out, and even many facts; because either I, or Stokes, or Broderip, felt that it was anticipating twenty or thirty years of the march of honest feeling to declare it undisguisedly. Nor did I dare come down to modern offenders. They themselves will be ashamed of seeing how they will look by-and-by in the page of history, if they ever get into it, which I doubt. You see that what between Steno, Hooke, Woodward, De Luc, and others, the modern deluge systems are all borrowed. Point out to the general reader that my floods, earthquakes, &c., are all very modern, also waste of cliffs; and that I request that people will multiply, by whatever time they think man has been on the earth, the sum of this modern observed change, and not form an opinion from what history has recorded. Fifty years from this, they will furnish facts for a better volume than mine. The changes in organic life, which I intend to be more generally entertaining than the inorganic, and more new, must be deferred to Vol. II. I will attend to your other requests immediately.

Very truly yours,
Charles Lyell.

P.S.—I have been very careful in my work in referring, where I have borrowed, to authors, and am not conscious of having ever done so without citing in a note. I doubt whether I have embodied sentences from any author so much as from you, and you will see that that is in great moderation.

I conceived the idea five or six years ago, that if ever the Mosaic geology could be set down without giving offence, it would be in an historical sketch, and you must abstract mine, in order to have as little to say as possible yourself. Let them feel it, and point the moral.

To POULETT SCROPE, ESQ.

London : June 20, 1830.

My dear Scrope,—I go on Tuesday morning to Havre, for the Pyrenees. I trust you will agree with me in what I have said in the winding up of the chapter on the Calabrian earthquake of 1783 as to excavation of valleys. I have never seen this subject touched on in all the theories of valleys— compare my allusion to the land-slips in the Batavia shock in 1699. Forests floating on the sea in Jamaica speaks also volumes as to effect on valleys. Whewell tells me he has just been at Oxford, and with a cavalcade of forty horsemen went out for Buckland's field lecture. He held forth more than ever about the impotence of modern causes. Had he and Conybeare been engaged to review me, they might positively have done some harm in *retarding* the good cause, and much more in stopping the sale of a certain number of copies.

In the volcanic causes, the only material thing for which I am indebted to Von Hoff is the tracing out the volcanic regions more fully than is found in other books. As to the philosophy or geology of volcanoes, he gives nothing that I have thought worth borrowing, and he attempts but little. Daubeny had seen it, but profited less than he should have done by it.

Whether my concluding theory of excess of subsidence will do, is more than I can predict. Like many others, it is my own, and thought out after turning the subject over in every way. I have not had time to study Elie de Beaumont's dashing theory of Epochs of Elevation, but have seen enough to be sure that much of it will not stand. The review of his notions will come properly in my next volume, for it is by fossil zoology alone that all dates of the relative upheaving of mountain chains must be decided, or guessed at. It is preposterous to say that the Alleghanies were raised when a certain European chain was, *because* they are parallel. He runs riot in these extravagant generalisations, just as he compared the Dauphiny Alps to the mountains in the moon, in sober earnest. E. de B. does not, as far as I can see,

entertain any suspicion that before laying down as an axiom
that 'parallelism of axis of elevation indicates contemporary
elevation,' some respect should be had to what may now be
going on ; or if we know too little as yet, at least let us wait.
I expect to come into collision with his doctrines, for he
seems to me to be embarked just as Von Buch was, in his
' Elevation Crater' system, on the plan of speculating on ancient times without caring about modern causes. Think of
his saying that *the Deluge* may have been caused by the
sudden rise of S. America! I respect him as one of the best
of the young Frenchmen. He has done much good, but
three-fourths of his theory won't stand five years. I am
more anxious than I can tell you that you should hit it off
well for Q. R. Of such an article as many reviews would
jump at, there is no fear; but if Murray has to push my vols.,
and you wield the geology of the Q. R., we shall be able in a
short time to work an entire change in public opinion.

Ever very truly yours,
CHARLES LYELL.

To POULETT SCROPE, ESQ.

Havre, France : June 25, 1830.

My dear Scrope,—Not being able to cross over to Honfleur to-day, because of tide not serving, I have seen all the
geology which the cliffs afford. Chalk capping, under which
Upper Green Sand just as in the back of the Isle of Wight,
and blue marl at bottom raised in same manner by under-
mining—tide rising 22 feet, and sweeping away land 500 feet
high. Having done this, I took up Elie de Beaumont's new
work, published in ' Ann. des Sciences.' Now I am somewhat
glad to find how much my views differ from his. In his
memoir entitled ' Recherches sur quelques-unes des Révolu-
tions de la surface du Globe,' &c., he begins by saying, that
whatever be the causes which have raised mountain chains
and put the beds composing them into their present position,
it must have been ' un évènement d'un ordre différent de
ceux dont nous sommes journellement les témoins.' In my
comments on the Huttonian theory, I throw out that there
have been, in regard to separate countries, *alternate* periods

of disturbance and repose, yet earthquakes may have been always uniform, and I show or hint how the interval of time done, may make the passage appear abrupt, violent, conclusive, and revolutionary. Now E. de Beaumont's reigning notion evidently is, that because 'le redressement des couches s'arrête brusquement à tel ou tel terme de la série des couches de sédiment (in different chains), et affecte avec une égale intensité toutes les couches précédentes, le phénomène de redressement n'a pas été continu et progressif, mais brusque, et de peu de durée,' &c. Now it is undoubtedly brusque as far as the solution of continents, but why therefore the 'convulsion' of short duration? I do not find anywhere in him the notion of restoring in imagination the geographical features of Europe of N. Hemisphere by causing the land to sink, which has been since upraised—as to change of climate he has nothing to do with it. I told you before, that I have, as far as I, and others whom I have consulted know, to answer for all the sins committed in that new theory. Before I dared hazard it, I went to Stokes, who has often condemned my follies to the flames, but he said 'he was sure there was much in it, and might be as much as I thought. At all events, it would do much good to try it, and would set others to work.' E. de B. gets hold of three or four chains, proves that one has been elevated since the Jura limestone, another parallel since the chalk, and then infers that they were upraised nearly about the same time, and, as they are parallel, he thinks it *very probable* that these accidents of the surface, which have a community of direction, 'ont été formés pour ainsi dire du même coup, et sont les traces d'une seule et même commotion.' 'Ann. des Sciences Nat.' tome 18. Whatever may be said of my theories, there are sweeping conclusions in E. de Beaumont, for which I would not be answerable for a good deal. At the same time his work is in many parts, as far as I can judge without being able to test the facts, of great merit. It is difficult to follow him, and for so new a subject he has been I think unpardonably concise, and therefore obscure in laying down his views. I do not half comprehend what he would be at, yet I have talked much with him on geology in general. My confidence is shaken by knowing that he is an unflinching

follower of Von Buch's, in theory of ' Elevation Craters.' I
mean to make almost straight to Olot in Catalonia, but my
companion, Captain Cooke, must go and see the Druidical
monument at Carnac in Normandy. This out-of-the-way
line will cost some extra days, but you shall hear in five or
six weeks of the volcanic region. I shall be surprised if Olot
was destroyed by lava, or an eruption—probably an earthquake, but if lava, so much the better. I will see what the
river has done in three centuries.

    With remembrances to Mrs. Scrope,

        Believe me ever truly yours,

            CHARLES LYELL.

To HIS SISTER.

            Toulouse : July 9, 1830.

My dear Marianne,—I arrived here yesterday after having traversed from Honfleur to Bordeaux, a great part of
Normandy, Brittany, and La Vendée, on the most detestable
roads, and through some of the finest countries I ever saw.
It is common to see waggons with nine horses unable to
move the machine out of deep ruts. In the diligence twenty
persons had to turn out several times in the mud, during
heavy storms, to let the carriage be dragged out. My companion, Captain Cooke, took me this round-about, to see the
great Druidical monument of Carnac, near Quiberon Bay,
which is a fine thing, though we have paid dearly for it in
time, &c. However, I learnt of course something of the
geology of the countries we went through. Cooke is a com-
mander R.N., son of a country gentleman near Newcastle,
well informed, a good linguist, knows Spain, a botanist, and
gets on in geology. He is a little too fond of lagging a day
for rest, as here where nothing is to be done. But such
peccadilloes I pardon for his manifold virtues. At Bordeaux
we learnt a great deal; the place abounds in geologists, full
of zeal, and their museums and conversation were very instructive. Here we have a great gun of the old Wernerian
school, D'Aubuisson, who amuses me much. He thinks the
interest of the subject greatly destroyed by our new innovation, especially our having almost cut mineralogy and turned
it into a zoological science. In short, like all men, he dis-

        T 2

likes that which destroys his early and youthful associations, and he has too much to do as an engineer to keep up with the subject. It is what they call cold unseasonable weather in the south of France, but just what we like. The verdure is splendid and probably very unusual here. Fruit half ripe, vines promising no crop for want of sun. Instead of suffering by heat, they tell us we shall be too cold in the Pyrenees. The line of huge blocks at Carnac are nearly as big as Stonehenge in parts, and, forming ten rows, reaches four or five miles in a direct line E. and W. How they got them up is marvellous; some are twenty feet high. We go tomorrow towards Ax, meaning to make a short excursion to a volcanic district round Olot in Catalonia, where Cooke has already been as a botanist; we then return to Bagnères de Luchon, and then to Pau—future movements in the Pyrenees will be governed by what we find in our first excursion. I hope to receive news at Bagnères de Luchon, and hear whether my book comes out or not. I shall be glad to hear your honest opinion of the work, regarded as one comprehensible to the uninitiated. I am afraid that what delights my friend Scrope more than all—the honest history of the Mosaico-geological system—will hurt the sale. D'Aubuisson said this morning : ' We *Catholic* geologists flatter ourselves that we have kept clear of the mixing of things sacred and profane, but the three great Protestants, De Luc, Cuvier, and Buckland, have not done so; have they done good to science or to religion ?—No ; but some say they have to themselves by it. Pray, gentlemen, is it true that Oxford is a most orthodox university ? ' Certainly. ' Well then, I make allowances for a professor there, dividing events into ante and post-diluvian : perhaps he could get no audience by other means.'

This attack against Buckland convinces me that the French Institute chose Conybeare before Buckland, because they considered the latter as trading in humbug, which I am sorry to say is notoriously true of Cuvier, but not of Buckland, for although I am convinced he does not believe his own theory now, to its full extent, yet he believed it when he first started it.

The quiet and perfect order and calmness that reigned

at Bourbon, Vendée, and Bordeaux and Toulouse during the
heat of the elections, afford a noble example to us—never
were people in a greater state of excitement on political
grounds than the French at this moment, yet never in our
country towns were Assizes conducted with more serious-
ness and quiet. There is no occasion to make the rabble
drunk. All the voters of the little colleges are of the rank
of shopkeepers at least, those of the highest are gentlemen,
only 20,000 of them out of the 30 millions of French. They
are too many for such jobbing as in a Scotch county, and too
independent and rich to have the feelings of a mob. This
city has elected all seven ministerial men—a rare case. The
little colleges have in general elected four out of five opposition, the great colleges four out of five ministerial. The
former are the democratic body, yeomen, merchants, &c.
The landed interest is in the greater college. The number
of royalists here is attributed to the personal influence of
the ex-minister Villèle. A gentleman told me at dinner to
day : ' Villèle is the only man Charles X. can choose. Polignac must yield. Villèle lost his place not by the censor-
ship of the Press, nor by the law of sacrilege, nor by any of
his unpopular ecclesiastical measures, however impolitic
some of these may have been, but he lost it by his bill of
primogeniture, a necessary measure for the salvation of
France, but so hostile to the feelings of the people, that it
is at the bottom of much of the opposition now manifested.
Every family is in arms about this measure ; yet we shall be
ruined if this partitioning goes on for ever.' I have been
amused at their telling me that Polignac, when in England,
became an ultra-aristocrat. They say that Chateaubriand
made a stand against Almack's and the exclusives, but
that Prince Polignac gave into it, and as the Stewarts lost
their crown by the monarchical lessons taught in France, so
P. will lose his premiership by the ultra-aristocratical
notions taught by the D. of Devonshire. Think of a Frenchman reasoning in this manner !

Love to all. Your affectionate brother,

CHARLES LYELL.

Prades, near Perpignan, Valley of the Tet : July 21, 1830.

My dear Caroline,—I believe it is an age since I last wrote, which I believe was addressed to Marianne from Toulouse. Thence we went by a baddish diligence to Foix on the Ariége, one of the most picturesque places on the northern flank of the Pyrenees. From thence we went to Ax, one of the great watering-places. I got two famous days' geology on that side of the mountains, and we filled a small box full of good fossils, which will throw light on the chain, of which I got a clear idea, as to structure.

There is something in the general character of the deep valleys of the Pyrenees, different from what I remember in other chains, or rather the characters which belong elsewhere to minor valleys and those in rocks of a distinct nature, happen here to acquire a grand and Alpine scale. After Ax, where we found a number of French strangers, not like our loungers at watering-places, but really invalids, we ascended from the region of oaks and beeches to that of pines. A forest of silver firs was covered with the lichen, which I think we used to call in the New Forest ' old man's beard,' some of them being several feet long. I admired them so much, that Cooke was amused not a little. He is a great fancier of pines, and has discovered two or three new kinds, and looked on the lichen as a loathsome disease. He has rather a fancy to begin insect-hunting, and I could hardly persuade him that if we set up a net, there would be an end of geologising. When we got into the centre of the chain, into the granitic region, I was glad to relieve *ennui* by joining him in catching those which were polite enough to offer themselves without trouble, and I have put them up in paper as Alpine kinds. We travelled on mules over the central chain, and then entered French Cerdagne by the valley of Carol. We went to Bourg Madame, as it is now christened, a village burnt by the Spaniards during the late war. On this frontier, close to Puycerda, we discovered a freshwater deposit unknown before, of considerable extent, filling an enormous valley, and containing animal, or at least testaceous remains, and wood. Although an old affair, and with a valley 500 feet and more cut through it, the nests of planorbis and lymnea and the whole aspect of some beds are singularly like the Loch of Kinnordy. We heard from good authority an unfavourable account of the state of Catalonia, where there are three parties it seems—the constitutionalists, who think King Ferdinand too despotic; the royalists, who consider him as the best of all possible kings ; and the Carlists —I fear the strongest party—who consider him as not half despotic enough, and who would re-establish the Inquisition in full force. We were dissuaded from entering at present, and as Cooke feared to descend to the eastern extremity of the Pyrenees because of the heat, he kept upon the hill tops among his favourite pines, intending to visit the valley of Andorra, and then to go to Bagnères de Luchon, where our letters await us, and where I am to join him. My work was clearly in this region, and to-morrow morning I shall be at Perpignan. This cool season, the heat is positively a mere trifle even down here in comparison to what even Murchison worked regularly in, in Auvergne, Vivarais, and the Vicentin. As for me, I have only been just warm hitherto. My present scheme is to see the valley of the Tech, Tet, and La Gly, on the eastern extremity of this chain, and the newer formations at their foot near the sea, then to see the flanks of the chain, always the most interesting point to a geologist who is not too fond of mere mineral hunting, and so to get to Bagnères de Luchon in seven, eight, or more probably ten or twelve days—all which time I shall be without news from England since my departure. I have got through part of the Spanish grammar, probably to small purpose, except at some future time. I must get a cram in pronunciation from Tom. My companion has been very agreeable, but, as you may suppose, not quite so far gone in geology as to be inclined to keep it up with as much spirit as I do, and I doubt not I shall make great play now I have got sufficiently interested in the study of this magnificent chain, to care less for the birth of my book than I did *en voyage*; but in writing to you, I am curious again to know whether it is out, which I shall not learn for some two or three weeks, or possibly more. What I saw of the Spanish peasants in the valley of Puycerda, pleased me much, with their red bonnets, fine sunburnt countenances, swarthy almost, and active mountaineer walk. They make capital guides, and will beat a horse in a day's journey in the hills. I need hardly say that my pure Castilian did not aid me much in interpreting their language, but Italian did much. Yesterday I was among pine woods; to-day, olives, vines, wild pomegranate, &c.

With my best love, believe me, your affectionate brother,

CHARLES LYELL.

To HIS SISTER.

Perpignan : August 8, 1830.

My dear Fanny,—I have just returned from a successful expedition into Catalonia. I believe I mentioned in my last that Cooke and I separated, after finding it impracticable to get into Spain, and after losing three days in the vain attempt. Encouraged by some Catalonian gentlemen here, I went in two days by diligence to Barcelona. A glorious view of the Pyrenees and of the delicious country bordering the Mediterranean to the south. The heat so great, that five Cuba merchants, my companions, pronounced it beyond that of the tropics. It did me no harm. We were fanned by machines at breakfast and dinner as in India, went without coats, drank incessantly, perspired while sitting still as if in full exercise; but there was so much novelty in all to me, that I never felt my energy flag. As I entered Barcelona on a *fête* day, I was struck with the gentlemanlike appearance of the numerous priests, a fine, venerable-looking set in general, respected by the people, so different from the poor, sneaking clergy of France, who look as if they had no business in the country. The garrison, 8,000 strong, was in full review in honour of the Queen's birthday, before the Conde d'Espagne, an officer who distinguished himself under the Duke of Wellington in Spain, and who is looked up to as the main instrument of ridding Spain of the last French invasion, by outdoing the French in diplomacy. He was covered with orders. About 2,000 of the Royal Guard were among his troops—a mark of the high favour he is in.

The next morning, July 24, I called on the Count at the Palace, with Stephens, an Englishman, a great favourite at Court. He, d'Espagne, is obliged to be very inaccessible, or he would be mobbed, but his regard for the English ensures them a polite reception if recommended to him. As Stephens was intimate with Vyvyan and other brother Oxonians, and as I was the companion of Captain Cooke, a friend of d'Espagne, I met with great attention from the Count, who is Captain-general of Catalonia, and has more power than an Irish viceroy in that great and populous province. He called the next morning and left an invitation to dinner. I dined with him both that day and the next, and received from him a special passport signed by him, and letters to Governors, Abbots, &c., and was pressed to take with me one of his body-guard. 'I can ensure you respect from the authorities, but no magistrate can guarantee you against the brigands. You will have heard that we are not troubled with many now, but in the mountains, a district of smugglers, who can say what may happen? For my sake, and that I may feel at rest, do not refuse.' I accepted, but when the Captain of the Guard came with the 'Mucho,' I declined, as I felt no insecurity, and as not one of the guard could speak French, nor serve as interpreter. With a mule and a guide on foot, who knew a few scraps of French, together with a few sentences of broken *patois*, and Italian of my own, I got on without one misadventure through a glorious country for a geologist. Saw Monserrat, the salt mines of Cardona, Vich, and the volcanic district round Olot, and returned across the Pyrenees by Massanet to Ceret. A naturalist at Olot, who had written on the volcanos, talked French well. I re-entered France, August 7, and find that in my absence a great revolution was accomplished. The tricolor flag hoisted on all the churches, and the utmost order prevailing, without any apparent apprehension even on the part of the royalists as to security of person and property. I am not a little glad that I saw what I have seen in Spain before this news reached them. For whether there be a convulsion or not there, the apprehension which the authorities entertained had got to such a pitch even before I quitted, that but for the Captain-general I might not have got out again so soon;

delays to travellers will at least be multiplied. I must tell you my adventures and what I saw in Spain in future letters, for it was a tour full of novelty. The sort of feeling one has there of the government and laws and state of the nation is such, that the topsy-turvying of all things here seems a trifle to me in comparison. Here you know everything, but in Spain there is a darkness and mystery, and an evident sense of danger which makes you uneasy; and had I not been excited unusually by the scene and the important geological facts I learnt there, I should not have enjoyed it as I really did throughout the whole fortnight. I have got so interested now about the Pyrenees, that even the seeing in the newspapers how a dynasty had been upset, or rather how they had upset themselves, seemed a milk-and-water affair in comparison. I hope to keep in this mood till in a few weeks I have run along the flank of this fine chain to Bordeaux.

Although the Prefect, Sub-Prefect, and almost all the lower officers here have been already dismissed and replaced, I got my passport this morning with such promptitude, that I cannot but admire them. They seem determined that the machine shall work well in the midst of so complete an over-turning of the old hands, or out-turning. The outs even do not pretend that they are in any apprehension here, and other travellers are moving about with perfect confidence. No one has been even scratched here, and the troops are in perfect good-humour, though they kept up the old flag in the citadel, or at least refused four days to put up another until ministerial orders came, expressly commanding it.

The only popular act of violence here has been the pulling down the crucifix erected by the congregation. I expect to hear they have done this almost throughout the country; it was the banner of a party which has chiefly ruined the old *régime.*

With my love to all, believe me your affectionate brother,

CHARLES LYELL.

CHAPTER XII.

AUGUST–NOVEMBER 1830.

VOLCANOS OF OLOT—EARTHQUAKES—COUNT D'ESPAGNE—CORK-TREES ON
THE FRENCH SIDE OF THE PYRENEES—PINE-TREES—POLITICAL STATE
OF THE COUNTRY—EXPEDITION TO MONT PERDU—LETTER TO SCROPE ON
GEOLOGY—RIPPLE-MARK—PARIS—LOUIS PHILIPPE—POLITICAL AGITA-
TION IN BELGIUM AND FRANCE—STUDYING SHELLS WITH DESHAYES—DIS-
TURBANCE BY THE MOB—RETURNS TO LONDON—SCROPE'S REVIEW IN
' QUARTERLY '—REMOVES TO RAYMOND BUILDINGS.

CORRESPONDENCE.

To POULETT SCROPE, ESQ.

Bagnères de Luchon, Dépt. de la Haute Garonne : August 10, 1830.

My dear Scrope,—As I have just returned from the
Spanish side of the Pyrenees, I will not delay thanking you
for your suggestion that the Olot volcano should be visited,
a hint which has turned out fertile in good results to me,
not only in regard to igneous rocks. After working a little
on the Ariége from the plains of Toulouse, up to the source
of that river, from the tertiary at the base of the chain to
the granite which separates France from Spain, I descended
with my companion Captain Cooke to Puycerda. . . . But I
must now tell you something of the volcanos of Catalonia.
Like those of the Vivarais, they are all both cones and
craters, subsequent to the existence of the actual hills and
dales, or in other words no alteration of previously existing
levels accompanied or has followed the introduction of the
volcanic matter, except such as the matter erupted necessarily occasioned. The cones, at least fourteen of them
mostly with craters, stand like Monpezat, and as perfect, the

currents flow down where the rivers would be if not displaced. But here, as in the Vivarais, deep sections have been cut through the lava by streams much smaller in general, and at certain points the lava is fairly cut through, and even in two or three cases the subjacent rock. Thus at Castel Follet, a great current near its termination is cut through, and 80 or 90 feet of columnar basalt laid open, resting on an old alluvium, not containing volcanic pebbles, and below that, nummulitic limestone is eroded to the depth of 25 feet, the river being about 35 feet lower than when the lava flowed, though most of the old valley is still occupied by the lava current. There are about fourteen or perhaps twenty points of eruption without craters. In all cases they burst through secondary limestone and sandstone, no altered rocks thrown up, as far as I could learn, not a dike exposed. A linear direction in the cones and points of eruption from north to south. Until some remains of quadrupeds are found, or other organic medals found, no guess can be made as to their geological date, unless anyone will undertake to say when the valleys of that district were excavated. As to historical dates, that is all a fudge. I found out the man who provided Maclure with all his antiquarian information, and after reading it, I can assure you that there never was an eruption within memory of man. There was an earthquake from south of Olot to Perpignan about four centuries ago, of which the exact date and all the circumstances are known. It did not do serious damage except at Olot, where the whole town, save one house, *fell in*. So it may do again, if such a shock as rent every house in Genoa in 1829, and shook the Apennines and Piedmont, should return, for there are underground passages, and subterranean rivers, and grottos, and hollow mountains, in which the 'rimbombo' is remarkable in the suburbs and environs of Olot. I say *fell in*, for houses and monasteries sunk *entire* into the ground, and have been dug out, their roofs on a level with the soil, within the time of men now living there. No shock or even tremor has been felt since, so that although Olot is far from the great line of fire, I do not think that earthquakes invalidate the rule, that at a distance from the great line, the shocks are *rarer* and less violent. The hollow foundation was an accident owing to volcanic eruption perhaps of another epoch, certainly very ancient.

Believe me, dear Scrope, yours ever most truly,

CHARLES LYELL.

To HIS SISTER.

St. Gaudens, Haute Garonne: August 16, 1830.

My dear Eleanor,—Before I forget my Spanish adventures, I will tell you a few of them before they get too stale. Since I re-entered France I have heard nothing but abuse of the Count d'Espagne, from whom I received so much attention. On my arrival at Perpignan, I found that the Count's cousin-german, Baron Raimond, Prefect of the Department, had fled in disguise to Barcelona. He has left behind him the character of a man of business and talent, but a tyrannical instrument of the late ministry. He was most detested for having thrown seven Spaniards—refugees—into prison, where they had remained half-a-year, unable either to obtain a trial or accusation, or any notice of their offence. All declare this was done at d'Espagne's instance. The prisoners were released when Baron R. fled, and I got acquainted with one, a gentlemanlike well-read man, both in French and English literature, apparently rich, a thorough Spaniard, and proud of his country. He declared that d'Espagne was a sanguinary partisan without ability, that he was a Frenchman, never more than a colonel when the Duke of Wellington was in Spain, and that he owes his elevation to accident, &c. My own opinion, after conversing on many subjects with him (d'Espagne), was, that he was well-informed, but far from a man of first-rate talents, and if he is comparatively so in Spain, it would lead me to suspect that *dans le pays des aveugles les borgnes sont rois.* In Catalonia they give him credit for getting rid of the French by his diplomacy, and for having shot indiscriminately the *priests* as well as laymen, who were murdered in an insurrection which the King sent him to quell. To the fear inspired by his defiance of the ecclesiastics, many impute the tranquillity which reigned afterwards in his province. The Carlists or high-priest party hate him, but they fear him

much. To their dislike, however, I rather impute the failure
of our entry by Puycerda. I believe I never told you the particulars. When Cooke and I reached the frontier, we went
immediately to the Alcalde of Puycerda, to get our passports
regularly signed. This Mayor or Justice of Peace (for I
believe he serves in both capacities in Spain) kept us two
hours waiting, and then said he could not sign till next
morning, though he could not make any objection to our
passports, which were quite *en règle.* Next day we waited
as before, and were then examined and cross-examined, as
to our objects, &c. The slow and dignified air of the
worthy magistrate were so diverting, that Cooke, who was
not so keen as I to employ the time otherwise, thought
himself well repaid. We were remanded for a third interview, but it was agreed that Cooke should go, while I geologised, and had a good day's work. On that day Cooke
found a priest with an immense shovel hat on the bench
with the Alcalde, who cross-examined in a most inquisitorial
way, and said that our objects seemed suspicious. ' That he,
a captain in the British navy, should travel for stones and
plants, neither he nor his companion being *Médecin,* or *Agent
des Mines*—that a society was organised in London for dis-
seminating liberal principles of government throughout
Europe—how was he to know we were not Commissioners ?
They had orders from Madrid,' &c. &c. Cooke showed an
old passport from d'Espagne, and offered to send an express
to Barcelona, but was told we must both appear before them
the day after, *i.e.* the fourth day ! The director of the
French Douane now entreated us not to go. ' You risk your
personal safety ; they often throw foreigners into prison, and
when it is found that they are innocent, they let them out to
be sure, but no other redress. I will go to the commandant
of Puycerda, who is indignant at their treatment of you,
and he will guarantee your personal safety and your return
to the French frontier well and good.' The military commandant, a man of family, and very gentlemanlike, said he
would rather decline a squabble with so troublesome a
parvenu as the Alcalde. So we demanded our passports to
be returned, but the devil a bit would he give them up.
' As to the Conde d'Espagne,' said the great man, ' I receive

my commands from a higher source, *from the King himself !* '
When this was afterwards told d'Espagne, he said, and the
saying soon reached Puycerda, ' I wonder it never occurred
to him that as I happen to be on the top of the ladder, and
he at the bottom, I may kick him off.' The fact is the
Mayor was a Carlist, an inquisitionist, and his hatred got
the better of his prudence. We went without our national
passports, and, but for Baron Raimond's politeness at Perpignan, I should never have got into Spain. After such a
warning, I made a complete study of the police, and took
every precaution, by getting d'Espagne's special passport,
&c. Once every day a signature must be got, which is often
very awkward, as when the rivers were impassable, and I was
sheltered in a miserable hovel far from any mayor or magistrate. But the most annoying event was on my return. I
had taken pains to get not only my passport and the guides,
but also the passport of Mons. le Mulet, *en règle* at Olot, but
was told that the latter would require another signature on
entering France. When I got to Massart I was told I could
not pass unless I got another horse. I went to the Alcalde in
no small rage. Luckily he could speak French well. I showed
him all my letters from d'Espagne, pass, &c., and said that
the Mayor of Olot had insured me against any delay. Like
all Spanish authorities, he first was very dignified, made me
sit down, and then wrote a letter and gave orders about
other matters. This I am told they always do, lest by
attending to you at once, they should let themselves down.
But at last he stirred in right earnest, insisted on my
guide taking a day's less pay, and got with some trouble
another beast, who happened to be provided with a proper
passport. ' I could sign your passport,' said he, ' but have
not authority to sign your mule's. There is no magistrate
nearer than S. Lorenzo (half-a-day) who could give you
that ! ' What a glorious state of political economy! yet they
are quite right. The priest and Mayor of Puycerda frankly
told the French Douanier that ' the fewer English and
French who travel in Spain the better,' and so it is for their
views. By prohibiting books, travelling, internal circulation,
of commerce, &c., they may retain their absolute dominion
some years longer, and I shall not pity them if they are

then hurled down from their place like the *Parti prêtre* in France. I must defer a curious religious ceremony which I saw till next letter.

My love to all. Believe me your affectionate brother,
CHARLES LYELL.

To HIS FATHER.

Bagnères de Luchon : August 17, 1830.

My dear Father,—I arrived here last night, and found that Captain Cooke had had the patience to wait here. Had he known, he says, how much I should accomplish in Spain, he should have braved the heat. We are now going to work at the Central and Western Pyrenees. The excursion to Spain I prize the more, as the gates are now closed upon us; at least no prudent man will now venture across the frontier. There is such excitement in Andalusia, that whether it pass off or not, the facilities of travelling are at an end. On my way from Perpignan to Barcelona, on the French side of the Pyrenees, I passed for many miles through a great forest of cork-trees, with some ilex intermixed, and I came back through a much larger wood of cork in returning from Olot to Cerat in the valley of the Tech. Some were as large as good-sized oaks, and more like old olives than I had imagined. I was surprised that the bark was only red when the outer bark had been taken off for cork, so that the red colour spoke of in Southey's 'Don Roderick' is due to laceration. I suspect that Mr. S. got his information second-hand, and was not aware that it was no characteristic of natural scenery, that the white bark is more picturesque, and that the partial removal of so thick a rind has a very bad effect in the landscape. The old trees are quite disfigured by it; he has been guilty of singing the praises of the mere result of commercial industry, and might be annoyed if he knew how unpoetic and vulgar a feature he has seized upon. It was a very fine country I went through, and I had several days of wild rides from Cardona to Olot, by mountain pathways through endless woods of pines and olives mixed and some fine ilex. It looks much in the state of Sicily, but perhaps farther advanced. The heat is shown by the fact that there are machines for fanning you and driving off flies as you eat;

milk, butter, coffee, or tea to be got anywhere, but strange
to say you can get a better cup of chocolate than you could
often meet with in France, in a poor peasant's house, for a
small sum, and made in five minutes. Their brown bread is
not so black and bad as in Sicily, the wine generally good
and cheap, the meat very bad, cost of living about the same as
in France, but less fair dealing in small places. Although the
French journals pretend the contrary, their national prejudices, especially against them, are very strong in all parties,
high and low. In fact they are ignorant, and immense pains
are taken to keep them so. The gravity with which after
consultation, my dictionary and Spanish grammar were
allowed to pass, the only books I had, was diverting. Yet
the arts of writing and reading appear to be pretty generally
spread among the poor in Catalonia, and they are advancing.
As I knew d'Espagne might feel it as a slight if I declined
one of his guards, I determined to see if there was among
them a good interpreter, in which case it might have been
worth while. As Catalonia is on the frontier, and the French
were settled there longer than in any other part of Spain, I
fancied it impossible that out of the hundred men one should
not be found to speak French. The barrack was a curious scene,
a swarthy race of most active-looking armed mountaineers.
They carry a small rifle and a cord with a loop, which they
throw over the head of any one they pursue, much like the
lasso of South America. This singular weapon of the police
gives them immense advantage in pursuing an enemy up hill.
As each group was asked if any one of them could speak
French, they replied with a kind of shout, *Somos Catalanes*, a
little indignation being felt at the very idea of such an
unnational acquirement. Even in the enlightened city of
Barcelona, a *Castilian* is regarded as so far a foreigner that
genteel families from Castile, settled there, find the *national*
prejudices much against them in society! The abbé of
Monserrat moaned over the splendid monastery which the
French had levelled to the ground by their cannon *in bello
independentiæ* (he talked Latin to me), and over the reduction
of his friars to the small number of three hundred! And
how many had you before? Nine hundred! On the top

of a hill 3,000 feet high, singing morning, noon, and night, in
an empty church! They have rebuilt the convent in a vulgar
square house, like an enormous manufactory. At Olot I
arrived on the *fête* day of the Patron Saint, and saw the
procession well from the balcony of my friend the apothecary,
Bolos—what a good name for a pharmacien! His daughters,
even the youngest child, seemed to be so tired of such sights,
that their only amusement was the interest I took in it. The
standards of twelve saints of the town came first, with an
image at the top of each. When the procession stopped, to
chant before different altars in the streets, these standards
were made to rest on the ground erect, by means of ropes
connected, like the shrouds to a mast, with the top of the
pole, and these ropes held by men and boys who balanced it
in this manner. The children held small flags, and could
not resist using them to fan themselves as the heat was
great. When laughed at for this, they often made a drive at
the boys round them with the sacred banners, then came a
regular English scuffle. The combatants got beyond the
length of their tether, the saint was tugged on one side, and
the image of St. Stephen, after three swings, came within a
few inches of my head in the balcony. The older men did
not seem to scold the youngsters for this, as it seems a sort
of saturnalia, and I saw several other similar rows. Before
a long line of Carmelitans who were chanting, was a girl of
about seventeen years, dressed in white, and with two immense
white paper wings on her shoulders. She carried a great
silver cross. 'Look out for the angel!' exclaimed Bolos's
children to me, and sure enough there never was a better
figure in any pantomime. A cherub who preceded the
Capucins was nothing in comparison. One part of the
procession was really imposing: it consisted of 'the
devout,' private individuals, chiefly peasants, who volunteered attending, each with a long torch and with a
brown cloak thrown over them. They went two by two,
so that there was a long line of flame as far as the eye could
reach in either direction. I observed to Bolos that there
was much religion throughout Catalonia, though the French
say there is less than in other parts of Spain. His reply
was, 'C'est un pays de montagnes, monsieur.' Want of

space must cut short my Spanish journal till another opportunity.

With love to all, believe me, your affectionate son,

CHARLES LYELL.

To HIS SISTER.

Bagnères de Luchon : August 18, 1830.

My dear Caroline,—I have been writing a long letter to
Scrope this morning with an account of the Olot volcanos, in
what they agree, and in what they differ from those of ten
other volcanic regions which we have both visited between
Clermont and Capo Passaro. I was in Spain yesterday,
having crossed the frontier a few miles on the summit of the
ridge. All is quiet in Aragon, the French news universally
known. Many of the people give the present king's ministry
the credit of having wished several reforms, but of having
been prevented by the ultra-priest-and-king-party at Paris.
I wish this may be true. The climate here is almost cold, a
sudden change from Barcelona, where the thermometer was
for three days at 94° Faht. At Olot 90° and above. My mule
was furnished with a pair of thick wooden stirrups, clumsy
things which I fancied would be close and hot, but they were
luxurious as keeping the sun from striking on my boots,
which felt as if they must crack when exposed to his rays.
Cooke tells me that the great forests of moderate-sized pines,
mixed with Olives and Ilex, which I passed through in Catalonia, were all of *Pinus halepensis*. It is not so handsome as
Pinus pinea, but resembles it. The old forests of the Pyrenees have suffered terribly by the axe, and will soon be
annihilated. They have been magnificent, and are still fine
here. Beech, silver-fir, and high up *Pinus uncinata*.
D'Espagne is now near us here, having marched as great a
force as he can muster to the frontier, for fear of an explosion at so critical a conjuncture. For my part I expect the
troops will be as ready as any to set up a new order of things,
the officers being very generally in favour of diminishing the
power and revenues of the church.

What you mention about the election and the anecdotes
of the mob-rule which I see daily in the papers, make me

ashamed of our system. The calm manner and the rational behaviour of the French in their last elections, when greatly excited, forms a proud contrast for them. Much public spirit was shown in placemen resigning or braving the threats of displacement, when ordered by the Prefects to vote for ministers. The south of France is more monarchical than the north, and they are relieved beyond measure at finding that at Paris the republican party did not muster stronger. They forget, however, that it was not Paris, but the Deputies who conducted the revolution. I have heard furious disputes. One day a fierce republican, a collector of the taxes at Ceret, was nearly silenced, and at last one of his opponents said, ' But in point of fact, sir, do you not anticipate the nomination of the Duke of Orleans to be king ? ' ' C'est probable, monsieur, car nous sommes toujours les singes des Anglais.' Did I tell you a Frenchman's comment on the removal of the cross of the mission from the Cathedral at Perpignan ? ' Chacun a son tour, le bon Dieu a eu le sien.' An American here tells me that he was opposite the hotel of the Prefect at Bordeaux, when the mob nearly killed him. He kept his drawn sword in his hand the whole time. They threw his carriage and four fine horses into the Garonne. The latter of course were drowned. Not one person was killed at Toulouse, and the reports of deaths were all fables. I heard a dispute this morning in which a friend of Charles X. said, ' You stand up for the responsibility of ministers—well and good—how then can the king be responsible ? If you punish his advisers, you must admit the principle " that the king can do no wrong." ' The answer was severe enough. ' Monsieur, dans nos jours ce n'est que la canaille qui violent leurs serments en France.'

Believe me, your affectionate brother,

CHARLES LYELL.

To His Mother.

St. Sauveur, Pyrenees : August 25, 1830

My dear Mother,—Since I last wrote to Caroline from Bagnères de Luchon I have made with Cooke a successful expedition to near the summit of Mont Perdu, which is within a few hundred feet of being the highest of this chain. As the last letter from Kinnordy told me you were reading my book without being pozed or wearied, I ought to tell you that besides affording the most picturesque scenery of these mountains, this celebrated point is of the highest geological interest, from its affording an exception to the general rule. In other chains the loftiest and central parts consist of rocks which do not enclose any remains of shells and corals, and you must go to the flanks or to the low grounds at their base in order to collect such objects. Here, on the contrary, you find at an elevation of between nine and eleven thousand feet, a profusion of sandstones and limestones, in the very middle of the Pyrenees, full of plants, shells, and zoophytes, many in so perfect a state, and in such thick beds, that you cannot doubt for a moment that you see the bottom of an old sea, now covered by glaciers, or so high that it supports no vegetation. There is one scene, the celebrated Cirque de Gavarnie, an amphitheatre of rocks with a high cascade on one side, which is quite unlike anything in Switzerland. On my arrival at Gavarnie, I had a fine view of this in a clear sky. The next morning we started for Mont Perdu. Having learnt that our whole time would be occupied with going and returning without leaving any time for work if we did not sleep on the mountain in a shepherd's cot, we resolved to do so, as other geologists had done. First we climbed up by steep precipices, and over much snow to a great rent called 'Orlando's breach.' Just before you arrive there, you must cross a glacier which is very steep, and overhangs a deep precipice. The guides had hatchets to cut holes, by which alone you can stir a step ; in addition crampons are fixed to your shoes, and you have the usual alpine pole with an iron pike at the end. Six smugglers carrying burdens on their heads, and armed like ourselves with pikes, followed in our train, and reminded me exactly of my father's plate of the ascent of Mont Blanc of Saussure and his party. Roland's breach is an opening in a ridge or wall of limestone, which runs down from the summit of Mont Perdu about seven or eight hundred feet in height. You cross this portal from France to Spain in about forty steps, and you are carried at once to a new world as it were, a grand mountain view with

six distinct distances opening on the Spanish side. The
foreground is a remarkable scene of desolation. Although
not covered with snow, the rocks are bare, with scarcely
a patch of green, not even on the most level parts. The
anatomy of the hills is exposed in a manner very unusual,
and very satisfactory to a geologist. You see that the strata,
once horizontal, have not been carried up 1,000 feet without
getting twisted and thrown about in the most confused
manner. After a long day's hard work, and very improving,
we crept into the Spanish shepherd boy's cabin. A small
flock is sent up to feed on a few patches of green for about
two months in the year. On the side of a precipice a hovel
has been made, covered with turf in so rude a manner, that
when we came to it, and it was pointed out to us, we looked
for a long time and could not guess where the habitation
was. You crawl in on all fours, as into a cave ; the inside
was ten feet by six, and about five feet in its greatest height.
In this was made a fire of wood, the smoke escaping through
chinks in the wall, and by the hole of the entrance, which is
never closed even by a stone. The inside of this hut was
smoke-dried quite black. We had each taken up a blanket,
and kept up a fire all night, but unluckily there was not
room for us, especially my companion, who is a head taller,
to lie at full length, so we passed an uneasy night enough,
and were not a little delighted when a second fine day
enabled us to crawl out at sunrise. I am quite sure that had
I taken two blankets and bivouacked instead, I should have
slept sound after my fatigue, and another time I would
always do this, for had we had a bed instead of hard stones
to lie on, the being unable to turn and lie at full length
would still have been painful for so many hours. The poor
lad gave me a bag of hard oats to put under my head,
' because,' he said, ' I might find a stone (which was there for
a pillow) not soft enough ! ' We of course had taken up
provisions, for he had not even a drop of milk, and was
obliged to go half an hour's distance for the nearest water,
and three hours for a single stick of fuel. He passes two
months with a child (his brother), is very happy, congratulating himself that he has not to sleep out like the smugglers, with the risk of imprisonment for life. It was matter

of no small regret to us that the present severity and delays of Spanish passports forced us to abandon our proposed continuation of the same line of observation down to the base of the Spanish side of the chain. When we returned over the glaciers it was necessary to make a new road, all the deep holes being filled with water which had frozen in the night. We brought back a famous quantity of fossil shells.

With love to all, believe me,

Your affectionate son,

CHARLES LYELL.

To HIS SISTER.

Bayonne : September 10, 1830.

My dear Marianne,—At Orthez we collected a good many fossil shells, and had a present made us of some by a French *propriétaire,* an amusing character. He took us to his marl pits on the summit of some hills covered with oak, fern, heath, &c., in short quite New Forest scenery, which reminded me of some similar marl pits near Stony Cross, where I used to hunt for shells. He had bought a good quantity of land very cheap, out of the savings as we afterwards learnt of a place of no less than 200*l.* sterling per annum, an immense thing in France, which he had got in the tax office from the minister Villèle, whose physician his father had been. He has turned agriculturist, and improved the estate very much. But he lives in hourly dread of losing his post, which we were afterwards assured he certainly would do, and his gloomy anticipations made him see all the late changes in a light which may be taken as a good example of what thousands of the ' outs' now feel in this country. We gave him a dinner at the hotel, and as he never suspected we should learn anything of his real history, he launched out without reserve, declaring that France had within her, in an unexampled degree, all the elements of happiness, but when things were well they could *never leave them alone,* that a *pure despotism* was the only form really suited to the French, and that Villèle was the greatest minister they ever had, that the deprivation of the Villèle peerages was not only illegal but *une sottise;*—

that persons now in place had not resisted the late changes, because on former turns out they had not changed more than the chiefs, but had it been known that all were to be upset, the revolution would have been withstood. To be quiet, it was supposed would have brought its reward with it. But the new order cannot last, &c. Then he uttered a furious tirade against place-hunting. ' Every shop-keeper has left Orthez for Paris, and each thinks he shall return a Prefect, or sub-Prefect at least. To be only a mayor, is what a cobbler would not condescend now to accept, if he is but a Liberal. As Lafitte is from Orthez, they all go up to him, to beg that so and so may be turned out to make room for them.' If clever men could talk like our little friend, it was high time that the demoralising effect of Villèle's system should cease. As for place-hunting, while men can buy estates out of the savings of a few years, there will be enough of that. If they want to cure the disease in part they must economise and reduce salaries.

Your affectionate brother,

CHARLES LYELL.

To POULETT SCROPE, ESQ.

Bayonne : September 11, 1830.

My dear Scrope,—I trust that neither the length of the article which the spirit has moved you to pen, nor the strength of the doctrines will prevent the editor from interposing it between me and that powerful party among us, who are nervous to a degree on certain subjects, and of whom even some of the moderates, have already hinted to members of our family, that my work, though certainly ' creditable to the author's talents, contains opinions that may well cause some alarm !' The decision of the Q. R. would settle the doubts of many of those good people, who only want to be told from authority what they *may* think. How I wish I could profit by the perusal of the slips, particularly on the part relating to the laws of running water, before I enter upon an explanation of experiments which I wish you to make, and which I should like to make the subject of a paper by Scrope and Lyell for the G. S., refer-

ring to it, if not concluded before, in my next vol., as about
to appear! It will require patience to obtain all the results,
but I am now more sure than ever that we may out of
small and apparently trifling observations, if not construct,
at least overturn many much-famed theories. I have for a
long time been making minute drawings of the lamination
and stratification of beds, in formations of very different
ages, first with a view to prove to demonstration that at
every epoch the same identical causes were in operation. I
was next led in Scotland to a suspicion since confirmed, that
all the minute regularities and irregularities of stratification
and lamination were preserved in primary clay-slate, mica-
slate, gneiss, &c., showing that before their transformation,
they had been subjected to the same general and even acci-
dental circumstances attending the sedimentary accumulation of secondary and fossil-bearing formations. Lastly, I
came to find out that all these various characters were identical with those presented by the bars, deltas, &c., of existing
rivers, estuaries, &c. Now, in my present tour, I have more
in support of these positions than it would be possible in a
sheet to explain, but I will just put down the facts and
theories as they occur, *currente calamo*, and I do it, in full
confidence that you will soon begin the experiments they
will suggest. I have *igneous* experiments to suggest at
another time. On the flanks of the Pyrenees is an extensive
formation of vertical, curved, and contorted beds, of thinly
laminated sandstone, sand, clay, slatey marl, slatey limestone,
&c., all thinly bedded, as well as thinly laminated. The only
organic remains in a thickness of many hundred feet are
numerous and beautiful impressions of fuci. The sandstone
slabs have almost all the ripple-mark, visibly and exactly
preserved. The fuci are spread out conformably to the
planes of lamination, bending with the undulations of the
ripple-mark. The ripple-mark is strongest in the sands,
and indicated by layers of mica. A section longitudinal to
the furrows, and ridges, gives a slightly undulatory lamination : transverse, gives a wavy and curved line, and other
figures common in primary and secondary rocks. An inspection whether of the surface or of the transverse section
of the laminæ of a ripple-marked slab, shows at once that

such a disposition of laminæ is the effect not of the action of running water upon matter deposited, but on matter depositing. The general notion, I believe, has been that the furrows have been scooped out; it is not so, the ridges are made by addition, not the grooves by subtraction. Hence a difficulty may be got over.—How could 600 feet and more of ripple-marked beds be formed in succession? how could the ripple influence the bottom at such a depth? Answer, It arranged the sediment in that way near the surface, which fell through an undisturbed medium hundreds of feet in that form, and fell in that order on the bottom. Qy. If sand and mud be spread through a river, going at such a rate as to hold it in suspension, and a wind blows against it, and raises a ripple, does not this cause lines of retardation which cause matter to fall from the upper part in long coils? Try experiments. The ripple cannot penetrate many fathoms deep, may not its superficial effect or sediment be the cause? The ripple arrangement is seen in sandstone on very different scales of magnitude, from minute curves of a few lines to waves a foot and a half large each. I observed a strong ripple-mark on the sand of the river Gave de Pau, parallel to the bank at Orthez. I destroyed it in certain parts. Next day it was not renewed in those parts, but remained where I had not touched it. I effaced it on a sand-bar at the confluence of a mill-dam. In the evening, when the sluice was open, a muddy stream came down. Next day the sluice being shut, I found the new surface of the bar as deeply ripple-marked as the sea-sands ever are— and although it was soft sand and the water only 2 *inches* deep, a strong ripple of the river, washing over, did not in the least blunt or obliterate the undulations, yet loose grains of sand thrown into the hollow were immediately washed away. I threw sand and small pieces of dead leaves water-logged into a creek, the sands of which were ripple-washed, the depth of water being about six inches. The fresh matter was washed to and fro, sometimes all in suspension, sometimes thrown down irregularly, but every now and then all the vegetable matter arranged itself along the axis of the ridges, and waved backwards and forwards like sea-weed, as if growing in lines upon the top of the ridges. After a whole hour of this movement there was no tendency of the newly-injected matter either to increase the ripple-mark, or to diminish it by settling in the hollows. It is evident that the preservation of the ripple-mark is consistent with considerable agitation of the bottom. In sandstones this peculiar disposition of laminæ is more marked and more general than in mud-stones. But if mud be deposited on an uneven surface, it will preserve the form of the mould. When shale and marl alternates with sandstone, they must be ripple-marked. Gneiss, as Macculloch truly says, does not in Scotland present in hand-specimens those small undulations so common in laminæ of mica-schist. May not this be that the granitic schists (gneiss) in which are much felspar, came from mud, aluminous clayey beds, &c.; those in which are only quartz and mica (mica-schist) from ripple-marked micaceous sandstone? This is thrown out as a guess.—I once saw the sand-dunes of Calais, which are from one to two hundred feet high, ripple-marked for acres, as distinctly as a sea-beach. This theory must apply to air, therefore, as well as water. It was fine-blown sand, and no vegetation on those parts to cause or interfere with the wind's motion. The sands had been moved recently at Ussat. I stirred up the fine sand of the Ariége. It rose in clouds, and the mica was seen suspended in thin broadish plates all parallel, and reflecting the sun quite in a blaze. They fell on a steep bank at last, which glittered with them. One should have lots of mica for experiments. I will get a box full from Scotland. Meantime pound up mussel shells, freshwater, which will do perhaps as well. A large and deep trough, with gently slanting sides, might enable us to experiment. Get a paddle-wheel which will turn with the hand, and make a ripple *ad libitum,* and sand and mud of different kinds to be deposited. Then we will afterwards mix matter in chemical solution. After a due series of failures, blunders, wrong guesses, &c., we will establish a firm theory.— Have you read Dr. Young's experiments on the arrangement of sand upon boards vibrating by different notes of stringed instruments? The symmetrical forms obtained are wonderful.   Believe me ever yours,

CHARLES LYELL.

To His Sister.

Bordeaux : September 20, 1830.

My dear Caroline,—I have received your letter on my arrival to-day, besides two epistles, one from Broderip, and another from Scrope. The latter informs me that Lockhart had allowed him three sheets for his article on me, and that he had printed and revised nearly *four*, an unconscionable length, which may probably incur the pruning of which he is very apprehensive ; and though I expect a good article written *con amore*, I am much afraid there are parts which it may be prudent for the editor of the ' Quarterly Review ' to omit, lest he should excite hostility from the Oxonians and others. There are some of my doctrines to which he demurs, though in general with me. As to my original receipt for heating up or cooling down our planet to the required degree of temperature, he was very animated against it in his first letter, but at last he fairly owns that he gives up the attack, ' that he shall not grapple with it at present, and though perhaps not true, it is certainly most ingenious,' &c. Broderip says he is making marginal notes for the second edition, of which therefore he makes sure. I am glad that even in vacation time they contrived to give Cuvier a public dinner,[1] at which Fitton was President, and Broderip vice-president, and thirty men, including all the eminent naturalists then in London, were there. In addition to his higher claims to such a distinction, Cuvier is always most attentive and hospitable to the English at Paris.

Cooke is gone into Spain, I suppose, this day, the very one fixed on by 17,000 Spaniards to enter, as they declare here, sword in hand. Ferdinand the beloved has just published, I am told, an *ordonnance* ordering any one they catch to be shot immediately, which is a good set-off against the late deeds of the brother Bourbon. ' The Spaniards,' says Cooke, ' conduct a war and even a revolution, with such slow and deliberate measures, that you have only to pass into the next province, and it will be weeks or months before the mischief is there.' I must say I would not go in myself, unless to

[1] At the Geological Society.

take part with the Liberals, to look close at them, and be able to give a true account of what I expect will be a serious business; but Cooke positively means to make a scientific tour, and he ought from experience to know how far it is practicable. I got him out three out of four wet days, along the cliffs south of Bayonne. On the last a tremendous surf, in addition to the heavy rain, cooled considerably what little courage was left in him; and although he soon revived again, I shall get on much better alone. Murray of Simprim said to me, 'Take care and get a companion, if you have one at all, on a short lease, renewable at pleasure.' I am happy to say we parted excellent friends, but I did not do the most geology in a given time when with him. Since I left Bayonne I found at Dax some famous French provincial workmen, and a very pleasant society. Delightful old oak forests, as wild as ours in Hants, interspersed with beautiful heaths, and here and there a tract of sand covered with pines, *pinus maritima*, the only one which will grow there. I think I never saw such fine old Roman walls and towers as at Dax. I made many acquaintances, and a geological discovery, with which I will not trouble you now.

I ought to have begun with the Bayonne cliffs, although the Parisians do not appear yet to have explained them well. There is the key to the Pyrenees, and God grant a revolution in Spain, that half of it may not be closed to geologists. Had I been alone I should hardly have made such a blunder as to begin in the interior, when there was a coast at hand exposed to the tides.

With love to all, believe me,

Your affectionate brother,

CHARLES LYELL.

To HIS SISTER.

Paris: October 2, 1830.

My dear Marianne,—I found all the world here, as you may easily suppose, full of politics, and on no other subject can they think or talk. Prévost, when I saw him first, had just been dining with the King, in the dress of a private of the National Guard, having been invited as the principal member

who happened to be in Paris, when the Geological Society have
thought fit to go up with an address which Prévost penned.
He sent a copy in to the *roi très citoyen* about an hour
before. The reply, which was delivered *viva voce*, was very
neat, and was in allusion to each part of the address, and in
the same order. If Louis Philippe had not the gift of making
these speeches, with the ease which he does, and always to
the point, and just such as please the Liberals, his labour
would have been insufferable, for the number he has had to
make, and the variety of the subjects, is extraordinary. All
agree that he plays his part as if born for the peculiar situation in which he now is. The difficulty of finding lodgings
here now in hotels is considerable. The confluence consists
in no small measure of place-hunters from the country, quite
confirmatory of all I had heard in different towns of the departure of innumerable expectants of new promotions. This
hotel is full of military men, called, as it were from the dead,
the forgotten and proscribed campaigners of Napoleon, who
are reinstated to a certain extent, those who were not too
old, and who are now in part employed to drill the National
Guard. A great many cannot get employment, and are
praying night and day for the entry of the Prussians into
Belgium, in the hope of the French being drawn into the
affair. A finer opportunity, they say, could not have happened
for 'resuming our natural limits on the Rhine ; ' but then
they have a pusillanimous cabinet, and unless the people rise
again, as they ought to do, there is great fear that the Belgians may be put down, and that the ex-ministers will not
be condemned. Fine constitutional notions ! The affair of
Brussels has excited great interest here, and they contrive to
persuade themselves that it is much more a counterpart of
their own than it really is. The effect of these stirring news
every hour has been that none of the French naturalists have
thought of geology, and Prévost has made but little progress
even in reading my book, which ere this he was to have half
translated. He says moreover that he finds himself too little
of an English scholar to be able without great labour to comprehend *les périodes soutenues* of the introductory essays. I
found Deshayes excusing himself for having made no progress in naming my Sicilian and other shells, by his necessity

of working for his daily bread for encyclopædias, &c.; and having ascertained that his situation was very bad, I came to an agreement to take him off all other work in order to give me a private course of fossil conchology, in which he is to give me all his time for a month, towards the zoological part of Vol. 2, also to give me two months' additional work when I am gone, providing me with the results derivable from his great collection and that of others at Paris, relating to the newer formations, which he intends to publish in a Manual of Conchology a year and a half hence. I shall thus be giving the subject a decided push, by rendering the greater wealth of the French collectors available in illustrating the greater experience of the English geologists in actual observation, for here they sit still, and buy shells, and work indoors, as much as we travel. I am nearly sure now that my grand theory of temperature will carry the day, and at least I have had the satisfaction hitherto of finding no one dispute its entire novelty. I will treat our geologists with a theory for the newer deposits in next vol., which, although not half so original, will perhaps surprise them more. I am prepared to hear from Murray that the political hubbub has injured the sale, for all the Natural History and Medical booksellers here declare they shall be ruined; they are literally reduced to sticking up political pamphlets in their windows.

October 3.—A *soirée* at Baron Férussac's last night. Instead of natural history, he was full of an approaching election contest, for a place for which he is to stand in the south. He got in by three votes as a Charles X. man, a short time ago, but now he has hoisted the tricolor flag. I shall have to work for three weeks or more here, and shall be glad to hear from you. The best geologists are not yet returned from their summer campaign to Paris. One of ours, Weaver, is here. Great disappointment among the officers to-day, because the Prussian invasion of Belgium is contradicted.

With love to all, believe me yours affectionately,

CHARLES LYELL.

To Poulett Scrope, Esq.

My dear Scrope,—I must not delay any more writing for fear you should put off your proposed expedition to the sea-coast, in expectation of my joining you, which I shall be unable to do, although nothing would really be to me so great a treat. I must, however, be satisfied to run one hare at a time, and I am flying at high game just now. I am going through a course of fossil conchology, particularly with a view of satisfying myself as to the identification either of recent with fossil tertiary species or of those of different basins with each other; and I am having, at considerable expense, tables made out of more than 3,000 species of ter-tiary shells in the collection of Deshayes and others in Paris, of which the localities are well determined, and have been almost all visited by myself. The results are already wonderful as confirming the successive formation of different basins, and the gradual approximation of the present order of things, and will settle, I hope for ever, the question whe-ther species come in all at a batch, or are always going out and coming in. A multitude of other surprising conclusions arise, and the harmony of the whole with the phenomena of earthquakes and volcanos, of geological super-position, of the position of land-quadrupeds (fossil), of contents of fresh-water lake formations, &c., is very beautiful. I am too busy to enter into the ripple theory. Prévost has been working at it here, for in the Calcaire-grossier it is a common pheno-menon. He thinks with you it is connected with undulations of inferior currents, &c.

Make experiments on waves in an estuary. It is to me a most puzzling phenomenon. The wind was blowing up the Gironde one day, and as I thought the waves rolling up, although the tide ran down several miles an hour. I threw some pieces of paper into the river (we were at anchor), and was astonished to see them first fly out of my hand to the south, or up the stream carried by wind; but instantly on touching the water float down very swiftly, passing over the waves as if there were no wind against them. At last the wind got up so as to make the waves curl over and break in foam upon themselves. Of course, when pieces of paper came to one of these breakers the falling wave stopped it for a moment, but the instant afterwards it was seen far below, floating on the superficial water. It was clear, then, that the waves were tidal waves; and I then remarked that on them there was ripple which the wind caused. This small ripple was all the wind could raise. Now on watching this ripple I saw it often remain quite abreast of me, the great waves passing down, just as a person in a long barge going down a canal might keep close to you as you stood on the bank, by running from the prow to the stern at the same rate as the barge was carried down. Here I apprehend was a complete equilibrium in part of the upper stratum, while the ripple retained its place. I found just within the mouth of the Gironde, in the sands at low water, a large tract with a beautiful double ripple-mark, a larger one parallel to the shore, and the other at right angles, consisting of small ones. The great ridges were 5 inches or so high, the smaller ones 1½ about. They were like a plan in relief of longitudinal and lateral valleys in a mountain-chain; but the summit of the larger and higher ridges had not been disturbed by the cause which added the lateral ones. I am afraid I shall not be intelligible without a longer letter.

Deshayes has found at St. Mihiel, south of Verdun on the Meuse, not far from where you mention, and in that same deep valley, four old needles of limestone like those of chalk in the Isle of Wight, one called the Devil's Table (do you remember?); they are composed of old remnants of the white crystalline limestone of highly inclined strata. They have three distinct *horizontal* lines of perforations like the columns of the Temple of Serapis, the hollows, sometimes empty, but *thousands* of them filled with the saxicavas, *now*. He has shown us some; they go *all* round the Needles. The same are seen on the neighbouring precipice at similar heights. The shells get petrified *in situ*, and so are imperishable; species cannot be determined. You will at once draw all the consequences. All that the *river* has done, is at best up to the highest mark of the alluvion. And here, in the middle of the continent, at such an elevation, we have

preserved the sides of an old frith, and proofs of the successive rise of the old rocks out of the sea, for they are worn
by the wind and water mark at the three places perforated.
Believe me, ever most truly yours,
CHARLES LYELL.

To GIDEON MANTELL, ESQ.

Paris: October 10, 1830.

My dear Friend,—I find from Pentland you have had the
disagreeable job of dunning Deshayes again. The Sieur
D., as Murchison calls him, is undoubtedly one of the most
' imperturbable ' characters in these respects I have met with,
nor can I defend his conduct in this, and some other small
matters. It does not, however, arise from anything dis-
honourable in his conduct or intentions. He was, you must
know, for ten years a medical man, in considerable practice
for a beginner, in one of the unhealthy and therefore not
well (or wealthy) inhabited quarters. He had about fifteen
patients a day, who kept him constantly at work, and never
paid. At last he determined to cut the affair, and try to live on
natural history, which, although his passion, he had abandoned. He has earned his bread with great difficulty by writing for encyclopædias, &c., and spent every sixpence on very
expensive works, on fossils and recent conchology, between
30*l.* and 40*l.* worth, and on shells. Just as I obtained the
crag shells from you and Taylor, Rozet, who has published a
small work on Geology of no originality or value whatever,
engaged him for five hundred francs to write a small concho-
logical appendix, and this entirely prevented his doing anything for me. Had I known his situation, it would not have
happened, but to tell the truth, his bearing was so independent, that I should as soon thought of asking Broderip to
name my shells for *argent comptant* as Deshayes. Now I am
better informed, I see that he is not justified in giving me a
day without pay, for he has sacrificed his existence to make
himself, for the benefit of science, the first fossil conchologist
in Europe. I have now engaged three months of his time,
to enable him to teach me conchology, and to construct

tables, which I have planned for my second volume of tertiary shells, and to name all my Pyrenean specimens, &c. At the same time, I am enabling him to cut inferior work, and to use the same materials for a Manual of Conchology. It will make a deep cut into the small sum which Murray is to pay me for vol. ii., and will indeed consume all which the amanuensis and the extraordinary expenses had not eaten up, but I find already that it will pay me in the satisfaction of giving an essential push to our favourite science. Already the results of Deshayes' collection are yielding fruits unexpected by himself, and very confirmatory of the order of succession of tertiary formations which I had arrived at, from purely geological observations. The crag though probably older than almost all the tertiary formations of Sicily, is still a formation containing a decided preponderance of living species, and between it and the London clay you will see how magnificent a series of events I will describe. As your fossils could not be properly compared till we come to those genera, for I am going through the whole comparing with Deshayes, you must still let us keep them a little, but I will not leave Paris without them. Cuvier last night spoke with great pleasure of having made your acquaintance, and hopes you will visit Paris. He is not in spirits about political affairs, and consequently I got him for the first time to talk about fossil anatomy freely. Remember I have heard no geological news for three months, save the dinner to Cuvier, and a note from Murchison saying that Conybeare had fired a shot at me in 'Annals of Phil.' What is that about? Politics absorb all the thoughts of geologists here. There are croakers enough about the state of France, but I see no ground for it, and I believe if left to themselves they will get on. To do without some odious taxes, with an increased army, and with commercial bankruptcies innumerable, is the difficulty for the moment.

Ever most truly yours,
Charles Lyell.

To HIS SISTER.

Paris : October 19, 1830.

My dear Caroline,—I must stay here a week or two more in order to make myself fully master of the details from which my great tables of shells are to be drawn up. I have gone through a great part of Deshayes' collection systematically, following Lamarck. I had no idea that shells so studied could be so interesting, or that the arrangement was so philosophical, being founded on the *animals* which inhabit the shells. With an immense collection before you of recent and fossil shells (35,000 individuals and 8,500 species), all well arranged in their natural groups, and with a clever conchologist at your elbow, you may in a short time make yourself master of more than I ever expected to know. I am determined to begin a collection in earnest, having already a fine start in all my Italian and Sicilian shells now named, and in eight hundred named fossil species, which I have bought of Deshayes. No one in England has an idea of the results to be obtained from a good collection of fossils, or the value and accuracy of these medals when well collected, but I trust in my next volume to make them know it. This morning all my Etna shells were examined ; out of sixty-three only three species not known to inhabit the Mediterranean, yet the whole volcano nearly is subsequent to them, and rests on them. They lived on a moderate computation 100,000 years ago, and after so many generations are quite unchanged in form. It must therefore have required a good time for Ourang-Outangs to become men on Lamarckian principles.

The night before last we were alarmed here by a mob, not for ourselves, but for the inmates of the Palais Royal, because they were quite unprepared, and the guard weak. The people think, as John Bull would have done under similar circumstances, that the proposal to abolish the punishment of death is prejudging the case of the ex-ministers, and many moderate *bourgeois* even are become violent against what they call a *ruse.* The lower class, who suffered most in the fight, look on the ex-ministers as their prisoners

of war, and are exasperated at the notion of their escaping
scot-free. So they came as they say, to serenade ' Le citoyen
Philippe, propriétaire, No. 200, Rue St.-Honoré.' Several
thousand of them sung the Marseillaise and Parisienne
alternately, with such energy, that when two diligences
came down our narrow street, and when we should hardly
have heard a musket under the window, we heard every note.
Anything but cannon would have been drowned. It was like
the roar of several tempests bellowing in correct time.
There was a ferocity in it that was more horrid than anything I ever could have conceived, yet we were six times the
distance of the king's rooms. After keeping up this for
three or four hours, there was a call of ' Vincennes,' the
prison where the ex-ministers are. Instantly they all went
off, rushing up all the side streets from the Rue St. Honoré
so fast, that we thought the cavalry were charging. Luckily
the general at Vincennes was prepared for them, and they
could do nothing. At half past two o'clock they returned
here, and sung again. The first shout awoke us all like a
clap of thunder in the dead of night. Since that, all is quiet,
the national guard can be relied on, and are 80,000 strong.
The king was a good deal alarmed, but acted with spirit and
prudence, and no one fears now, and a reaction in favour of
the king has been produced. A small mob last night, who
collected there, went away cheering him, upon his coming
out on foot, and saying ' that they should have *justice*, but
not *vengeance.*'

With my love to all, believe me, your affectionate brother,

CHARLES LYELL.

To G. POULETT SCROPE, ESQ., *Castle Combe, Chippenham.*

9 Crown Office Row, London : November 9, 1830.

My dear Scrope,—I received several letters in Paris
before I read your article congratulating me upon having
fallen in with a critic who was capable of handling the
subject in an able manner, 'and in whose review a friend's
spirit pervaded the whole, though he finds objections to those
parts most generally controverted.' Among others, 1 was
delighted at Murchison having been taken in, and attribut-

ing it to Sedgwick, whom he had not then seen on his return.
Perhaps you know that he idolises even more than the
Cantabs 'the first of men,' as Adam is usually styled there,
and although he really has talent enough to appreciate most
fully the merits of a good thing when he reads it, yet the
eulogy was somewhat more from the soul than might have
been poured forth on one younger than himself, and who was
neither entitled as president or Professor to show him the
way. Murray, among others, wrote to say that the sale of
the first three months before the ' Quarterly Review ' appeared
was 650 copies, and he had no doubt that an article acknowledged ' to be at once *masterly* and *popular*,' would soon help
them to dispose of many more. For myself, I assure you without reserve, that I am quite sure of, what I am sure will be the
universal opinion, that it is incomparably the best thing you
ever wrote, and in point of style as well as originality, a great
stride beyond any former composition on a scientific subject.
I include those parts where you oppose me, even when I still
differ very considerably. The suggestion that the Mediterranean would sink if separated from the ocean, is so self-
evident when stated, yet so new, and opens such magnificent
views, and explains so much, that it would have redeemed
fifty pages of prosing. The explanation, which I had
endeavoured to hit off in vain, of the sandbanks opposite the
indentations in river channels, is so clear, that like most
discoveries, all will wonder how they missed it. It is your
own I conclude *the attributing it to the momentum.* These
hints will enable me greatly to improve my second edition if
I have one, as is expected. Such a broad-side will do far
more than my book to sink the diluvialists, and in short all
the theological sophists. Conybeare I am told has published
in the ' Annals ' ' an explosion ' against me. I shall not
answer it, but thank heaven he cannot turn the battery of
the ' Quarterly Review ' against me now. Basil Hall at the
first meeting of the Geographical last night, which I just
cut into from the Dover coach, said to me, ' Well, I will tell
you fairly, I did not think you could have written such a
book.' And so, I know, almost all of our council will say of
your article, but neither I nor Lockhart are of that mind,
and I am gratified that you have by an anonymous article

surprised many with an admission of your capability of doing
what I have frequently endeavoured in vain to assert you
might do, when regretting that you had cut in some measure
science, for political pamphlets.

Murray ought to send you 70*l.* for the good you have
done the review,[2] to say nothing of probable aid to his book.
I am going on now with vol. ii. It would sell, even if
stupid (which it shall not be), so great is the excitement for
and against, which vol. i. and your article have together
caused. I am in all the mess of unpacking, so must conclude, and with kind remembrance to Mrs. Scrope, believe me
ever most truly yours,

CHARLES LYELL.

To DR. FLEMING.

9 Crown Office Row, Temple, London : November 12, 1830.

Dear Dr. Fleming,—I am just returned from Paris,
where I spent five weeks, steadily applying myself to the
study of recent and fossil conchology, in the rooms of M.
Deshayes, whose collection contains 8,000. Although I have
thus late entered in earnest on this branch, I hope some years
hence to possess as fine a collection, and perhaps as much information respecting it as any one here ; and I am building a
cabinet and have purchased considerably in good tertiary shells.
It might perhaps be agreeable to you, as it would be most
serviceable to me, to exchange duplicates of French tertiaries
for carboniferous fossils, and might induce you to collect
these zealously. I shall always be ready to send you what
you may wish in return in this way, as I know no point
where so much has been done in this, as by you in Fife.
Your letter on my book was very kind, and at a time when
in Paris, I could find no one whose mind was not occupied
with politics, and when I had received no news of the progress my book was making, it was most consoling. Every
one agrees with you that the size of the volume and the
type has been much against it, and Murray is of that opinion
too. But as your friendly anticipations have been already
realised, I shall be able to correct this, in a second edition,

[2] Note by Mr. Scrope : 'He gave me 100*l.*'—G. P. S.

at no distant period. It seems that, for the first three months, they sold fifty copies a week, and since that I have heard no bulletin, except that the demand is constant. Murray declared, that if I could give him vol. ii. in six months, before half the purchasers of the first had either forgotten it, lost it, gone abroad, died, &c., he was convinced that in twelve months from the present time not a single copy of either volume would be unsold. With this encouragement I shall be glad to persevere, and benefit by any criticisms either in letters or in print, which you or others may come out with.

It would be doing me a service, now that I am obliged to spend money on secondary fossils with hard-dealing merchants, if you would exchange them against books which you may want; a proposition so common in France, and I think so fair between naturalists, who must make sacrifices, in order to form their libraries, that I propose it, *sans cérémonie,* and it might justify you in seeking more keenly for Fife shells and corals, than you would otherwise feel disposed to do. Believe me, ever truly yours,

CHARLES LYELL.

To HIS SISTER.

London : November 14, 1830.

My dear Marianne,—I have pretty good news to tell you about my volumes. . . . The booksellers assure me that if the latter part of my work is as popular and readable as the first, it will prove an annuity to me. Whewell of Cambridge has done me no small service by giving out at his University that I have discovered a new set of powers in Nature which might be termed 'Geological Dynamics.' He is head tutor at Trinity, and has more influence than any individual, unless it be Sedgwick. The young baronet [3] who franks this says that my new theory of deriving salt from inland seas, like the Mediterranean, went the round of the newspapers in his county (Cheshire) with great applause, as they felt a local interest in the explanation. He himself swears by it. He is very fond of conchology, and promises to send me

[3] Sir Phillip de Malpas Grey Egerton.

immediately a complete set of the N. of England freshwater shells. I have fixed on new rooms in Raymond Buildings, Gray's Inn; very light, healthy, and good. I suppose in ten days I shall have shifted quarters.

I shall send you soon some exceedingly entertaining German books, and hope when you find them likely to be useful, and make some progress, we may read them together in the spring, as soon as I have got rid of vol. ii. Lieutenant Graves is arrived, and has, at this most opportune moment, brought me several boxes of shells from S. America. I shall be in a bustle for ten days I fear, getting into my new rooms. They are on the same staircase as Broderip, whose library and great collection of recent shells, worth some thousand pounds, will enable me to dispense with laying out money, some of which would have been necessary otherwise. I shall also be very near Stokes. Brown the botanist has just given me some good information about a mummified Egyptian plant, and I am going to get a wood-cut executed. Several friends have asked me to dine out, and as I see I must make a firm stand, I have refused all, and shall only go to a club dinner once a week. There is scarcely any steering a middle course. All my friends who are in practice, do this all the year, and every year, and I do not see why I should not be privileged, now that I have a moral certainty of earning a small but honourable independence, if I labour as hard for the next ten years as during the last three. I never was in better health, rarely so good, and after so long a fallow, I feel that a good crop will be yielded, and that I am in good train for composition. How I wish we could be fellow-labourers! Yesterday Milman the poet, whom I scarcely know, sent me a long and curious extract from a German book about some theory of the followers of Zoroaster, to help me for the second edition. I have sent for a translation from the Miss Somervilles. If you were near this smoky hole, which after my late stay in Paris, appears dirtier than ever, I would gratify your desire to be useful to its full extent, but since we are both pinned like shrubs to the spots we are planted in, we must be satisfied. I think I may be out in six months, poz; Murray says five, but that I cannot. I am very confident that the matter in

vol. ii. will be more generally popular than in vol. i. if I
succeed in making as much of it as I see it admits of. I
shall only go to the end of the tertiary period. It will
include the history of the globe as far back as the time
when the first of the existing species came in. If I have
succeeded so well with inanimate matter, surely I shall
make a lively thing when I have chiefly to talk of living
beings!

I shall send the shuttlecocks and battledores, but don't
quite wear them out in the winter, as I may be with you
before out-door exercise is always within reach, if your
account of ordinary springs be true.

With love to all, your affectionate brother,

CHARLES LYELL.

CHAPTER XIII.

FEBRUARY–SEPTEMBER, 1831.

APPOINTED PROFESSOR OF GEOLOGY AT KING'S COLLEGE—VISIT TO CAM-
BRIDGE—JOURNAL TO MISS HORNER—MRS. SOMERVILLE'S 'MECHANISM
OF THE HEAVENS'—PARTY SPIRIT ON THE REFORM BILL—ON EDUCATION
OF BOYS—MADAME DE STAËL AND HER WRITINGS—SIR WILLIAM NAPIER
GOES TO SCOTLAND—EDINBURGH—LORD COCKBURN—DR. GREVILLE—
KINNORDY.

[Mr. Lyell made a tour of some weeks during the summer of 1831 to the Eifel, a volcanic district between the Rhine and Moselle rivers.

He was appointed Professor of Geology in King's College, London, and gave a course of lectures in 1832 and 1833. After which, finding that the duties would interfere with his schemes of travelling and original research, he resigned the office.

He was appointed Deputy-Lieutenant of the County of Forfar in 1831.]

CORRESPONDENCE.

To DR. FLEMING.

2 Raymond Buildings, Gray's Inn : February 7, 1831.

My dear Dr. Fleming,—I hope you are proceeding successfully with your volume on organic remains, of which, now that I am writing on the secondary strata, I should be very glad. If you were printing it I should crave a sight of the sheets done, in order that I might benefit. We expect you will throw much light on the organic remains of the carboniferous era. I am finishing the last two chapters of a somewhat longer volume than that of which you have, I hope, received a copy from Kinnordy, and I mean to lecture from the slips, as far as that can be done, having magnified all the sections a hundredfold, which are to adorn my book, for the lecture-room. If the course of twelve lectures should prevent my being out with the volume before my marriage, in other words, before a year or more, I shall sadly regret having accepted of the chair. But I hope to do both. The King's College is coming on gradually, the number of students having augmented by more than one-third, both in the junior and senior departments since Christmas, and promising to increase again in the same ratio at Easter next. Sedgwick wishes to make your fish scales a means of identifying his Caithness schists, and all their fish and the trionyx, with a member of the lowest 'old red.' I am sure I have no *objection*, for I would as lief start with vertebrated animals and freshwater, as with a universal ocean, and the simplest forms of animal life. Only I require rather more proof for identification than the Welsh Captain did, when he made out that his hero and Philip of Macedon were the same, 'for there is a river at Monmouth, and likewise also there is a river in Macedon, and there is salmon in both.' If the *species* of salmon and of plants could be proved to be 'as like as my fingers to my fingers,' then I think it would do well enough. In one of your last letters you mention having reptiles from the carboniferous era in Scotland—this is delightful news to me. I expect to prove that there are three distinct mammifers in Stonesfield slate. Vernon admits that his single saurian vertebra was only in alluvium, but that alluvium entirely composed of mountain limestone.

Ever most truly yours,
CHARLES LYELL.

To GIDEON MANTELL, ESQ.

March, 1831.

My dear Mantell,—I have been within this last week talked of and invited to be professor of geology at King's College, an appointment in the hands entirely of the Bishop of London, Archbishop of Canterbury, Bishop of Llandaff, and two strictly orthodox doctors, D'Oyley and Lonsdale. Llandaff alone demurred, but as Conybeare sent him (volunteered) a declaration most warm and cordial in favour of me, as safe and orthodox, he must give in, or be in a minority of one. The prelates declared 'that they considered some of

my doctrines startling enough, but could not find that they
were come by otherwise than in a straightforward manner,
and (as *I* appeared to think) logically deducible from the
facts, so that whether the facts were true or not, or my
conclusions logical or otherwise, there was no reason to
infer that I had made my theory from any hostile feeling
towards revelation.' Such were nearly their words, yet
Featherstonhaugh tells Murchison in a letter, that in the
United States he should hardly dare in a review to approve
of my doctrines, such a storm would the orthodox raise
against him! So much for toleration of Church Establishment and no Church Establishment countries. It is, however, merely a proof of the comparative degree of scientific
knowledge diffused. Pray be so kind as to give me the
earthquakes. A shock in Sicily which threw down Melazzo,
seems to have occurred nearly on, if not on the same day as
Dover. Another just announced in China has killed, they
say, a million of men, all in favour of modern causes;—it is an
ill wind, &c. The young Prince George of Cumberland told
me the other day of you and the great lizard, which last has
taken much hold of his imagination. 'Tis clear, as Aber-
nethy said, you will ride on that beast. Don't throw away
any great big specimens, for if I lecture, I shall be as greedy
for them as I have hitherto been shy of them. I will get a
scene-painter to put Etna and Auvergne on scenes as large
as in a theatre, on canvas, from Scrope's and my sketches.
Scrope writes, ' If the news be true, and your opinions are to be
taken at once into the bosom of the Church, instead of contending against that party for half a century, then, indeed,
shall we make a step at once of fifty years in the science—
in such a miracle will I believe when I see it performed.'

Ever yours,

CHARLES LYELL.

To HIS SISTER.

London : April 7, 1831.

My dear Marianne,—I almost forget where you are in
regard to my affair of *the chair*, but I think I said I had
indited an epistle to Bishop Copleston (the minority of

one) to tell him among other pleasant news, that I meant to
do my best to show in volume ii. that no deluge has swamped
Europe within 4,000 (I might have said 40,000) years. On
his other points I was fully agreed with him, and on this I
told him I had no objection to his drowning as many people
as he pleased on such parts as can be shown to have been
inhabited in the days of Noah. In his answer he thanked
me for my candid statement. Although some of my friends
assure me I shall lose by this concern, I do not; and most
of them agree with me that it places a man who wishes to
devote himself steadily to a branch of a science, in a much
more agreeable, influential, and respectable situation, and it
is not like taking orders, a step without a retreat.

Conybeare has taken his last shot at me. Daubeny has
fired off one at me, in a timid sort of a discharge, and full of
puffs about ' Oxford being proud of me, though I am not of
the Oxford school of Geology.' Sedgwick's attack is the
severest, and I shall put forth my strength against him in
volume ii.

Your affectionate brother,

CHARLES LYELL.

To GIDEON MANTELL, ESQ.

Athenæum, London : June 22, 1831.

My dear Mantell,—I take the largest sheet I see near
me at the Athenæum to write an apology for my remissness.
A visit to Cambridge of a week with Buckland, Conybeare,
Daubeny, and other Oxonians, who were returning a visit
which the Cantabs made to them, was the first cause of
silence.

We were lionised with a vengeance—lectures, experiments (optics, polarisation), feasting, geologising, and evening-
party going, and nocturnal smoking and cigars, and by way
of *finale*, Conybeare and I took our *ad eundem* degrees, and
were admitted M.A.s of Cambridge. Then came an arrear
of work here, and my father and brother in turn gave me
lots of interruption, notwithstanding all which, as also much
lionising of Conybeare, who wanted to be introduced to
divers persons, I have reached p. 110 of printing volume ii.

The four great slabs are two of them worth nothing, but the
other two are as magnificent specimens of the forms of
ripple as I ever saw. Many thanks. Murchison and his
wife are gone to make a tour in Wales, where a certain
Trimmer has found near Snowdon 'crag' shells at the
height of 1,000 feet, which Buckland and he convey thither
by the deluge.

You heard of Fitton's accident? Changing his residence
as usual; going from his country seat near Sevenoaks to a
new place eleven miles north of town, taking a maid-servant
to Harley Street in a gig, horse ran away in Regent's Park,
dashed against a gate. Fitton's arm said to be broken high
up, but Brodie can't make out where ; feared that the blade-
bone was injured, but hopes not, as Dr. F. is doing so well.
Lonsdale says that the great femur of which Trotter has
given us a cast, was of an animal that had paddles. What
is the largest paddle-bone you have? No room to talk of
age of reptiles.

Ever most truly yours,
Charles Lyell.

JOURNAL.

[Extracts from letters and journal to Miss Horner, eldest daughter
of Mr. Leonard Horner, to whom he was engaged, and who was with
her family at Bonn on the Rhine.]

Rotterdam : August 4, 1831.

My dearest Mary,—I tried, as soon as the steamer passed
Bonn, to compose my thoughts by steadily reading Macculloch's vol. ii., and succeeded tolerably, till I landed at
Cologne, and then walked to the Cathedral. It was eleven
years since I had been in it, with my father, on our way to
Italy. I remember well the splendid painted windows, and
that singular semicircular aisle behind the altar. Although
there is no such architecture in Italy, this gigantic fragment of a church reminds me of Italy. There is nothing
there so striking as the innumerable monuments of the
designs of architects and priests, which outran all reasonable

calculation of means. It is rare to see an Italian church finished. But I suppose if Catholicism had remained in full force five centuries more, Cologne would have been finished, for the secret of our own grand gothic buildings having been finished, was their having been added to, during many successive centuries, by 'corporations that never died.' After I had left the Cathedral I heard a bell begin to toll, the finest sound I thought I had ever heard. I returned, and found about sixteen men or more pulling at one rope in the great tower, or to as many strings attached to one main rope. This huge bell, they told me, was only rung on a few occasions in the course of the year, not on Sundays and ordinary *fête* days. This was the eve of the King's birthday. They say it is the finest bell in Germany. I was never so much struck with any, even at Rome. I waited a quarter of an hour to hear it in the great tower, ' swinging slow with sullen roar,' and then tried to find my way to the town hall, but lost it ; fortunately, for I blundered on a street where I saw a mark in a stone of a great flood of the Rhine in February 7, 1784, when its waters were seven or eight feet higher than they have ever been since. I cannot understand this, since in February there could not have been a great melting of Alpine snows, but that for half a century since, nothing like such a flood has occurred, is a good fact ; the effects of *rare* combinations of the most ordinary causes, are scarcely ever reckoned upon enough in geology, and I could preach for a page on this text, if I did not fear I should put you to sleep. We have had to-day the certain news of the Prince of Orange having entered the Antwerp territory, and so many stories of that city having been attacked, that I fear something must have happened. As the Dutch have the best organised army, and are more united, I think they do well to show a little spirit, and unless a general war comes on, they will get better terms by it. Here the war is popular, and the going out and return home of the regiment quartered here, to exercise, is quite a triumphal procession, morning and evening. Tell your mother, with my love, if she finds her thoughts fatigue her, I recommend her to read, if she can get it, Madame de Staël's chapter on the efficacy of study in such cases to calm the mind. It is in her work, ' L'Influence des Passions sur le Bonheur des Nations et des Individus,' one of the most splendid productions of the age. The sun has just made an effort to break through the clouds, so we shall have no rain I think.

August 7.—Sent my Rotterdam-arranged MS. to feed the hungry devils, and corrected a sheet and a half of text, of that set up in my absence, the fruit of three hours' hard, steady work, aided by my steam-engine Hall, for in this stage of the business more than half of the work is mechanical. I hope to receive from the binder my old copy of the work of De Staël, of which I spoke, that you may have it. Be careful not to erase my marks. I shall re-read it with delight; it is the work of a mind which had been agitated strongly by almost every passion, which had felt intensely, and therefore wrote eloquently—her rules and precepts are often inexplicable to ordinary affairs and common minds, but it is a splendid production. I recollect the chapter on study (though I have forgotten its details) made a lasting impression on me, and taught me among other things to take delight when engaged in composition, in improving, enriching, and embellishing thoughts, instead of deeming that drudgery; just as an artist retouches a painting. My copy of De Staël is just returned; I have been dipping into it. It is the more striking as being so exaggerated. The balance against the chances of happiness in many of the most natural and best passions, is not so unfavourable as she paints it, but such would the world be, if all were as romantic, sensitive, and imaginative as she was.

August 11. —Went to Paganini's concert, Prandi having sent me a ticket. Paganini can play both beautifully and wonderfully, but I would rather have half an opera another time, than all his performance.

August 12.—Boué, the French geologist, called, just arrived with his wife from Paris. Am to see him again to-morrow, and to show him my Sicilian volcanic specimens. At Athenæum saw with pleasure that the Dutch had routed the Belgians.

August 13.—Before I go to rest, a few words of a day of

steady work, and few incidents—wrote to Mantell, dispatched nearly half the printed Tables to Deshayes—walked over the rooms at King's College, which promise to be handsome and well lighted. Found the table of Athenæum this evening covered with new books—in this respect a club is a real economy—Whately's new lectures on Political Economy. In the first of those which I skimmed, is a long combat against those who object to the Rev. Professor's science, as opposed to the Bible. He says that the cry is louder 'than against geology,' because people will admit that the sacred writings were not to teach us *physics*, but say that a science connected with *human* concerns should be in accordance even with the letter.

August 14.—Webster brought the corrected drawing from my sketch of the Val del Bove on Etna, one of the grandest scenes in Europe. Although a poor representation, it will give some idea, and be useful geologically ; I shall go to Chelsea, and hope to find the Somervilles.[1] Arrived at five o'clock, and found that dinner was at seven : was lucky enough to find Mrs. Somerville in the garden, told her of my expedition to Eifel, &c. She showed me her volume,[2] all done save the index. The preliminary discourse will be popular, dedicated to Lord Brougham, as he suggested it ; no word of flattery to him. The Murchisons have been staying with the Conybeares in Wales, which country, said Dr. Somerville, 'they have now done.'

August 16.—Mrs. Somerville said yesterday that she never knew politics in London so embitter people as in this Reform Bill : the fact is there never were such private interests at stake before. Few bills for the confiscation of property ever passed in any country which smote individuals so hard. When a man has several brace of borough appoint-

---

[1] Mrs. Somerville, b. 1780, d. 1872. Author of *Mechanism of the Heavens*, from La Place; *Connection of the Physical Sciences, Physical Geography*, &c.

[2] *Mechanism of the Heavens*. 'Mrs. Somerville's volume is ready all but the index. The preliminary discourse into which I dipped will be popular and of high interest; it touches on all the grandest subjects. Speculations, for example, on cavernous structure of the interior of the earth ; what the total number of comets may be ; what the climate of other planets, and in which of them our plants, &c., could live. How the sun and other bodies look from each of them, and an immense variety of other topics.'—Extract of letter from Mr. Lyell to his father, August 15, 1831.

ments, each worth forty or fifty thousand pounds in the
market a year ago, he may well cry out, and may easily
persuade himself that the good derivable to the country is
problematical. The Liberals say that the cutting up counties will make each district a close borough, and so a
new aristocratical influence, as bad as the old, will arise. I
doubt whether it will be as bad, but the new House will be
aristocratical enough. A letter from my mother to-day announces her recovery. She is going up to the hills, to a shooting quarter of ours. Some day or other, we will go up to
that shieling in the midst of the heather, where there is a
fine air, and where many a time I have enjoyed the glorious
view of the Grampians on one side, and on the other of
Strathmore and its lakes, with the Sidlaws and the Lomonds
of Fife in the distance. Although the game was scarce, I
used to like it better than Lord Airlie's preserves, as he
has no scenery, though thirty brace for one bird on our
ground.

August 17.—The best day's work, at least the greatest
number of hours at composition since Rotterdam—it will
require a multitude more, before I have done this volume.
The facts which are given in a few sentences, require weeks
of reading to obtain. A letter from my aunt Heathcote,
one of the most amiable creatures in existence, and moreover
an elegant style of person, inviting me to pay her a visit.
How lucky that the very day I got it, she must have received
mine to say I meant to come! Lonsdale has just told me
that one of the principal reviews, as he heard from Fitton,
has just held up Sedgwick's speech, and my first volume, as
models of style for all English scientific writers. I must
learn where this is, for as Messieurs les Critiques give one
some hard rubs, it is fair to seek out their commendations.
It is a reward for doing a thing with care. Did not Miss
Parker [3] remark that the gravel of the Rotherberg was as free
from scoriæ and volcanic rocks as that above Friesdorf?
What are volcanic rocks? she would say. I answer, Any of
those which the said Rotherberg in general, or Rolandseck,
are made of. Webster tells me, he spends all his time in

---

[3] For seven years the valued friend and governess in Mr. Horner's family.

teaching ladies geology, and remarks that they have more liberal curiosity about science than young men. The fact is our boys are worse educated now than girls ; they spend ten years and more at school and college, in learning Latin and Greek ; two weapons which nine out of ten never use. They are mere linguists of dead tongues—they get *learning* without *knowledge*, and without the taste of acquiring a knowledge of things. A system which would make our youths not scholars, but lovers of knowledge, would be a grand reform. It would be easy, for it is natural in youth to love it.

August 23.—Made good progress in my MS. I forget whether I told you that Mrs. Somerville's book was shown me in boards, but without the index, at least a hundred pages longer than my first volume, and about the same dimensions in other respects. Mine even was deemed too corpulent, as well as having too much crowded into a page, which in the second volume we mean to correct, though using the same type. About 500 pages of Mrs. S. is algebraic, x, y, z, and sealed save to those deeply initiated. The introductory hundred will be popular, and I trust serve as sails and winds to waft the heavy cargo on through these unpromising times. While I was working here at four o'clock Lord Northampton came in, and sat cozing three quarters of an hour, about fossil insects, shells, and how he should vote on the Reform Bill. After supporting second reading he is to go to Rome, and *not* leave his proxy. ' They will alter it so that I cannot say what the third reading will be, and I am not a thick and thin man that can trust another with my vote.' I have just received your letter. Do not suppose that my marks of Madame de Staël's sentiments are always of approbation—they are like the hear, hear, of M.P.s of something which strikes, often as true, often untrue, or doubtful. I look upon her as a phenomenon—a giantess almost rivalling any male giant of her age. Her imagination was vivid and poetic, and tempered by judgment, and she had a philosophical mind also. Had she written first-rate poetry, she would have shown that the very best power of mind which can be developed in man may belong to her sex —which nevertheless would be unfair, since they have many

excellences which men cannot have. I think her work the
more interesting, as showing how a powerful mind, placed
under the circumstances that a woman's must be, would
think and write. Hereafter her writings will stand out in
relief more prominently, among those of her contemporaries,
than they do now ; very few as boldly, and this is saying
much, when the general standard is so high, and so many
minds are highly cultivated. But when an equally great
vigour of genius is given to some woman, who has more of
the perfect qualities, the peculiar attributes of her sex, we
shall have something more splendid and characteristic and
original. Like most great characters, she was a mixture of
good and bad. From much that I have heard of her, I can
believe her speech about her daughter, that she was *bête
comme son père*, although her husband was much above par,
and if he had not been, she would have fancied it, had she
been amiable. But her extravagant admiration of her
father, whom she overrated, in a great degree from filial
feeling, was a redeeming quality—as regards feeling, what
in this case overpowered her better judgment. Had our friend
Mrs. Somerville been married to La Place, or some mathematician, we should never have heard of her work. She
would have merged it in her husband's, and passed it off as
his. Not so De Staël, she would have been a jealous rival.
What she says of Nature never having given great talents,
without accompanying the gift with a love of displaying
them—a desire to make them tell—is most true, but much
that she says of the incompatibility of even moderate ambition and love, is false and curious only, as showing that she
had found her desire of fame to be that of one placed in a false
position. A man may desire fame, reputation, and even
glory, for the sake of sharing it with one he loves. A woman
cannot share it with her husband, it will be the utmost she
can do not to make him of less importance by it. I am
sure that it is very common in ambitious men to look upon
whatever they acquire, whether of fortune or fame, as
chiefly valuable for the sake of others. I could, and should
feel this ; for a wife, children, brothers and sisters, that the
sharing it, constituted its chief and almost only value—that
the mere personal aggrandisement was of itself a shallow

pleasure. As for love being an episode, &c., it is true both
to man and woman in the sense of that romantic sort of
love which the Authoress of ' Corinne ' would try to persuade
herself should run through life, but what never could nor
ought to do, or to be more than an episode in any person's
life—but speaking of love in a more sober sense, I do not
believe it, and could point out proofs of the contrary.

To reap the full benefit of what the Germans are doing
and thinking would require no moderate intimacy with the
tongue, and its idioms. I am sure you will work at it with
more zeal if you believe you can help me by it, as I labour
with greater spirit, now that I regard myself as employed
for you as well as for myself. Not that I am at all sanguine
about the pecuniary profits that I shall ever reap, but I feel
that if I could have fair play for the next ten years, I could
gain a reputation that would make a moderate income for
the latter part of my life, yield me a command of society,
and a respect that would entitle me to rest a little on my
oars, and enable me to help somewhat those I love. I met
Smith, F.R.S., to-day, coming to ask me to join Wallich the
botanist and Sir Robert Inglis at R.S. club dinner. Smith
poured out, ' I am in the middle, and full of astonishment
and admiration at your having, among other things, been
the first to take the bull by the horns, about the antiquity
of the Earth, and contrived to do it in so inoffensive a
way,' &c.

August 26.—The Chevalier Ricasoli came in to receive
the second packet of Tables, which he is to take to Deshayes
to Paris. He caught sight of my plate, intended for a fron-
tispiece,—from a sketch of my own, a view in Catalonia in
Spain, the volcanos of Olot. Prandi[4] was delighted. ' Ah !
that is Olot indeed. Well I fought there for thirteen days,
five were we encamped on that hill by that church ; and so
the hollow was the crater, and the cone a volcano. How
odd, I never dreamt of that ! at its foot we killed 400 of the
factions ! but from that hill (another volcano), as we passed
to enter that valley, a discharge of shot thinned our ranks.
We could not return it with any effect, we were on that flat
plain, and on the slant of the cone, concealed in the vines.

[4] An Italian refugee, who resided many years in London.

There is the defile, through which Milano brought his troops from Gerona, and this is Olot, with its villanous population, of 30,000, all factions.' 'What do you mean by *factions*?' 'Why those were the times of the established constitution, they were rebels, *now* the legitimates.'

August 27.—An hour and a half good work before breakfast, ditto after, then Hall reading De la Bêche, while I shaved, half hour's recreation. At one o'clock walked, met Judge Patteson, a coze with him. Called on Spiller, librarian of the House of Commons. Davies Gilbert in the library, very curious to hear about the Eifel volcanos and the gravel of Rotherberg. Stepped into a neighbouring *café* and took a cup of coffee; on the table was the 'Westminster Review'—read a splendid extract from Napier's 'Peninsular War.' It must be a noble work. One day last spring, when I called on Sedgwick, he read me a whole chapter of his friend's book, an animated and graphic sketch of a skirmish in Portugal. Scott in his best historical novels never made you more completely present on the spot. Returned, and worked two hours; have made great way to-day, and yet a day of much enjoyment withal; seven hours' work, and quite fresh. I shall take a walk to the Athenæum. Is it not strange, after what I have said above, that at the Athenæum the first man I sat down beside was Colonel Napier?[5] I was struck with his conversation, addressed to Stokes, and on asking who he was, found it was he whose third volume I had been admiring; a fine-looking man to boot. Stokes showed him a five-franc piece of Henry V., for the Bourbons are carrying on a parody on the exiled Stewarts to the last. There are coins which Stokes has of Henry Ninth, King of England. Did you ever see Stokes's toy of the man whose head you can cut through with a penknife? Napier remarked, 'In these times such a gift would serve a king better than such dollars.' In the library found Greenough and Decimus Burton. Greenough turned up to a note he had made on the Rotherberg in 1816. It struck him in every respect as it did me—no trap in the gravel, which therefore is not posterior. He speaks of a section by the walnut-tree near the farm-house in the crater, which I missed.

[5] Sir William Napier, K.C.B.

To Dr. Fleming.

London : August 29, 1831.

My dear Dr. Fleming,—I was very glad to hear from you, on my return from a tour of four or five weeks to the Eifel, that tract between the Rhine and Moselle, in which abundance of extinct volcanos have burst through a country which is composed of strata much like the Sidlaw (where it is free from trap), the same sandstone and shale, and similar fucoidal plants, connected in some few parts with trilobite limestone, and with genuine *old* coal.   Through these the supra-marine volcanos burst, and strange holes have thus blown through the mountains, each eruption having been almost invariably at some new point.   I daresay you may have heard of a certain case of diluvium of the Rotherberg, *i.e.* gravel on one side of the crater of one of these modern (tertiary) volcanos.   The fact is, it proves nothing as to the post-diluvial, or rather post-alluvial origin of that volcano. The gravel was there before the eruption, and contains no volcanic matter.   I never saw so clear a case, and it is one of ten thousand proofs of the incubus that the Mosaic deluge has been, and is I fear long destined to be on our science. Now I am fully determined to open my strongest fire against the new diluvial theory of swamping our continents by waves raised by paroxysmal earthquakes.   I can prove by reference to cones (hundreds of uninjured cones), of loose volcanic scoriæ and ashes, of various and some of great antiquity (as proved by associated organic remains), that no such general waves have swept over Europe during the tertiary era—cones at almost every height from near the sea, to thousands of feet above it.   I mean when lamenting the rash introduction of such hypothetical agency, to hint at the unfairness of gaining votes for new 'guesses' by the temptation of a promise to explain the Mosaic deluge, if they will consent to adopt these new views.

At the same time, I intend to allude to the debt due to *you*, for having been the first who had the courage to maintain the untenable notion of a non-exploded dogma, and of having faced the obloquy which others are at this day no longer in danger of sharing.

I don't lecture till April, and unless I see the Institution promising very well, shall only give a very short course of nine lectures at the most, on a branch of the subject. Some of my friends think it was a foolish fancy. Others, with Whewell, that it will serve ' to give an anchorage to my thoughts, and keep me more steadily to our great subject.'

Remember me kindly to Mrs. Fleming, and believe me, ever most truly yours,

CHARLES LYELL.

To G. MANTELL, ESQ.

London, August 30, 1831.

My dear Mantell,—I hope you got my letter after my return from Germany? Since that I have been detained here much longer than I expected, and am now on my way to Scotland; first to Edinburgh, then to Forfarshire. I wish much to hear from you, as your silence has made me uneasy. Write to me here, and your letters will be forwarded free to me wherever I may happen to be at the time in the country. Barrow has sent me a box of specimens from the new island [6] thrown up off Sicily in a spot where I did not at all look for a submarine eruption, but am nevertheless well satisfied therewith. Chocolate-coloured sands and scoriæ of the same hue. The ' Britannia ' man-of-war passed over the spot some months ago, and feeling her bottom struck as if by a rock (slight earthquake), she sounded, and found eighty fathoms. Now the isle is 200 feet above water, and is still growing. Here is a hill 680 feet with hope of more, and the probability of much having been done before the ' Britannia ' sounded. I congratulate you, one of the first of my twelve apostles, at Nature having in so come-at-able a part of the Mediterranean thus testified her approbation of the advocates of modern causes. Was the cross which Constantine saw in the heavens a more clear indication of the approaching conversion of a wavering world? more especially as the first box of specimens from the new isle came through the post office by the Mediterranean steam packet, and was

---

[6] Graham Island, which rose August 1831 in the Mediterranean, and towards the close of October was nearly levelled with the surface of the sea.

presented by Barrow to me before he had opened them
himself, eight or nine days after they had been thrown up
hissing hot.

Ever most truly yours,

CHARLES LYELL.

To MISS HORNER.

Steamboat 'Soho,' Orford Ness : September 1, 1831.

I have been profitably employed this morning since breakfast in reading Conybeare's last little work, and several
chapters of Omalius d'Halloy's new ' Elements of Geology.'
What a proof of the interest now excited by the science,
that since my first volume appeared we have three works,
Macculloch, two volumes, De la Bêche, and this, on Geology,
besides innumerable memoirs !

September 3.—' Soho,' off Bass Rock, three o'clock. We are
at last in calmish water again in the Forth after a rough, rainy,
and very slow passage, against a north wind. All ill, and I
was obliged to lie in my bed nearly all day yesterday. I had
an agreeable compliment paid me in conversation, which I
overheard. One of the party was describing the washing away
of the small port of Carse on the Solway Firth, near Dumfries ;
which led another to say, ' Have you read Lyell's account of
the ravages of the sea of the E. and S.E. coast of England? '
To which a blind old talkative gentleman, who I learnt was
a Dr. Milligan, replied, ' I have ; it's by far the first work on
that science, and the writer, ye ken, is a countryman of ours.'
On the score of his being aware of the latter circumstance, I
deduct one half from the value of the eulogium, and am
satisfied with the remainder.

Edinburgh : *September* 4.—I have just ten minutes to
spare, before dining in Charlotte Square, and I doubt not
you will like to hear all that passes between me and your
old friends. Macbean took me to call on Mr. Thomson,[7] with
whom I was much pleased. He told me he had heard
Chalmers [8] preach, and make a somewhat wire-drawn dis-

---

[7] Thomas Thomson, advocate. ' Formidable in dignity and in antiquarian
learning.'—*Lord Cockburn's Memorials.*

[8] The celebrated Scotch divine.

tinction between an ungodly man and a sinner. ' Wonder
what he would make of an ungodly sinner?' said I. Finding
the celebrated preacher was to perform service in the afternoon, I determined to go. I went to the kirk, and for a
shilling got an excellent place, which is a great point, as
there are many words in Chalmers' sentences which one may
miss if not very near. I have a great prejudice against
popular preachers in general, and almost all I have heard
have strengthened my original impressions. But not so in
this instance. It was a very long discourse, but admirable.
The subject was ' repentance,' a hackneyed one enough. You
are to repent to-day, not because it may be too late to-morrow,
as you may die; but because, though you are to have ever so
long a life, your case will be more hopeless should you resist
this warning. Your mind will be hardened by the habit of
resisting. He then explained the effect of habit, and its
increasing power over the mind, as a law of our nature, with
as much clearness and as philosophically as he could have
done had he been explaining the doctrine to a class of
university students in a lecture on the philosophy of the
human mind. But then the practical application was enforced
by a strain of real eloquence, of a very energetic, natural, and
striking description. I hardly think any critic could have
detected any expression or image that sinned against good
taste, although he was very bold and figurative. But unfortunately, every here and there he seemed to feel that he
was sinning against some of the Calvinistic doctrines of his
school, and all at once there was some dexterous pleading
about ' original sin,' which interfered a little with the free
current of the discourse, and gave me an idea, that in order
to be popular with a Scotch congregation, a preacher must
deal a good deal too much in the unintelligible doctrinal
points of theology. Upon the whole, however, judging from
this single specimen, I think I would sooner hear him again
than any preacher I ever heard, Reginald Heber not
excepted.

September 5.—When I got to Mr. Maitland's this morning,
I found him with an excellent newspaper account of the new
volcano in the Mediterranean, prepared for me to read.

The Solicitor [9] did not arrive till half-past eight o'clock. You would have been amused with the first interview. Maitland, having been employed to draw up for the new Reform Bill the boundaries of the borough of Musselboro', was showing his map to me ; so when he introduced me, Cockburn gave me a moderately good reception, a bow, and, if I remember right, a shake of the hand. We all began to move to the breakfast room, when the Solicitor exclaimed (the first sentence I heard from him), ' Is it *Mr. Lyell?* Lord! I thought it was some d ——d surveyor come about your map.' Then he shook my hand in right good earnest. On my telling Cockburn how much I was struck with Chalmers, he gave me a most favourable sketch of his various acquirements, private character, &c., and said that Chalmers was now studying so hard, that he had told him (the Solicitor) that he had been *idle* all his life before. On my observing that I could not agree with Macbean, in wondering at the small number of successful popular preachers, because they had such trammels to contend with in the pulpit, so many religious dogmas to steer clear of, &c., Cockburn said, ' Besides, how many of the most popular weapons of oratory are utterly prohibited—sarcasm, irony, ridicule ! ' I agreed to go to Bonaly [1] to-morrow. I then sat down, and took a good hour and a half's work at composition of the MS. chapter of my second volume, which I left off at in town. I delight in thus making the business of acquaintance-making and calling on people the mere recreation necessary between hours of work. After this set-to, I found, thanks to the lawyers' early hours here, that it was only twelve o'clock. Called on T. Allen, not in town, then at Jamieson's. I then went to Hibbert's, at home, glad of it, as I wanted a good confab about the Eifel, to which he had laid a six months' siege, whereas I had only attempted to take it by storm in as many days.

Kinnordy : *September* 8.—On my arrival here, I found every one quite well, and two letters from you. Mrs. Hibbert's

---

[9] Solicitor-General for Scotland, afterwards Lord Cockburn, author of *Life of Jeffrey, Memorials of his Time,* an old and valued friend of Miss Horner's family.

[1] Lord Cockburn's country-seat near the Pentland Hills, Edinburgh.

drawings of the Eifel are really splendid, most of them worth
publishing, yet when she began the tour, she had hardly attempted to sketch from nature. She has caught that style of
Scrope which gives you a sort of map-picture of a country, a
half panorama taken from high points. A view, for example,
from a hill looking down upon the Laacher See, and taking in
the whole lake, and the hills surrounding it, and part of the
adjoining country. Never having heard of that beautiful
map which Von Oeynhausen lent me, they lost much time in
constructing a map, which Hibbert did trigonometrically.
They have really collected both rocks and shells, and worked
and sketched to great effect. In the 'loess' descending
from Kruft to Andernach, they found a vast number of land
shells, which seem to indicate a modern date to that formation. About nine o'clock Dr. Greville [2] came in, and insisted
on my breakfasting next morning, promising as a bribe that
he would get ready, if I came, a drawing of the new volcano
in the Mediterranean, copied from one an officer has sent
him, as a present for the 'Sky Parlour' [3] here. At ten
o'clock I went to Mr. Thomson's—a capital supper—grouse,
&c., and all good things, and the guests just sitting down as
I entered—Mr. Maitland, Professor Napier, [4] Professor Pillans, [5]
and one or two more. Pillans has been with a friend along
our Angus coast, geologising in an amateur way for his
health. Next morning early I went to call on Maclaren,
that furious radical as Editor of 'Scotsman,' and a mild,
amiable, simple-minded man in private, and a great reader
and lover of geology. Greville's drawing of the volcano was
quite beautiful—his talent of rapid drawing and colouring
is truly enviable. Hibbert was there, and quite delighted
with a piece of antiquarian geology which T. Thomson had
given me at supper. He has fished out an old Latin legend,
I believe in Pinkerton's 'Lives of the Saints,' of St. Niniano
(or St. Ringen in corrupt English), evidently founded on those
impressions of the feet of animals which have recently made
such a noise in the geological world, preserved in the Dum-

---

[2] Botanist, author of *Cryptogamic Flora*, &c.
[3] Museum at Kinnordy.
[4] Macvey Napier, editor of *Edinburgh Review*.
[5] Professor of Humanity Chair in Edinburgh.

friesshire red sandstone. The saint had collected all the cattle of that country, to sprinkle them with holy water and bless them. A thief thought it a good opportunity to steal a few, but a bull ran at him and kicked him dead, and the saint softened the rocks, so that like wax they received every impression of the furious animal's hoofs as he plunged about, and then they were hardened in attestation of the miracle. Hibbert offered to show me a gun of the Spanish Armada, encrusted with shells, fished up from the sea, and other antiquarian remains, bearing on a chapter I was then writing, canoes, &c., so we went to see them. Macbean showed me Williams' original paintings of scenes in Greece. What a glorious country their poets had to write about, as far as natural scenery was concerned! He was a fine artist. When we arrived at Bonaly, the Solicitor-General was waiting, ready for a walk; showed me the bowling-green, then to the terrace, which I admired much, though the distant view was lost in the haze The white pony came to receive his caresses. At the end of the walk I was shown the statues of Robert Bruce and another, ' which the mason had painted white, and Horner cleaned by whole bottles of acid.' After ascending to the top of Graham's Dell, he showed me the marsh surrounded by rocks where the Covenanters used to preach, then over the height, and down to the *Linn* by the single elm tree. ' Here I have seen Frances [6] stand with an umbrella, which keeps off the water longer than you would think, picking wild flowers.' Then we went over Mushroom Hill, where we filled our pockets with mushrooms, and back by Torphin. He talked to me much of Brougham, how treacherous as a friend, and how gigantic his power as a servant of the public. He has cleared off nearly all arrears, and if he does what he proposes here (all the forty appeals, Scotch), he will do what has not been done for centuries. I talked to him of the state of France, when I was there last year, their new peerage bill; of Spain, of which he asked me much; of Buckland, with whom at Craig Crook he was much delighted; of Etna, and I made him understand the difference between the volcanic and other rocks of the Pentlands in the dell. This led him to talk of Playfair, and he cited an

---

[6] Miss Horner's next sister, now Lady Bunbury.

elegant passage of his about Hutton. On these and fifty other subjects did we enlarge, and I think exchanged more ideas than I have often done with men with whom I have been acquainted for years. There is a delightful simplicity of character, an entire absence of ceremony, without any unpleasant familiarity, and without inviting it, a disposition to be amused and informed, and a wish to be thoroughly idle, when he is not working, that is delightful in him; and I observed that he could in an instant, on his return, sit down to severe work, for he has a dreadfully difficult case about patents submitted to his arbitration, which takes up much time. This decision, whatever it be, must ruin one or two very rich men, for the law is in a state of disgusting confusion and ambiguity. Mrs. Cockburn had returned from Edinburgh, where she had been spending her morning with her sister. I thought her very kind, and what is no small merit, always looking so happy and cheerful. On my return to the drawing-room, I found a carriage arrived with Napier and T. Thomson. We sallied out again for a walk in the garden; Cockburn with Thomson, and I with Napier. From the latter I learnt many particulars respecting Playfair's style of composing and amending his MS., which amused me, and some good anecdotes of Dugald Stewart, of whose style of writing I have always been a great admirer. At dinner, the principal fun was a general attack on Napier, for having been shooting three whole days, and not having yet taken out a license. At first Thomson told him with such gravity how concerned he was that an information had been laid against him, that he was literally alarmed, and began to guess who had done it, or instigated the informer. When he began to hope it was a joke, Cockburn said, 'How can you expound the law to a class after thus violating it? Why the very act is an implied forfeiture of your chair!' Cockburn told us of a curious action which was brought against the Siamese twins, who had been out shooting, and had returned blows on being assaulted by a Colonel. This led me to talk of what I saw of these youths, and when I mentioned how freely they conversed with others, and yet never to one another, he asked me why? 'I suppose they are each acquainted with so precisely the same facts, that they have no

ideas to exchange. Do you not remark that people of the
same family, who always live together, sometimes sit silent,
although the best of friends, and yet are talkative to
strangers?' 'Yes,' he replied, 'and I have often thought
that scarcely any companionship, scarcely any conviviality
is so delightful, as that same *social silence.*' He laid a deep
emphasis on these last words, which brought forcibly to my
recollection many an occasion when I had felt the same,
and when I would have preferred to walk or sit in a room
with one I loved without speaking, to any conversation with
an indifferent person. Cockburn asked me about the Zoological Gardens in town, and on my describing the prehensile
tails of the 'spider monkeys,' declared he had often thought
that if men had had tails, what volumes would have been
written on the peculiar beauty of such an appendage to the
human form, and of its utility, &c. Then he went on with
speculations on the sort of caudal ornament he should like
for himself. He put Napier on the defence of the last No.
of the 'Edinburgh Review,' protesting against it, as the
dullest, heaviest, and most unreadable one ever published;
but all in joke, and the editor took it well, and really fought
a famous battle—much better than about the license. It
appears that Sir William Hamilton wrote that stirring article
about Oxford. In short we had a most agreeable merry
party, and when I went away Cockburn made me promise to
call on him the moment I returned to Edinburgh.

I left Edinburgh for Kinnordy, and met Marianne and
Caroline at Glamis, and in an hour afterwards was at
Kinnordy, having spent only three days most agreeably, and
you will allow not idly at Edinburgh. The drive from
Kinross to Perth through Glen Farg is beautiful, and not
having been able to get an inside place, I enjoyed a seat on
the outside on a sunny day very much. I found the place
here in great beauty, the more so after the uncommonly
sunny summer which they have enjoyed, and which has
still left all things green, but with a warm look upon them.
I like the change in the garden very much, the turning part
of the flower borders into the English style, and a magnificent rank and file of dahlias now in full blow.

September 9.—I must return once again to Bonaly. I

forgot to say, that when we rejoined the ladies after dinner, there was an animated conversation about Dugald Stewart's monument, which is now rising on the Calton Hill, and I think promises to be beautiful. Turner, R.A., when in Edinburgh this year, augured ill of its effect, but I foretell it will be no failure. Cockburn wishes to have Dugald Stewart's bones removed to the spot. Some one (Napier I think) declared that the relatives would not have them taken from consecrated to unconsecrated ground, but Thomson denied that they had any objection. ' It is not for the sake of the bones being really there,' said Cockburn, ' but if people could be made to fancy it, their associations would be stronger of the monument belonging to the man. Well, I will grant you, if you please, that this is mere illusion, and what is more, one which I do not require to aid my feelings, but others will; and a procession would not be enough, unless the bones were carried there, or some of the bones.'

Kinnordy, September 12, 1831.—After Hall's arrival (the day after I came here), I set to work as of old in the ' India paper room,' a most cheerful drawing-room and a fine height. The paper, covered with birds, &c., was made by order for the room, and is not faded in thirty-five years. Walked with Marianne with baskets for mushrooms, which we filled, so that they were heavy to carry to the hill above Kirriemuir. A glorious view and day.

Letters from Donald Ogilvie (Lord Airlie's brother) canvassing us for the county, Maule, M.P., having become Baron Panmure. Others from Hon. D. Halliburton (brother of Lord Aboyne) who opposes on Maule's interests. I had risen early to secure two hours before breakfast, after which went to the Episcopal chapel with a large party. I then rode to Cortachie Castle, and apologised to Lord Airlie at never having acknowledged the letter of his agent announcing my nomination as Deputy-Lieutenant of the county.

I am getting on very steadily with my work, and really believe as well as in town ; which I like to believe, as nothing can be more cheerful than the whole party here at present, my mother being in very tolerable health, and all the rest as well and happy as the day is long.

September 15, 1831.—Cockburn mentioned to me in

my walk with him at Bonaly, that the best thing Playfair
ever wrote in point of style, was a controversial pamphlet on
Dr. Reid's eligibility to a Professorship at Edinburgh, notwithstanding certain alleged unorthodox opinions which the
ministers of the Kirk charged him with entertaining. As it
is precisely the kind of topic which I have been obliged to
touch upon in my history of the progress of geology, and
may have again, I should much wish to read it, but cannot
buy it, as it is out of print. I will try and think of any work
on Switzerland or the Alps for you to read. I wish you could
get up the geography of the country of the Alps and neighbourhood, which E. de Beaumont has made the foundation
of his ' successive elevation theory.' I must take things in
their turn, and shall be unable to think of our tour much for
some time. As to geology having *half* of my heart, I hope
I shall be able to give my *whole* soul to it, with that enthusiasm by which alone any advance can be made in any
science, or indeed in any profession.

September 17.—On Thursday the 14th, by way of a lark
(to use a slang term), we went, seven of us, with a wheelbarrow to the hill of Kerry, and filled it entirely full of mushrooms, besides five baskets full. Much joking about setting
up a ketchup manufactory, and chalking the walls of Kerry
with ' Lyell's Matchless Ketchup.' To-day walked with
Marianne to seek for ferns and other plants, which Stokes
wants for his herbarium. Dined at Baldovie: went in the
open carriage, and returned at night by a fine moonlight.
Laing Mason and several other lairds at dinner.

CHAPTER XIV.

SEPTEMBER–DECEMBER 1831.

NEW VOLCANIC ISLAND IN THE MEDITERRANEAN—LIFE AT KINNORDY WITH HIS FAMILY—ELECTION EXCITEMENT—MURRAY OF SIMPRIM—RETURNS TO LONDON—DINNER AT THE LINNÆAN CLUB—CUMMING THE SAIL-MAKER —GEOLOGICAL SOCIETY MEETING—DUKE OF SUSSEX AT THE ROYAL SOCIETY—VISIT TO JERMYNS—THINKS OF GIVING UP HIS PROFESSORSHIP AT KING'S COLLEGE.

JOURNAL.

Continuation of Journal to Miss Horner.

Kinnordy : September 20, 1831.—A frank brought me a Malta gazette, containing an excellent account of Graham's Island, the new volcano in the Mediterranean. The island is 160 feet high, a mile in circumference, and twenty-five miles from Sciacca in Sicily. It has been landed upon, but is expected to be washed away in time by the sea, which has already made perpendicular cliffs since the eruption. One fragment of limestone has been found on it unburnt, thrown up from below. The submarine eruption is still going on with great violence.

After my usual morning's work on Monday, I went to see Fanny's [1] bee-hive. Yesterday she tried the stratagem of taking away the honey, and then uniting the bees with those of another hive, instead of killing them. Strange to say, it is believed that in consequence of the additional heat generated by a double number of bees, they eat so much less ; that the consumption of the two hives when united is, during winter, no more than that of one when separate. One hive is put under the other in the night, and the bees go up to the higher one, which is full of honey. When the swarm there

---

[1] His sister.

z 2

is sleeping, and during the night, they acquire the same smell, and in the morning are recognised as part of the same fraternity. Nearly all entered; but five or six dozen did not get in, but tried to enter next day, when I saw them. They were detected as strangers, and much fighting ensued, and lots of dead bodies were cast out of the entrance, until nearly all who had been excluded, and probably some of the others, were killed, as also one queen bee, for they never suffer two in any hive.

One of the great amusements here this summer seems to have been to watch the bees defending the entrance of their hive against a tinea, a species of moth of the same genus, and nearly the same size as the clothes moth, which happened to swarm by thousands every night. The guard at the door was trebled while the pest lasted, and numbers were employed in dragging out the moths which entered to plunder the honey. Tender as these tinea appear, they rarely are killed. The wasps are great depredators, and swarm also; they are usually pulled out by five bees, but are scarcely ever killed. Caroline says, 'A wasp looks so like a thief hustled out by a party of constables!' I afterwards rode with Eleanor a most beautiful ride, across the Prosen, and a ford of the Esk, to Cortachie Castle, and then to Downie Park.

Sir James and Lady Ramsay came early to dinner. Ramsay wanted me to go to Banff, but as I had declined Captain Farquharson's tempting invitation to go with him to Invercauld, I could fight off with a good grace. They always make me a speech here about 'knowing how fully my time is employed,' &c., which would annoy some people who like to be thought to produce books without any effort; but as I have no such vanity, I am content to be thought privileged to do as I like. I took Ramsay to the top of the Tower,[2] and to see the Canoe:[3] afterwards walked with Caroline and Eleanor[4] to a 'den' on the river Carity, to collect ferns for

[2] In the garden at Kinnordy, containing a museum of curiosities.

[3] An ancient canoe taken out of the marl in the Loch of Kinnordy, in the summer of 1820. See *Transactions of the Geological Society*, 2nd series, vol. ii. plate x.

[4] His sisters.

Stokes, and got five species on the rocks. My door is open, and I hear Lizzy,[5] who is the best performer, playing ' the Bonnie House of Airlie,' of which she has a beautiful set. The two parrots which Tom brought from Africa are chatting away in the passage, but neither of them are given to screaming now, as they were apt to do at first. To-day the Mudies of Pitmuir, cousins of ours, but not very near, came to make a call from six miles the other side of Forfar.

September 24.—I went a long ride with Eleanor to Balentore, went over the shooting quarter, which is a suite of ' laird's rooms ' attached to those of the farmer. We went hunting a little for ferns and lycopodiaceæ, and with tolerable success, and returned by a different road. Balentore is our northernmost farm, fairly in the Grampians, and among the grouse.

If you are not frightened by De la Bêche, I think you are in a fair way to be a geologist; though it is in the field only that a person can really get to like the stiff part of it. Not that there really is anything in it that is not very easy, when put into plainer language than scientific writers choose often unnecessarily to employ.

My father, Maria,[6] and Fanny went to make a long round of calls in the open carriage, to Tealing—Mr. and Mrs. Fotheringham Scrymgeour; Sir John Ogilvy's, Baldoven House, near Dundee; Miss Graham of Balmure, &c. Sir John's family was in times of old, as tradition says, a branch of the Airlie stock before they were titled, and they (Sir J.'s ancestors) possessed the feudal castle of Inverquharity, which now belongs to my father, about four miles from here. It is in the valley of the S. Esk, and by far the best wooded and most beautiful part of our property. Although much fine timber has been felled there in the last twenty years, it would stand not a little cutting now. The river and park scenery about that place is so very superior to anything near Kirriemuir, that many think Kinnordy should have been built there. I rode with Eleanor and Maria to Pearsie, a beautiful highland place, and most comfortable house. When a boy, just before I went to Oxford, I thought it a perfect Paradise. When I visited it at that time it was a Mr. Wedderburn's, and Mrs.

[5] [6] His sisters.

Somerville, when Miss Fairfax, used to stay there, and as a girl came over from it to dine at Kinnordy, in my grandfather's time.

September 27.—To-day I went with Eleanor a long walk up the hills behind us on the north, and got two more plants for Stokes on the Clune and Castle Hills, where we started a great number of blackcock. On my return I found Dr. Fleming just arrived. He showed a strong interest about King's College, and I told him that as I had received my nomination in a manner that was highly complimentary, I should do my best for a short course. Upon this he said, ' If you lecture once a year for a short course, I am sure you will derive advantage from it. A short practice of lecturing is a rehearsal of what you may afterwards publish, and teaches you by the contact with pupils how to instruct, and in what you are obscure. A little of this will improve your power perhaps as an author. Then, as you are pursuing a path of original and purely independent discovery and observation, it increases much your public usefulness in a science so unavoidably controversial to have thrown over you the *moral protection* of being in a public and responsible situation, connected with a body like King's College. But then you must stipulate that you are to be free to travel, and must only be bound to give one short course annually.'

I perceive you are very busy and studious, which gives me great pleasure. Though I was born with a taste for natural history, I find that if I take up a new branch— botany for example, or conchology—I am obliged to do very little at a time, or I get a temporary distaste. It will never do to force these things, either on ourselves or others. Dr. Fleming has been spending much time in the Sky Parlour, selecting duplicate insects, some of them handsome things, which Eleanor brought from Jermyns [7] last spring, and which are not Scotch. He has no collector's box with him, which he says proves he came with honest intentions ; but he was greatly delighted to hear that ——, when here, had the inside of the crown of his hat *lined with cork !*

Kinnordy: September 30, 1831. — The papers of this

[7] His uncle Captain Heathcote's place, near Romsey, Hants.

morning were full of the dreadful hurricane of Barbadoes, in which 3,000 persons at least are said to have perished, and the whole island to be now a desert. This morning Dr. Fleming went away in spite of us, in our open carriage to Forfar, at a quarter before six o'clock, to meet the coach in a hard rain, which has continued all yesterday till to-day. I think I omitted yesterday to mention a letter from Tom to Eleanor from Portsmouth. 'We are under weigh for Cork, where we are to rejoin Admiral Codrington's fleet.' Sophia and Lizzy have discovered that an election is a good thing, as there is to be a ball on the 10th (a county ball) at Forfar, which has grown out of it. As we could not go out for the rain, we had a good game of battledore and shuttlecock, with some famous rackets which I brought from Paris last year.

October 3.—My father went to the election to-day at Forfar, and I have done a great day's work, nearly finishing my fourth chapter. I walked with Eleanor, and got a few ferns, and in the garden for apples. *Eight o'clock.*—Messenger arrived with the news that the Tories have triumphed, and Colonel Ogilvy is M.P. Sophia tells me it is quite right that Colonel Ogilvy should be member, because he will conduct the Forfar ball with more spirit.

October 4.—Yesterday I took a walk with Maria to Kirriemuir quarry, the largest in this country, one of my father's, which, although not a very valuable possession, is really a fine object. Lord Airlie called to-day in his carriage and four, and Captain William in his carriage, when I was out, a sort of formal way of paying a visit of thanks. But his manner was anything but formal, and he thanked my father as warmly as if he had put his brother in. When Eleanor and I called at a cottier's to-day, an old woman said, 'It has been sair work to get on, they Parliamenters,' alluding to the 14,000*l.* spent last year by the Airlies in trying to get into the borough. But this county election has not I suppose cost 300*l.* My brother Harry was to have been on the roll this year, but I am glad my father has stopped it, as certainly it would be throwing away money, for a reform in the Scotch system cannot be far off. I went to call on my old nurse Nanny (or Mrs. Yeates) to-day, who is not well. She talks away with the familiarity of one of the family, in which she takes a real interest.

October 8.—Mr. Blackadder the surveyor came from Glamis to consult about a map of the county, which he is to construct for a joint memoir on the Geology of Forfarshire, for which we have been preparing materials for years, to be published by the Geological Society.

Sunday.—Walked with Eleanor to chapel, where the carriage also went full. The principal talk was about a dire event, no less than the putting off of the county ball, which was to have been Sophy's and Lizzy's first Forfar *début.* And why? Because Mrs. General Hunter, the directress, had heard that the Forfarians were resolved to break every carriage to pieces of the Tories, if on Monday the news should arrive, as they anticipated, of the bill being thrown out by the Lords! Dresses and all for nothing! and the Miss Drummonds staying at Cortachie for nothing else, and a large party at Logie for the same. So Colonel Ogilvy, the director, was obliged to send round notice only two days before, that the ball was put off.

October 9.[8]—I hope the returns of this day, as long as my life lasts, will very rarely see us separate; never, I would willingly say, if I thought all my wishes could be realised.

I rode alone to Glamis, to choose some plants (fossil) which Blackadder's wife and daughter had collected, they having taken a fancy to fossilise. He rode back with me as far as Kirriemuir. Met Wedderburn Ogilvy of Ruthven, who rallied me on my republican principles—but in great good humour he said, 'As you are looking better than last year, I presume you have not been fagging so hard.' When I returned I found that my father, Fanny, and Sophy were gone to dine at Fotheringham, and to sleep.

Thursday.—The Fotheringham party returned. They met Sir Francis Walker, Lady Drummond of Hawthornden and daughters, Mr. Fotheringham Scrymgeour, Lord Airlie, &c. It was just an Angus set-to of the old *régime.* They arrived at half-past six o'clock, and waited dinner one hour. Gentlemen rejoined the ladies at half past twelve o'clock! They in the mean time had had tea, and a regular supper laid out

[8] Miss Horner's birthday.

in the drawing-room. After an hour with the ladies, they returned to the dining-room to supper at half-past one o'clock, and my father left them all at it at half-past two o'clock!

The ladies did not go to *this* supper. Cockburn remarked that Forfarshire was the last stronghold of those Scottish habits being maintained, but it is fast wearing out. I forgot to say the other day at Cortachie that the conversation turned on a trial now pending at Perth on Lord Mar, for shooting at a man with intent to kill and wounding the horse. Lord Airlie said, 'I was very lucky once. An old man and a youngster came to fish at Airlie Castle, and I warned them off in the morning, so they went; but when I came back in the evening I found them again, and broke the old man's rod, and on the other's making off, shot at him when 90 yards off, and peppered him, so that he was obliged to have the shots extracted. Fortunately for me he did not prosecute.' Erskine of Linlathen did not vote for Colonel Ogilvy, because he will not take any oaths. He (Erskine) has built a house 220 feet long in front, they say the largest in Angus.

Our parrots have acquired a nice accomplishment, that of imitating the squeak of two doors as they move on their hinges, and of a wheelbarrow of which the wheel squeaks. Whenever they see the passage door move, they set up this concert, which I am now enjoying, and when they see the gardener's man, we hear the barrow performed. This has replaced the dog whistle, with which they used to salute Tom. All the sisterhood (five of them) went this morning to Dundee, with a beautiful basket of ladies' work, which I hope will bring 8*l.* or more for Lady Ramsay's table. They had the large carriage, with a pair of horses besides our own, a nineteen miles' stage.

October 15, 1831.—Our five lassies returned from Dundee. They sold their ladies' work at high prices, and it seems to have been an amusing meeting. Many gentlemen, to show their gallantry, were buying at the tables of ladies of their acquaintance. As they came back through our village of Kerry at night, a mob of about 40 reformers *hissed*. They put down the windows lest stones should be thrown, but this it seems was not intended. They think that as they had four horses, it was taken for Lord Airlie's carriage, but perhaps it was in honour of my father's Toryism.

October 24.—The Murrays of Simprim arrived. Having a horror of the habits of a considerable set of the Angus lairds, Murray voted himself a Perthshire man, and would have nothing to say to them. Indeed, had he been in my father's place, I am persuaded that, like him, he would have emigrated. But Mrs. M.'s great friend, Mrs. Farquharson of Invercauld, persuaded her to visit her namesake of Baldovie, and there I met them some seven years ago. I soon found in Murray, and he in me, a congenial spirit, and he was always asking me to Meigle, about ten miles from here, in the direction of Perth, where his place is. I went occasionally, and found him an original character, with a great love and respect for science, and who had all his life cultivated literary society; full of anecdote, and who had the art both in the country and when in town of filling his house with an agreeable society. He used to go occasionally to the Royal and Geological Societies with me, and at last they called at Kinnordy, in spite of the rule they had laid down, and continue to visit here. Our party was Mr. and Mrs. M. and their youngest daughter, about twenty years or twenty-one; Mrs. Kinloch and her sister Lady Ogilvy (the dowager); Annie Kinloch; John K. (the laird of Logie); and Mr. James Ogilvy. Murray told me that his cousin, Sir Alexander Mackenzie, had refused to read my book when he, M., bored him so to do; but lately he fell in with it somewhere by accident, and has been most active recommending it to all his friends.

Kinnordy: October 25, 1831.—I see by the paper that Leslie, Brewster, Ivory, Herschel, and Charles Bell are knighted —Guelphic Order, I believe. I begin to think that all our acquaintances will soon be dubbed Knight-Bachelor.

Edinburgh: 27.—I had an inside place in the coach, and read with much pleasure and profit, Playfair's Dissertation on the Progress of Physical Science in supplement to the ' Encyclopædia Britannica,' which Mr. Maitland procured for me when I was last here. I think I can give a new definition of geology modelled on Playfair's definition of Physical

Astronomy, which I will give you to read to your father, as he thought my former definition open to criticism, and I shall be glad to know if he likes this. ' Geology is a science arising out of the comparison of the phenomena of *past* change, observed on the surface and interior of the earth's crust, with the laws of chan ge*now* regulating the terrestrial creation, whether animate or inanimate.' It is perhaps too abstract for general readers, yet may be useful in the body of the book. A definition, I suppose you know, is of all things the most difficult, and must in the present state of geology involve a theory.

Saturday.—I shall write a few words before I get into the steamboat, just to tranquillise my mind a little, after reading several controversial articles by Elie de Beaumont and others against my system. If I find myself growing too warm or annoyed at such hostile demonstrations, I shall always retreat to you. You will be my harbour of peace to retire to, and where I may forget the storm. I know that by persevering steadily I shall some years hence stand very differently from where I now am in science ; and my only danger is the being impatient, and tempted to waste my time in petty controversies and quarrels about the priority of discovery of this or that fact or theory. I am glad to see that E. de Beaumont has reiterated in more explicit terms all those general views which I expect to overthrow, intoxicated, perhaps, by Sedgwick's eulogy, who was carried away by the novelty and dashing assertions of E. de Beaumont.

Monday.—I have been writing a few sentences for my work, and reading ' Don Quixote,' which is provided by the librarian here. We shall have a third night of it in the river, which we must not complain of, as we have had still water the whole way. I am convinced that if I can but keep my mind moderately fixed to the publication of my volume, that I shall outdo expectation in this volume ; but as the part I am now about is of necessity, from the present state of the science, exceedingly controversial, and as I am contending at once seven or eight distinct points with no less persons than Sedgwick, Conybeare, E. de Beaumont, and three or four others, you may well suppose how very carefully I must weigh every word and opinion. I see by the new articles just out that they are beginning to differ from one another, as well as from me, and I shall make new schisms among them by this volume. I shall be as careful as possible not to controvert uncivilly.

London : November 2, 1831.—I have most to tell you when in town, but least time to spare for writing. I fled from the confusion of unpacking, to the dinner of the Linnæan Club, which is always amusing. In a few minutes I forgot I was just off the sea, and heard all the news. Old Lambert, Prince Cimitelli, old Caley, and myself were all that sat down; an unusually thin meeting, but a merry one. Adjourned to the Society's rooms, and heard a learned botanical paper by Robert Brown, on the fructification of the Orchidaceæ. I was glad to meet my friend Dr. Hooker the botanist, also Burchell the African traveller, whom I had not seen since his return from the Brazils. Bicheno, Wallich, Curtis, all had some news to tell me ; a week's reading of reviews and magazines in Angus, or at the Manse of Flisk,[9] would have given me less information about what is now going on, and *coming out,* in sciences connected with my department. And then, the excitement is so useful ; many inquiries what precise *page* I had arrived at in print. I was glad to see Curtis the entomologist, and to compliment the meritorious but ill-requited author of the most beautiful and scientific work on insects ever published in England, on the merits of his last number, which I had found on my table. The great news of the day among the *collectors,* was the arrival of the long expected *sail-maker,* Cuming, a man who, having made a little money in trade in Chili, some ten years ago, and fished up a few new shells, which brought in in London six or eight guineas each, took it into his head to build a small sloop, and sail along the American coast and isles of the Pacific, dredging for shells and corals, and observing their habits, as also crustacea and other classes, till he filled four hundred chests of things, numbers of them quite new.

On my return home, I found my chum Broderip glad to see me. After a morning's work, went with Broderip to the sail-maker's, and found not a few eagles gathered together around the carcass—Lambert, Gray, Children, &c. His

[9] Dr. Fleming's Manse in Fife.

treasures only half unpacked, but much that was interesting. Dinner at Club of Geological Society, as usual an agreeable meeting. Murchison in the chair, agreed to meet on Friday, and tell me lots of news. Greenough, Broderip, Pringle, Stokes, Basil Hall, and strangers. Basil Hall brought his newly purchased autograph of the 'Antiquary,' by Sir Walter Scott, and two pages of preface lately written by Scott at Portsmouth, in his own handwriting, giving his reasons to Basil Hall why he liked the ' Antiquary ' better than all his novels. A paper of Dr. Turnbull, Christie's, on Sicily, led Murchison to call me up, and I had the field much to myself. I told you that a temporary cloud came over me in Edinburgh at seeing the controversial storm gathering against me in the horizon; but I must say it was dispelled by this meeting, for never on the first meeting, after my first volume appeared, did I hear it so much spoken of. Fitton declared to the meeting his conviction that my theory of earthquakes would ultimately prevail. Greenough made one of his ultra-sceptical speeches, saying that once he only doubted what ' stratification ' meant, but now he was quite at a loss to conceive the meaning of the term ' sea ' in geology, and he thought no one could explain what ' the sea ' meant. Fitton gave him some hard hits for thus fighting shadows, and ' assisting the cause of darkness ' by doubting elementary truths. Basil Hall, with more humour, said as a sailor, that he felt alarmed at hearing the existence of a great mass of waters (for he would no longer give offence by talking of ' the sea ') called in question, &c. He then eulogised my book, and after the meeting two American gentlemen came up to be introduced, and poured out a most flattering comment on its popularity in the United States ; but what was much more to the purpose than this sort of incense, they gave me many facts bearing on my theories respecting part of North America. I did not get back till near one, and then a great fire in Holborn, which was too grand a sight to lose, detained me three-quarters of an hour.

2 *Raymond Buildings, Gray's Inn : November* 5, 1831.— After a good morning's work, I went with Lonsdale [1] to see

_____
[1] William Lonsdale, b. 1794, d. 1861, assistant secretary and curator of the Geological Society. ' To no one man has our body been more indebted,

the panoramas of Quebec and Bombay, and instead of re-creation got into a long argument with him. He took up the cudgels in favour of Elie de Beaumont and Sedgwick, but I made him abandon some points and doubt on all the rest. I returned to work, but after this went by invitation to drink tea with the Murchisons, and was thus kept to geology till half-past eleven o'clock at night, so that after my work this morning I found my head ache, and thought to relieve myself by going up to Dr. Horsfield,[2] who lives over me, to see his insects, which he studies every Saturday. Here I was excited again, for the Doctor was just re-turned from Leyden, and told me that he found Reinward (Professor of Zoology, I believe) just publishing a grand work in Dutch on the East Indian Archipelago, with a geological introduction, in which he declares his entire adherence to all my ' Principles,' and illustrates them by facts of recent elevation, &c., such as, Horsfield says, will startle all but me. Now you must know that Horsfield is one of those heretics of whose conversion I never had any hope, so I was amused at the pains he had taken to shake Reinward from the true faith, but all in vain. As Horsfield has a prodigious opinion of Reinward, and as he has that natural turn for philosophical pursuits that renders a person somewhat restless when in a state of doubt as to the explan-ation of grand phenomena, he has been reading Leonhard and other works to strengthen himself in his old faith, and declared himself still an unbeliever, though he wants amaz-ingly to read my second volume. You will be glad to hear that your knowledge of Dutch will be so useful to me by-and-by. Mrs. Murchison was in high spirits, and has evidently enjoyed their tour,[3] and above all the York meeting. Her

---

whether for his publications, his conduct of our affairs, or the zealous and disinterested labour he bestowed in aiding and improving the works of his associates.'—Leonard Horner, *Geological Society's Proceedings.*

[2] Dr. Thomas Horsfield, F.R.S., b. 1773, d. 1859. For some years keeper of the Museum of the East India Company. His contributions to entomology and other branches of zoology are well known.

[3] During this expedition Mr. Murchison discovered that the Upper Silurian rocks (then called Upper Grauwacke) were stratigraphically conformable with the lowest members of the Old Red Sandstone formation, and that its dif-ferent beds could be recognised from their containing peculiar fossils. Mur-chison here laid the foundation for his great work on the Silurian system.

real delight in conchology is of great use to her in taking away the tedium of long sojourns in out-of-the-way places, and she had a fair opportunity of collecting. They have found *recent* Irish Channel shells in abundance, 300 feet high on the Lancashire coast! I am delighted, because the identity of the living species of plants and animals in England and Ireland made me argue that some great changes in the distribution of land and sea had taken place in that part within the modern era. It is clear to me that the muster of nobles, dignitaries of the Church, and scientific men at York, must have done great good. A hundred and fifty ladies, and many of rank, at the evening discussions, must also have ' popularised ' scientific pursuits. Lord Morpeth seems to have acquitted himself with great ability. Murchison made a capital cruise in the geological way. It is cheering to see how unabated his ardour is.

On his hearing from me how much your father was geologising, he has determined to write to him some hints about the geology of Düsseldorf, Cologne, &c., and to point out some things to be cleared up. The satisfaction he expressed at hearing that your father was so employed, amused me. He said very truly that a succinct account of what the Germans have done about that country would be a most acceptable boon to us here.

November 12.—This morning I wrote to Whewell, to ask him if he would review me, as Lockhart wishes, in the ' Quarterly.' Murchison has read the first three chapters, and liked them much. Necker of Geneva, Murchison, and Lonsdale breakfasted here. Necker expressed great regard for your father, and inquired about you all. I wish much that we may be able to pay a visit to him (in Switzerland), and I think he would put me in the best way of seeing parts of the Alps, where he thinks Elie de Beaumont's grand theory may be put to the test, and where he thinks it fails most entirely. Chamouni, moreover, is the grandest thing in Europe, and I should like you to see that.

King's College is in good odour. My friend Dodd, author of ' Autumn on the Rhine,' who declined a chair of English literature as unprofessional, he being a barrister in practice, is trying to get a chair of common law made. I want the accession of a man like Dodd, who would strengthen the gentlemanly tone of the thing, as that is of more importance even than talent, to success.

Lord Lansdowne told Murchison yesterday that he had spelt my first volume from beginning to end in his late tour.

November 17.—We had yesterday an excellent meeting, both at the Club of the Geological Society and at Somerset House. A paper of Count Montlosier on Somma and Vesuvius excited much interest; we had a famous discussion. Buckland, Greenough, Fitton, De la Bêche, and I began, but Necker, on being invited, spoke half an hour most capitally. I never heard a better lecture. Otter, on seeing me at King's College yesterday, when I looked in after Geological Council, said, ' I am glad to see that you show your face here sometimes.' He looks very ill; he is probably too old for such a work as starting an university, though all has gone on quietly. There was a discussion at our end of the table about the costume of the King's College *students*, some saying that they ought not to have adopted the gown and cap, and others defending. As for the Professors, they all wear their own *gowns*, that is to say, according to the academical degree they may happen to have. Thus I shall have my Oxford Master of Arts gown; some have their Bachelor of Divinity; in short, the robes are as varied as at College of old.

Mrs. Somerville is returned. I have not seen her. Wollaston says she is looking ill, and attributes it to her anxiety about Greig,[4] who is recovering from a dangerous illness. Buckland is, I think, in pretty good spirits, though certainly very gloomy at times, and croaking about the state of the country. So are Stokes, Broderip, and many others. A feeling of insecurity of property has much lessened the demand for labour. Lockhart says that an additional stagnation of six months would make every bookseller bankrupt, except Longman and Murray; but this, I suppose, is a little too strong. As for public affairs, I have long left off troubling myself about them, as knowing that one engaged in scientific pursuits has as little to do with them in point of influencing

[4] Woronzow Greig, her son by her former marriage.

their career, as with the government of the hurricanes or
earthly motion; and if one becomes annoyed, there is an
end of steady work. Even Whewell is frightened about the
Reform Bill.

London : November 20.—We had a most agreeable muster
at Stokes's ordinary this morning, at which Buckland,
Brown, Basil Hall, Broderip and I were at breakfast. All
of a sudden I took it into my head to go and dine at Chelsea,
if I found Mrs. Somerville well enough. Seeing her book on
the table, led me to think of this, which I had before resolved not to do, till I knew she was better. Found Stokes
going to Athenæum, who agreed to go on to Chelsea, if I
would stop at the club ten minutes. Crawfurd there fastened
on me, and told me about a book on India, which he is
coming out with, and of a geological chapter he wishes to
write. On our way we looked in to see Chantrey's fine
statue of Canning. Is it not provoking, that just after
Chantrey has built a most splendid and truly elegant house,
Lord Grosvenor's agent has persisted in building just
opposite, in spite of remonstrances, a vile, tall steam-engine
chimney! 'Tis enough to sour any one for the rest of his
days, but Chantrey is luckily not atrabilious. I found Mrs.
Somerville quite glad to get me to dine there, and I spent a
most pleasant evening, and stayed very late. All kinds of
offers from her and the girls, to be useful to me in getting
up my lectures. Mrs. Somerville is looking oldish. I think
the book has done her no good, and when I see it, I am sure
that such a work must have been too great an effort for any
one in the time. It was a gigantic undertaking. Babbage
was offered knighthood, but declined, not as underrating it,
but because he had written on the necessity of titles being
given, &c., and it would look ill if it appeared that he had
done so for personal considerations. Martha and Mary S.
have made a beautiful collection of seaweeds at Bognor.

November 22.—I have been this evening with Necker, who
will, I hope, on his return from Aberdeen, visit Kinnordy.
He has been planning an expedition to Chamouni and the
Vallorsine for us; and as all the points, some of the most
picturesque, as well as geological of the Alps, which I ought
most to see, are the same to which he took his mother, an

oldish woman. last year, it is clear that you can go to them. Talking over Chamouni has revived my admiration for that region, to which all the rest of Europe which I have seen is poor in comparison, for grandeur and sublimity of scenery. Somerville was remarking that De la Bêche's book is unintelligible, and that I must write an introduction, which I told him I intended to do. Anticipating a small class, I have been thinking that the only way to ensure my not letting down my reputation in my lectures is to determine to write so many chapters for an easy introductory volume. You cannot think how much my spirits have risen, so far as the lectures are concerned, since I saw more clearly than formerly how easy they will be. Somerville's wine and coffee, and the excitement of walking home with Stokes, who kindly volunteered entering on what my introductory lecture ought to be upon, made my head ache, and I hardly got any sleep, and am almost good for nothing to-day. Mrs. Somerville begged me to send a copy of her book to Bonn, which is going in the parcel to Von Oeynhausen.

December 2.—I had a hard task yesterday to keep my resolutions of severe temperance, for Murchison was so unwell, that he asked me to preside at the Geological Society club dinner, and take his friend Colin Mackenzie to it; so I was obliged to push about the bottle, and to sham taking a fair quantity of wine. Our numbers were thinned by the Royal Society Anniversary and by the rain, but it went off well, and in the evening brilliantly. A sort of geographical paper, long and dull, and really of no value whatever, was to have been read by my travelling companion Captain Cooke, now again in Spain. We had staved it off till we had read out the others. But when your father's letter to the President came about the new isle north-east of Pantellaria,[5] we determined to shove Cooke's into the coal-hole without ceremony. So when Turner had read some passages, Murchison (for he was recovered enough to take the chair) said to the meeting ' that it would not be doing justice to Captain Cooke to go on, since there were certain maps and illustrations which ought to accompany, not yet arrived : that it was begun, just that it might keep its place, &c. I then

---

[5] Graham Island, visited by M. Hoffmann, the Prussian geologist.

officiated as secretary, Broderip being ill, and read Hoffman's paper,[6] which produced an animated discussion, into which I entered largely. The structure of Pantellaria, the wasting away of the new isle, the par-boiled fish, the floating cinders, the clefts at Sciacca, &c., all afforded subjects for Fitton, myself, De la Bêche, Dr. Babington, Murchison, and others to dilate upon. The President was knocked up by his extreme imprudence. He went to dine with thirty other F.R.S.s at Kensington, with H.R.H.,[7] and came home with Captain Smyth, who seduced him to a smoking bout till three o'clock next morning! punch I believe accompanying. He was half dead at the council. This morning he persuaded me to go in his carriage with him to see the Œningen freshwater tortoise (*achelonia*) which Bell, the dentist, has got over for his museum. After the paper of Hoffman's was over, old Whishaw came up to me and said ' he never heard one more interesting, and he thought Horner had drawn it up with great clearness and judgment.' He and Fitton mentioned having received letters from your father. I told Murchison long ago that I meant to dedicate vol. ii. to him, not only as President of the Geological Society, but because I use our joint and unpublished notes on Auvergne in the volume, and because we had together part of the best and longest geological tour which I ever made, my tour which made me what I am in theoretical geology, or completed my views on that. Scrope came here this morning, full of the frightful state of the country. For my part I am, in regard to politics, in the happy condition of Voltaire's gardener: ' Travailler sans raisonner, c'est le seul moyen d'être heureux.'

December 6, 1831.—A letter from Whewell, expressing his delight at my three hundred pages, and that he had actually begun a review of me for Q. R., excited me to work with spirit; and after two days thinking of the delay which the completing the subject in one volume more would occasion to the getting out the large part already done, I resolved to see if I could get it out as a volume by itself. Told Broderip I should dedicate it to him. He was more overjoyed than I

[6] Drawn up for the Geological Society by Mr. Horner.
[7] The Duke of Sussex, President of the Royal Society.

have seen him many a day, and the day after he declared
that if a legacy of 5,000*l.* had been left him, it would have
given him less pleasure. So I must explain to Murchison
that the third volume is the one he ought to have dedicated
to him, as it is the part in which he is strong, and where his
name will appear.

December 9, 1831.—I am just returned from a very late
sitting of the Royal Society, at which the President's (the
Duke of Sussex) speech was read. It touched on very
delicate ground, his election, and his conception of his duties,
and why he thought himself fit for the office. Yet there is
really not a word that one could wish altered. It raises him
much in our estimation, and I know few public men who in
so delicate and personal an affair could have acquitted themselves so well. His allusion to Herschel and his father was
excellent. 'Sir John Herschel, to whom I had the honour,
for I cannot call it the misfortune, to be opposed.' When I
get the annals or proceedings of R. S. I will send them.
You will read the Duke's speech with interest, if you have
ever heard of the feuds and schisms which our contested
election gave rise to. I was quite amused to find, from conversation with Whewell last night, how much his tone is
altered in regard to my anti-catastrophe system. I am sure
that, when volume three is out, there will be few of the
paroxysmal school left—comparatively few—and they will be
far more moderate.

December 12, 1831.—Dawson Turner was at Athenæum
last night, with whom I had some talk; also two good hours'
confab with Scrope, who has a most active intellectual mind
—alive to everything, politics, political economy, Hume,
Berkeley and Reid's metaphysics, geology, Irving, the
Roweites, and progress of fanaticism. I went this morning
to thank Otter for a volume of Horsley's sermons, one of
which on the discrepancy of Genesis with certain philosophical
discoveries is very striking, and perhaps the most straightforward, manly, and to-the-point declaration that any
eminent divine has made.

December 13.—The Athenæum was very entertaining last
night, so many members of the honourable House coming
there, after the new Reform Bill was moved, and giving their

opinions *pro* and *con.* I went to call on Mrs. Lockhart, who is expecting to lose her eldest boy, poor ' little John,' every day. I made a great blunder in missing the Geographical Society last night, through forgetfulness, where I should have got some sound information, instead of hearing politicians discuss the interminable bill. Young said, ' Scrope, nothing will please you, because you are a philosopher half a century in advance of your age, who wants to Owenise the representation, and divide the constituency of the country into parallelograms.' We all laughed at the sally of the old beau, for it hit my friend hard. You will like Scrope, I am sure, very much. Dined at the Geological Society club. Majendie, the famous French physiologist, sent by the Institute to Sunderland about cholera, dined with us. He complains of the public hospital. Agrees well with the opinion which you sent me about Mittau, that in general the less you do, the more you leave it to nature, the better. A pleasant club. Stokes, Greenough, Buckland, Lord Cole, Broderip, and a few more. Murchison was pheasant-shooting in the country, but cut in for the meeting. A short paper on Springs, and another by Mr. Hutton on the Whin Sill of Yorkshire, drew up Buckland, Greenough, Fitton, Murchison, and De la Bêche; and as they seemed much disposed to go on for ever, Buckland speaking five times, but not once too often, I was glad to sit quiet, having held forth at such length on the new isle, &c., last time. Mrs. Somerville has just sent me the most peremptory mandate I ever received from her, to come there next Sunday. I know not who is to be there. I was sorry to miss the Brahmin Ramahoun Roy last week. Went to Beaufort at Admiralty, to go over a new map, showing modern and ancient coast line of Adriatic.

Sunday.—Went with Stokes and Broderip to look over the splendid collection of South American shells brought home by Cuming.

Jermyns : [8] *December* 24, 1831.—I found them here all quite well yesterday, and Marianne and Caroline looking really quite blooming and happy and pretty. We are just going out for a walk together. I have announced to them my decided wish to get free of the professorial chair, and at the same

8 Near Romsey, Hants.

time recommended my aunt to send cousin Gilbert to King's
College, to prepare him for Cambridge. The examination of
the forty-one mathematical and classical students, and the
various lectures of Hall, Austin, and Otter, have given me
an exceedingly high idea of the general education system at
King's College, and I am certain that when once it is known
the increase will be very great. There is emulation, a
genteel set of youths, most of them going to Oxford and
Cambridge, and three elegant scholars and gentlemen.
Conybeare was at the Royal Society on Thursday. On my
hinting that I should probably get free from King's College
next spring (you remember the move he made towards my
getting the chair), he was almost sore, and said he trusted I
should think better of it, that it was a fine position, &c.
The fact is, Conybeare's notion of these things is what the
English public have not yet come up to; which, if they had,
the geological professorship in London would be a worthy
aim for any man's ambition, whereas it is now one that the
multitude would rather wonder at one's accepting.

Christmas Day.—Marianne, Carry, and I went to Hursley
Church, a long service, and it was a frost, and the church
not over warm. What you say of the Lutheran equality is
certainly a striking contrast to that most aristocratic form
of Christianity which has been established in England.
Gibbon Wakefield says in his pamphlet, amongst a long list
of causes which make the poor feel an aversion to the higher
classes : ' Even in church, where some of them preach that
all are equal, they sit on cushions, boarded, matted, and
sheltered from the wind and the vulgar gaze, whilst the
lower order must put up with a bare bench or a stone floor,
which is good enough for them.' We got out of the
carriage when we were half way back, about two miles, when
we got into fine forest scenery, so we had a delightful walk
on the fine gravel road, and a bright sunny frosty day. Just
as we were going to dine, the children raised a shout of
' Tom, I declare ! ' and a gig drove up. Great was my surprise
to find that it was my own brother Tom, whose ship had
come yesterday from the Downs to Portsmouth, who, finding
himself within three and a half hours' drive of Jermyns, had
just had time to get leave, boat it ashore, hire a gig, and

cut in for our five o'clock Christmas dinner. My aunt was quite
overjoyed, and we had a very merry, cheerful evening. I shall
stay to-morrow and return on Tuesday, and be thoroughly
idle all the time. Eleanor says ' Fanny had a cheerful letter
from Harry. He says he never had been in such good health
since he had been in India. The regiment was stationed at
Secrora, in the province of Oude, one of the very best stations
in India, and he was thinking of not taking his furlough till
they left that, which would make up his twelve years. We
are going a long walk, and Marianne has proposed taking
me on this fine frosty day to see where ' the chalk hills ' join
the sands, &c., which this house is on.

London: December 29, 1831.—A letter from Murchison,
concurring most warmly in the propriety of my cutting
King's College after the first course. 'I look to you as my
successor as President of the Geological Society. I wish
Phillips may become Professor of Geology at King's College.
Touching the book, I quite approve of your dedicating this
part to Broderip, but I shall be not a little mortified if any
delay, whether caused by *devils* or *angels*, should retard the
publication of the third part beyond my consulate, inas-
much as I had in the pride of my heart calculated (after what
you told me) on this dedication, as marking an epoch in my
life, of which I am justly proud, by the distinction of all
others most flattering to me, this tribute from my colleague
and friend.' Another letter was from Whewell, followed the
day after by his review of my book. Though it is not so
spirited in style, nor so rich in thought and original matter
as that in the ' British Critic,' [9] it is very good, enters completely into my conception of the subject, sees the bearing of
the whole on geology, and explains it in a clear way. Sir
Walter Scott's health improved on his voyage to Malta, but
they fêted him there too much, and he has rather relapsed.
Mrs. Lockhart is quite overcome at the loss of little John,
and sees no one.

December 31, 1831.—I have been walking back from
Hampstead on a fine frosty night. I feel quite at home at

---

[9] Review in *British Critic* on the *Principles of Geology*, vol. i., by the
Rev. William Whewell.

24 Church Row.[1]   Mrs. Mallet says, ' I am afraid if you do
not retain your Professor's chair you will not live in London
by-and-by.'   I said it was partly a fear of that, that first
induced me to court some official place, which should oblige
me to be in the metropolis of science, for I dreaded it becoming prudent to live elsewhere, which I knew would be giving
up my career for ever.   I talked with them already over the
pros and *cons* like old friends.   What a cheerful countenance hers is! there is such a happy look, as may gladden any
one who talks with her.

If I could secure a handsome profit in my work, I should
feel more free from all responsibility in cutting my cables at
King's College.   Do not think that my views in regard to
science are taking a money-making, mercantile turn.   What
I want is, to secure the power of commanding *time* to
advance my knowledge and fame, and at the same time to
feel that in so doing I am not abandoning the interests of
my family, and earning something more substantial than
fame.   I am never so happy as when at the end of a week I
feel I have employed every day in a manner that will tell to
the rest of my life, and last week will, I think, be one of
them.

[1] Residence of John Lewis Mallet, Esq.

CHAPTER XV.

JANUARY—MAY 1832.

[In January 1832 the second volume of the 'Principles' was published, and a new edition of the first volume, which was already out of print. In this year he married, and made a tour up the valley of the Rhine, and visited the Valorsine on his way home through Switzerland.

In 1833 he gave lectures at the Royal Institution, and revisited the Rhine and Bavaria.

He received the gold medal from the Royal Society for his work on the 'Principles of Geology.']

JOURNAL.

January 6, 1832.—I am writing in an unaccustomed place, the room in which Babbage's celebrated machine is constructing—Southwark; for having met him, he proposed to me to take a walk here, and has left me while he transacts business. He said, 'I am going to give a course of lectures at Cambridge.' 'How many?' 'Four.' 'I thought you told me it was loss of time.' 'So I did; but I am coming out with a pamphlet on the Political Economy of Machinery, and I mean to discharge it first in a lecture. I would not have gone through the work that these four have cost me, for many times my professional income.' He said that he would advise me to try a course of lectures before I judged of the degree of interruption they would give me—that he

believed the council of King's College would be glad to retain
me, even if I would only agree to give six lectures in the year.
As Fitton strongly recommended me not to give it up so
long as they permit me to do as I like, I incline at present to
give Otter notice by-and-by that I will not lecture next
autumn, and that I shall probably not lecture till after
Easter next year, and must be at liberty then to give as few
lectures as I like. My second volume is to be laid upon the
table of the Geological Society this evening.

I went by appointment to Greenough's, Regent's Park,
to get instructions respecting a recent formation examined
by his friend J. Burton on the shores of the Red Sea. On
my return I called on the Lockharts, and after a confab with
him, I went up to her. She was looking much worn by
anxiety, but I never saw her appear to greater advantage.
She brought down 'baby,' a child of four years old, who
broke her arm the other day—getting on very well. I heard
a long description of the accident. She showed me an ode
of ten lines written by Wordsworth on her father's leaving
England, in the highest strain of poetry and very original.
I shall try to get a copy some day, as it would delight you.
Another longer ode by the same, on the same subject, was
good, but in a commoner style and lighter strain. Just as I
was going, ' Conversation Sharp ' came in. He began to rail
against Clubs. ' Well,' said Mrs. Lockhart, ' I don't know
what I should do without the Athenæum. Lockhart is never
so entertaining as the days he looks in there, and that is five
times a week, but I don't let him stay there long.' I have
been getting on to-day pretty fairly. Babbage and Fitton
were so delighted at seeing my volume dedicated to a private
friend, instead of some great man, that in the effusion of
their feeling of friendship for Broderip, and their anti-aristo-
cratic spirit of independence, they came up, and each shook
me by the hand, saying, ' I am as much obliged to you as if
you had done me a favour.'

London: January 7, 1832.—I received a letter from Tom
this morning, sent when he was just sailing. I shall do all
I can steadily for him, but poor fellow he has small chance.
A better officer or man never lived, or who has roughed it
more, from being shipwrecked when nearly all were lost, to

every description of tyranny in the captain. His present one is so violent that he has broken more than one five-guinea telescope on the *heads* of the sailors, and sworn at his first lieutenant before the crew. His health suffered in such a manner on that vile African station, that I dread the West Indies.

I am very fully persuaded of one part of Mr. Cockburn's advice to you—to marry one who earned his bread by his labour. At least I am sure, that unless I can feel that I am working to some decided end, such as that of fame, money, or partly both, I cannot be quite happy, or cannot feel a stimulus to that strenuous application without which I should not remain content. I am told by most authors that, though slow work at first, yet, with a name, one can command the booksellers. I certainly feel that my power is increasing, though slowly. I fancy that if I am allowed to do little for several years, I shall by degrees get up a course of respectable length without much fatigue or loss of time ; and knowing, as I do, that it is a science that is only beginning to unfold its wonders, I hope to continue it always. If I can only earn some cash to enable us to travel, it will be a grand point for both of us. I propose this summer to go up the Rhine to Basle, see the Falls of Schaffhausen, then north of Switzerland or a peep at it, then Chamouni, and take a run over the Simplon, returning perhaps by Mont Cenis or some other pass.

Mill Hill, Hendon : Sunday.—I came by coach to this place after breakfast, passing through Hampstead. Dr. Fitton is residing here, at Mr. Wilberforce's house, a most delightful residence, eleven miles from London. Mrs. Fitton and children quite well ; six children. Conybeare and Babbage the only visitors ; most agreeable ; but not lying *fallow*. Fitton pronounces me to be rather thin. We have had great fun in laughing at Babbage, who unconsciously jokes and reasons in high mathematics, talks of the ' algebraic equation ' of such a one's character, in regard to the truth of his stories, &c. I remarked that the paint of Fitton's house would not stand, on which Babbage said, ' No ; painting a house outside is calculating by the Index minus one,' or some such phrase, which made us stare ; so

that he said gravely, by way of explanation, ' That is to say,
I am assuming *revenue* to be a *function.*' All this without
pedantry, and he bears being well quizzed for it. He says
that when the reform is carried he hopes to be 'the secu-
larised Bishop of Winchester.' They were speculating on
what we should do if we were suddenly put down upon the
planet Saturn. Babbage said, ' You, Mr. Lendon (the
clergyman there, and schoolmaster, and a scholar), would set
about persuading them that some language disused in Saturn
for 2,000 years was the only thing worth learning ; and you,
Conybeare, would try to bamboozle them into a belief that
it was to *their interest* to feed you for doing nothing ! '

Fitton's carriage brought us from Highwood House (the
correct name of his place) to within a mile of Hampstead,
and then Babbage and I got out and preferred walking.
Although most enjoyable, yet the staying up till half-past
one with three such men, and the continual pelting of new
ideas, was anything but a day of rest. We were disputing
sometimes on difficult scientific questions, sometimes on other
topics, Tom Moore's poetry, to wit. I cannot recollect Conybeare's favourite lines, but it is where Moore says to his
country that his songs on Ireland are not his, but, like ' the
breeze, they wake the magic that is all thine own.' Babbage
thought the Irish melodies superior to all the rest, in which
I agree, and Fitton cited these lines :—

When midst the gay I meet
Those gentle smiles of thine,
Though still on me they turn most sweet,
I scarce can call them mine.
But when to me alone
Your secret tears you show,
Oh, then I feel them all my own,
And claim them as they flow.

I am not the better for drinking hock or Rhine wine there
' to our friends on the Rhine.' I was as temperate as the rest,
out that is not saying much.

January 13, 1832.—Mrs. Mallet thinks your mother the
most exemplary person she ever met with, for making the
best of everything, and not giving way at difficulties ; and
admires much the courage with which she set out to Bonn

without Mr. Horner, in which I fully concur. I hope we shall both of us contrive to cultivate a disposition—which David Hume said was better than a fortune of 1,000*l.* a year — to look on the bright side of things. I think I shall, and believe you will.

Fitton tells me that they had serious thoughts of putting me out of the Council, in order to give me the Wollaston medal, but I am glad that this subterfuge is avoided. Besides, I would rather avoid at present anything so invidious as a prize, and it would be against the spirit of Wollaston's bequest. I went to Lockhart's, and on a solicitor coming in on business, I went up to Mrs. Lockhart. After an account of her visit last week with the children to the Horticultural Garden, I said : 'If you had not been there so lately, we might have gone together to-day.' 'Can you take me in on a Sunday ? ' 'Yes.' 'Well, then, I'll go again : it will be a nice walk this frosty day. Lockhart is going with Mr. Murchison to see some pictures, and we will go to the garden, and if you can take two ladies, I will ask Miss Alexander to accompany us.' So I marched off with my two ladies, and they were much pleased. Indeed the gardens are much improved and enlarged in the last five months. *Baby's* arm is going on well. At Chelsea I found Somerville just recovering from a touch of the gout. They were right glad to see me after three weeks' non-intercourse. I took in my pocket a volume of my work, but found that Dr. Somerville had read it right through, and made notes on it. She is looking well, and so are the girls. A walk home in a bright moonlight did me much good. I looked in at the Athenæum on my return, to see if any companion was there, and I found Stokes just ready to walk home with me. I forgot to mention, among my lucubrations, calling on Scharf and seeing a fine piece of scene-painting which he has done of my panorama of Val del Bove on Etna (nine feet long) for King's College lectures, and other scenic decorations.

Monday.—Whewell called, and is going greatly to improve the review. Murchison had pointed out that Omalius d'Halloy had based part of his new 'Elements of Geology' on the Lamarckian transmutation system, more than a justification of my having expended so much powder and shot upon it, though I did not see O.'s work till I had written my first half. As O. thinks man existed at the old coal epoch, and finds authority for it in Genesis, I expect Whewell will give him the lash.

Mrs. Somerville showed me an interesting letter from Ivory, written on receiving her book. He says astronomy has gone as far as it can in certain lines, till a new mode of calculating is invented, and that may be done.

London : Tuesday, January 17, 1832.—Sedgwick met me to-day, and carried me to his lodging, to show me his maps made last year, and his summer's work. 'After you and the Oxonians left us, I was kept in Cambridge by a parcel of old women, who came there on a visit. What think you! Phillips, R.A., wants me to sit for my picture. Maiden blushes suffuse my cheeks : very handsome, no doubt, to paint me for nothing, and he says I look best in a black stock. Very dry work geologising in Wales, all primary and old rocks, like rubbing yourself against a grinding stone.'

Henslow was at Linnæan. He sent us a beautiful paper on a hybrid foxglove, with plates published in Cambridge Transactions, the excellence of which cries shame to Oxford, which is so far behind, that they cannot attempt to get up Transactions on Physical Science.

Thursday, 19.—Went to the Council. Warburton, Sedgwick, Conybeare, Basil Hall, Murchison, Clift, Broderip, John Taylor, Turner, Fitton, Greenough. A grand discussion upon the Wollaston medal, and it ended by an admirable resolution that 'this year's dividend of the Wollaston fund should be paid to Lonsdale, to assist him in continuing his survey of the oolite of England.' We must force him to undertake field work, or his health will entirely give way, and he is so good an observer, that it is a sin to chain him to clerk's work and museum arranging. We had a very full meeting at the Geological Society, Henslow being there, and his famous paper on Anglesea incidentally alluded to, Conybeare got up, and quoted some lines from Pindar on the feats of the infant Hercules, as he observed that Henslow's first paper was such a masterpiece. At which 'the first of men' rose, and begged to observe, that as mythological allusions were introduced, he must inform them that

Henslow was making his first tour with *him*, his master in geology, when he went off on that expedition ; so that, like Minerva, he had sprung full-armed from the head of JOVE !

January 24.—A letter from Captain Cooke, R.N., from Madrid, dated December 1830. He has been geologising until laid up by a kick from his mule. Government there more civil to strangers, but robberies increase.

January 25.—I went to Royal Society last night, the fullest and best attended meeting since the Duke was President : he was not there. Conybeare introduced me to Dr. Pritchard, whom I have cited and eulogised. Chantrey said, ' Lyell, I've not seen you an age. I congratulate you on your new birth, and on being unlike so many of my friends in not looking older for your hard work.' Wollaston said he had read to Chapter Thirteen, and was going on most diligently. I have not had so many pats on the back I don't know when. I had a very good day's work yesterday. I think I never do so much as when I have fought a battle not to go out. The way to do much and not grow old, is to be moderate in going out, to work a few hours or half-hours at a time, to have nothing but a pot of porter and a chop the days you do not feast, and to go to bed at eleven o'clock. But, above all things, to keep out of the excitement of politics, which is now raging like a high fever here.

CORRESPONDENCE.

To GIDEON MANTELL, ESQ.

January 18, 1832.

My dear Friend,—Sedgwick is in town, and has been rather, *
I should say, wasting his giant strength on a barren primary district in Wales, which he owns was like ' rubbing himself against a grindstone.' Conybeare is here, and in good feather and spirits, but does not seem to have made much progress in his promised work on Geology. Henslow is to be at the meeting to-night, so I much wish we could see you amongst us. Buckland has also promised to come up.

They say he will give us the Cave book soon, but I shall believe it when I see it. Two or three pages of green sand fell through Fitton's hour-g'ass, and then to be positively

talking of giving it up till next vol.! after printing something. How lucky your chalk tables came out! Buckland
is reported to have said to his wife, when she asked him what
he should do for the Bridgewater prize of 1,000*l*, ' Why, my
dear, if I print my lectures with a sermon at the end, it will
be quite the thing.'

I am working hard in the hope of getting out vol. iii.
before I go abroad, for if I do not, I suppose it will not see
the light in a great hurry. Ever most truly yours,

CHARLES LYELL.

JOURNAL.

To MISS HORNER.

London: February 2, 1832.—At the Somervilles' on Tuesday.
They were remarking that it was odd that Chantrey and
Stokes, and Brown the botanist, and Troughton the optician,
and one or two more whom I forget, are all blind of one eye.
Mrs. Somerville had nearly finished another volume. I fear
that it can hardly sell, because, Sedgwick says, so few men
at Cambridge can go far enough to enter into it. He proposed to the Philosophical Society of Cambridge to elect
her, by acclamation, member on receipt of the book, but some
objected to the manner of doing it. Sedgwick says : ' It is
most decidedly the most remarkable work published by any
woman since the revival of learning.'

February 3.—Just returned from the Geological Society,
having dined at the club, and since taken an active part in
a debate, with Sedgwick, Conybeare, De la Bêche, Fitton,
Greenough, and others—which was prolonged unusually till
half-past ten o'clock—' on the old and new red sandstone,'
and other dryish subjects, which Adam made entertaining.
The paper was by him. The attendance was quite splendid,
very numerous, and all the best men there.

February 4.—To-morrow I expect an agreeable breakfast
at Stokes's. Captain Hewett is to go there with me, an R.N.
surveyor, who has been collecting according to my instructions in the North Sea, and wants a lesson from Stokes.
Murchison is to call here on his way, and to talk over some
sections for a chapter of mine on ' the crag,' one of which I

think of borrowing from his last year's note-book. Captain
Hewett called, and went with me to Stokes's, where we met
Brown, Murchison, and Captain King. Both the naval men
gave me many valuable facts in regard to tides, currents, &c.
Hewett has been reading all the *nautical* or hydrographical
parts of my first volume with a critical spirit, and will
help to correct and amplify for second edition. After
breakfast Murchison came to my rooms to talk over his
speech, and go over on a map my intended tour with you.
He stayed long, and I thought it impossible to get to Hampstead, but I thought I would try, and they had not sat down
to dinner. The party were Dr. and Miss Yellolay,
M. Prévost, Mr. Martin, Mr. Merivale. I thought Yellolay
very agreeable. He spoke with delight of the last meeting of
the Geological, and talked of the early days of the Geological
in Lincoln's Inn Fields, when he and Horner and Laird
were working together, and when the Medico-Chirurgical
Society shared apartments with the Geological. He supposed
that Mr. Horner was now cultivating his old hobby at Bonn,
desired to be particularly remembered to him, and recurred
with enthusiasm as we came home to the old times, and to
Francis Horner, &c. I was much pleased to hear Mrs.
Mallet talk with such amazing delight of the geological
lectures which some time ago Webster delivered at Hampstead. She tells me she could think of nothing else for a
time but geology, and that she has drawings now from
recollections made as notes to those lectures, which she is to
show me some day.

I have been so wishing that your father was within reach,
for I would give him the rough copy of my chapters in MS.,
and I should be so much benefited by his aid. Brown has
invited me to read some notes and corrections he has made :
this is all one can get from him. He will not help one to
any general views, but if you blunder at all in facts, he will
correct you.

February 10, 1832.—I am just returned from the Murchisons. My favourite, Mrs. Lockhart, was there. She
told me she had been desperately alarmed for two whole days,
the little dog Spicey having been missing. We are going
together on Friday with the children to the museum in

Bruton Street. Mr. and Mrs. Conybeare, I had not seen her for many years, ten I think. She is looking well. Mrs. Lockhart told me that Hogg was much fêted, and told her he was *dennered on* for the end of the month, if the grand *denners* did not kill him first. I took a very small box of insects taken last year at Kinnordy to the Linnæan Society, and you would have been amused at the sensation created by a pair of beetles from Catlaw, our highest Grampian near Kinnordy. They proved new, and belonged to a favourite family (the *carabidæ*) of a favourite order with all entomologists (the *coleoptera*). But I will not trouble you with a long account of this most *glorious* discovery, of which I have already written a full despatch to Eleanor. Mrs. Mallet has certainly a considerable turn for natural history, for she is reading my book with more eagerness than most of my friends. Have just returned from Royal Society, where a paper by my friend Captain Smyth was read, proving that the new isle did not rise on, or near *the shoal*, but in water one hundred fathoms deep. So that Prévost's elevation crater is in the highest degree improbable or impossible, but there is nothing strange in a volcano seven hundred or eight hundred feet high.

Friday morning.—Captain Smyth [1] called immediately after breakfast, and cut up half my morning, which would have tried my temper, but for listening to what he had to say about the Royal Society, and having renewed by a coze of two hours an old acquaintance which I much prize. His letters from France have disgusted him with the manner in which the eminent scientific men have thrown up their pursuits, and turned place-hunters, even Arago. Smyth has an ardent love for science for its own sake, in which his wife sympathises most strongly, and I know no people who have less worldly-mindedness and low ambition, and real happiness and contentment with a small income. Yet it proves that there is no great decline of science in England, that Smyth has hardly been a *loser* even in a pecuniary point of view.

[1] Admiral Smyth, b. 1788, d. 1865. Distinguished for his able survey of the Mediterranean, and later on for his devotion to astronomical science. He erected an observatory at Bedford, and the accuracy of his observations won for him the gold medal of the Astronomical Society.

He might have been hydrographer, I am told, and the Admiralty voted him a sum of 500*l.*, I believe, for his astronomical observations.

February 13, 1832.—Young Murray tells me Mrs. Somerville's book does not sell at all, and that Brougham had made great professions, both to him and his father, of what he would do to push it, whereas he has not taken the slightest pains to help it. The surprising and fine part of the work is, that everything which she there shows she so perfectly knows, was acquired for mere pleasure, just as others read a poem. The State might award her 5,000*l.* for the benefit conferred by a woman who could thus teach what Johnson justly called 'the most overbearing of all aristocracies, that of mathematicians,' how most of them can be equalled and surpassed by a lady who was merely reading for her amusement. It is true that there is not much display of *inventive* power in the book—not that kind of power, for example, which enabled Babbage to invent his machine, but there is some I understand.

I expect that in the course of the next ten years the seeds which I have sown in regard to the action of modern causes will come up in an abundant crop, and stifle all the school who delight in what they call 'the *poetic* mystery' of geological causes. For I find every day the hydrographers are coming to me for instructions. I have just drawn up some for Captain Fitzroy, who has my book, and is surveying in South America. Captains Hewett, Beaufort, King, Vedal, and others, are in continual communication. There has been a little talk of getting a bust from Chantrey of Mrs. Somerville, to be put up in the Royal Society at their expense.

Mrs. Mallet showed me her *geological* drawings and sections, which are very well copied, and I am sure I could soon make her a geologist. I was regretting you were not staying with her when she attended Webster's lectures. She observed, 'When one has for a long time been remarking and wondering at a thing, as I had, at chalk and chalk flints at Margate, the first gleam one gets on such a subject is so highly interesting.' Now the fact is that most people are as a friend of Stokes said, who was bothered with people wondering at the comet in 1811, and talking of nothing

else. 'Why,' said he, 'have these people thought of the
heavens for the first time? have they always *taken the stars
for granted?*' Most people would take a chalk flint *for granted,*
and if they had thought and doubted enough of what it
could be to be delighted when told it was often a silicified
sponge, it shows that they had minds apt for cultivating
Natural History, and prone to inquire into 'the causes of
things'—which people I believe are born with in very different degrees, not by any means in proportion to the talents
they may happen to have for other pursuits.

Did I tell you that a Belfast Professor of Divinity, Edgar,
has been denouncing an unfortunate Dr. Drummond, who
gave lectures on geology in the Royal College there, for
having declared that 'the changes proved with *tolerable certainty* by geology must have occupied time to which our
historical eras were as nothing;' and then the Professor wrote
to me, as Professor of King's College, to know if the lecturer
or he was right, expressing great alarm for Christianity if
such doctrines were true. I shall not go to Murchison's to-
night, for I am getting round again, and will be very quiet,
and to-morrow shall do a great day's work again.

February 19.—On Friday I went to the General Meeting
and the Anniversary Dinner of the Geological Society, at
the Crown and Anchor—a splendid meeting. Near the
President sat Lords Milton, Morpeth, Cavendish ; Sir John
Malcolm, Sir J. Herschel, Sir J. Johnston, M.P., Sir R.
Vyvyan, Sir C. Lemon, and other M.P.s. Literature represented by Hallam, Lockhart, Sotheby, &c. Geologicals—
Buckland, Conybeare, Fitton, Greenough, Sedgwick. Then
from Cambridge, Whewell, and many other good men. All
the best geological residents in town. Murchison was ill,
but got through the fatigue very respectably indeed. All the
speeches were short, and many of them able. I was glad
when mine was over, but Murchison made my reply easy,
by giving me something to say off-hand, as he told them he
should say nothing of my book till the evening ; and I told
them I should follow the President's example, in thinking
that they would have enough of me if they heard of me once
in a day. Hallam asked to be introduced, and talked of
Bonn friends. Ferguson of Raith asked me to walk with

him from the tavern to Somerset House, and asked after
you all. There were about one hundred and thirty at the
dinner, and a hundred of them picked men. When I came
up here on Wednesday morning, I found your father in town.
He drank tea with me in the evening. The Royal Society
has actually engaged Chantrey to make a bust of Mrs.
Somerville—a very gracious proceeding—and they sent a
deputation to communicate their wish that she would sit
for it. At Cambridge, too, I am told it has been decided
that it shall be a class-book. So I hope it will sell, though
slowly, and certainly it will now be useful.

I heard this morning that Senior means to give up not
only the chair of Political Economy at King's College, but
to abandon the course of lectures he had promised—nine in
number; his reason being that he is appointed Commis-
sioner for inquiring into Irish Tithes, &c. I shall consider
Senior's withdrawing as the loss of the man whom I had
most satisfaction of being a colleague with, as one of known
talent, and well known at Oxford.

March 16, 1832.—An excellent party at Murchison's last
night—many at dinner, and the rest came early, and evidently to a thing they enjoyed—very *select*. ' The first of
men, Adam,' *alias* Sedgwick ; Mrs. Somerville, the first of
women, not of the blue ; Schlegel, Brewster (*now* Sir David),
Babbage, Dr. Somerville, Conybeare ; Lords Morpeth and
Wentworth (Lord Milton's son), and Upton his tutor ; Sir
John Johnston, M.P. for Yorkshire, and his lady, who was a
Miss Vernon Harcourt ; a sister of hers, a very pretty lively
girl. Sedgwick as usual was in rhapsodies with her, and
talked to me of her all the way home.

I put my name down yesterday for two guineas to Mrs.
Somerville's bust. About 80*l.* subscribed—200*l.* required.
It will soon be filled up, and if not, Chantrey says he would
happily put down the rest himself.

Murchison is one who has worked at science chiefly for
the rewards, but not entirely, for if he had had no pleasure
in it he would have failed. Sedgwick and Conybeare for the
pleasure chiefly. What I shall always cherish, is a love for
science, rather than its rewards ; but I indulge the hope of
profit, as the best earnest of usefulness, and also of protec-

tion against its becoming a duty to accept some offer of an uncongenial situation.

Dr. Wollaston and Mrs. Somerville, of whom I have seen much, were very useful in ' mortifying' in me much of the vanity which I formerly had, and I think they were two kindred minds, and ought to have been united. She has done a real service to science, by showing many who are proud of inferior achievements what can be done by one of no pretensions. Of course there are some who are jealous of the homage paid to her. I hear every week of some one who says: ' What has Mrs. Somerville done in the way of *invention*? what is she but an expounder of La Place?' I cannot judge of her inventive talents as displayed in this work, and I don't suppose she has exercised them, but she has worked in all that has been done since La Place, and doing that is an enormous work. Undoubtedly the highest honours are due to original discoveries and inventions, and nothing is so rare as inventive talent, which Wollaston for one possessed ; but her introduction proves she has it, and how little hers has been called into action! At least there are not five persons in this country that could have written her book, if that number.

Sedgwick has declined a living of 1,100*l.* a year offered by Brougham, on the ground of not being able to retain his professorship with it. I hardly rejoice at it, except for the cause of science. As a votary of geology, Sedgwick of course must not exchange a fellowship of 430*l.* a year, and a chair of 200*l.*, for a cure of souls of 1,100*l.* It was merely a question of marriage : as a bachelor, he would be mad to exchange a college life, and his position in the University and in science, for becoming a country parson. Some will say that Sedgwick is old enough now to remain a bachelor. He is forty-seven, and from what he told me of his hopes and feelings on the subject of matrimony a year and a half ago, I believe he would be much happier, and would eventually do much more for geology. I warmly remonstrated with Murchison for advising Sedgwick to decline; and on giving my *matrimonial* reasons, got finely quizzed, as you may suppose, and they have been bantering me ever since. As to Sedgwick getting a stall, I hope he will. It should have been given before, but the Whigs may cease to reign, and he will not cease to grow older—and 1,100*l.* a year is not a bad thing to have to *give up* to a chancellor in exchange.

I gave Mrs. Mallet an invitation to my introductory lecture, with which she was much pleased, and pressed Mr. Mallet into the service immediately. I wish half my audience may be as intelligent listeners.

Tuesday.—Lord Milton was at Murchison's on Sunday, and it was lucky for Curtis the entomologist, for he was introduced, and Lord Milton also talked to him of his own insect-hunting as a boy, and how he knew the Yorkshire species now, and how fond his sons were of it, whom he would bring to see his collection. Sedgwick asked me to walk home with him. I found a gloom upon him, unusual and marked. I most carefully avoided all allusion to the rejected living, but now when the first excitement of the declining the boon is over, and that others have expressed their wonder at it, and that he finds himself left alone with his glory, he is dejected. He told me, Thursday last, that he wished before he left Cambridge to *do something.* ' Now if I take a living instead of going to Wales, I abandon my professorship, and cannot get out the volume on the primary rocks with Conybeare,' &c. Then he hinted that in a year, when this is done, he may retire on some living, and *marry.* But I know Sedgwick well enough to feel sure that the work won't be done in a year, nor perhaps in two; and then a living, &c., won't be just ready, and he is growing older. He has not the application necessary to make his splendid abilities tell in a work. Besides every one leads him astray. A man should have some severity of character, and be able to refuse invitations, &c. The fact is, that to become great in science, a man must be nearly as devoted as a lawyer, and must have more than mere talent.

As I was much amused with Power, in an Irish character, on the night I heard Miss Kemble, I went again to the Adelphi to hear Yates for a couple of hours, which is as long as I like to be at a play, and got home and in bed by eleven o'clock. I have enjoyed parties, and two plays this last month very much, because it was recreation stolen from work; but the difficulty in the country is, that on the contrary, one's hours of work are stolen from dissipation.

Wrote to your father on geological queries of Dr. Fitton —frank from D. Gilbert. His brother-in-law, Guillemard, came to ask me to come down a fortnight hence into the country with him. I refused with pain, but it is only by resolution of this kind that anything can be done. I always reflect that if I had a recognised profession, all the world would think me right, as called upon to be equally unfriendly; therefore, as I am determined to make science a profession, I will do it. In the end they will think me right.

Another day nearly gone and not through chapter i., so short a one too! But in it I am grappling not with the ordinary arm of flesh, but with principalities and powers, with Sedgwick, Whewell, and others, for my rules of philosophising, as contra-distinguished from them, and I must put on all my armour.

I found on my return three small volumes from Basil Hall—voyages, travels, second series; most diverting—not yet out, though reviewed three weeks! Stokes remarked to me that his enthusiasm about my first volume, and the manner in which he went about, when I was in the Pyrenees, crying it up as the finest thing ever written, was anything but that of an egotist entirely absorbed in his own literary works. He read it at the Athenæum till some one told him he kept it too long, and then he bought it, and they say thought and talked of nothing else but my geology for a month. I had met him at a relation's four years before in Angus, but was never intimate till my return from France. The first thing he said to me on my return was, 'Lyell, I never thought you had it in you to write such a book. I don't mean as a geologist, but as a writer. That passage about Vesta and Tivoli might do for the style of any English writer—nothing in Playfair beyond it,' &c. I was more pleased with a compliment as to style from Basil Hall,[2] than I should have been with an eulogy as to general views. In the former he excels, in the latter he is usually wrong; and

---

[2] Captain Basil Hall, R.N. Author of *Fragments of Voyages and Travels.* See Appendix A.

when he goes East, 'tis a sure indication that the true course is West.

Mantell was at Geological Society, and came home with me—in much better spirits, though overworked. I think he will soon flit to Brighton. He announces that a diploma for me from Philadelphia is soon to arrive.

Mantell is to be here in two hours to spend the evening, a great pleasure, for he is an old geological friend. I went and introduced myself to him on a tour *à cheval*, with a spare shirt or two in my pocket, in one of the Oxford long vacations; and having previously seen the Isle of Wight, ventured to question the whole arrangement of the Sussex beds below the chalk, which Greenough and Buckland, then the top authorities, had given Mantell, who looked up to them as infallible guides. Fitton afterwards confirmed my arrangement and acknowledged it in print, and I got great credit for it with Mantell, who being unable to travel and compare countries, but doing wonders in his small field of observation, was obliged to depend on others for more extensive generalisations. I have watched his gradual professional rise with much interest, and look forward to see him a great Brighton if not London surgeon.

March 31.—I am knocked up a little to day, and half meditate going to Hampstead and taking the Mallets by surprise; for Sundays are Murchison's evenings, and to-morrow Mrs. Somerville is there, and I don't like to go and come away late. I have paid 20*l.* for compounding to Zoological Society, so now I have compounded to all the Societies—a good thing over, instead of being bothered by annual payments.

April 7.—Last night at the Royal Society Sir J. Rennie told me that he and his engineers used my first volume as a text-book, and all I said of currents, &c., he had found true in examining harbours, estuaries, &c., professionally; and he offered to send me a MS. with facts bearing on power of running water, &c. Thus, he said, below old London Bridge the current descending through the arches has hollowed out the bed of the river to the depth of thirty-seven feet at low water (the ordinary depth of the Thames elsewhere is only seven feet), and this for sixty yards below bridge. I have

been so long off the hydrographical part of my book, that
this news was very cheering.

Went to Loddige's garden yesterday with Stokes and
Broderip, to see 100,000*l.* worth of Camellia Japonica in full
blow—a wonderful sight. Belgian affairs hurt their sale.
All the naturalists here are much grieved about Holland,
as they say the king did so much for natural history, far
more than any other sovereign.

April 10.—Yesterday morning I ran out to Kingston Hill,
where the farm of the Zoological Society is situated, to try
by walking about there to get rid of a slight headache. I
took Hall,[3] who wanted much some country air, with me,
putting him on the outside of the coach at nine o'clock, while
I went in. At ten o'clock we were in the farm of thirty-five
acres. Beautiful ground, a view like that from Richmond
Hill, down into the rich vale. Saw the Kangaroos, Quaggas,
Emus, and other birds and beasts, which are there breeding.
After walking an hour, went to the adjoining Combe Wood,
classical ground to entomologists, of which I had often read,
but never before visited. Saw some insects there, but as the
old keeper whom we met in the wood (once himself a collector
and dealer) said, ' London is walking out into Combe Wood,
and two houses (villas) building in the middle, and all the
underwood cut,' &c.

You asked me in one of your letters how I liked
Schlegel—so little, that I avoided him. I met him three
times, and exchanged some words each time. He is full of
conceit, talks incessantly and of everything, not like Humboldt, whose loquacity bored some people, but never me, because unmixed with self-conceit, like Schlegel's. He called
at Chelsea and annoyed Mrs. Somerville. He wanted to be
pressed, he said, to lecture at the Royal Institution, and
wished to know if *he* could be seen from all parts of the
theatre, because *the change* of the expression of his countenance would add great effect to his delivery of certain passages; and, ' I will lecture in French, for although I read and
speak English *well*, I should be more at home in French.'

April 14.—The old keeper, formerly a regular entomological collector, who I mentioned having met with in Combe

³ His clerk.

Wood last Sunday, was quite a character. 'When there were but a few gentlemen,' said he, ' who cared for insects, they gave any price for rarities. I remember a red beetle with long horns that sold for two guineas; and once at the bottom of an old stump, just by that copse, sir, I found a quantity of them, and *there was an end of they.*' That is to say, their market value was destroyed. Now that entomologists have multiplied, who really care for things, not because rare, but for the study of them, they scarcely ever buy, but exchange with one another. In another part of the wood we met with two youths who had purchased a regular instrument made for digging for chrysalises at the roots of trees. Think of an invention for that purpose! One of the youngsters said to me, ' What have you seen to-day ? ' I said, ' Urticæ,' the common tortoiseshell butterfly ; to which he replied, ' I have seen Gonepteryx rhamni,' the yellow brimstone papilio. I was rather pleased with this adventure, because it was a cold ungenial day, and it required all this to make me believe that it was really the Combe Wood that I had read of as a boy. We found some black beetles, and I found that Hall remembered the exact names—it is wonderful how fond he is of entomology. I should never make him take to geology in the same way. He has planned himself a nice wire net-hoop, which goes into his hat, and takes out and fits into a walking-stick. To-day Lt. Graves dined with Broderip, and I joined them at dinner, sending, in old college fashion, wine down to Broderip's rooms. He is a thorough good sailor, and a naturalist; sent by the Admiralty to survey Loch Neagh in Ireland, and going out to the Grecian Archipelago.

It was curious enough that I should have met Spedding just as I was turning into King's College Chapel, because about two months ago we were saying what a complete failure that same chapel was of Smirke's, not being even so much like a place of religious worship as most conventicles. Spedding and I had been accustomed to those two beautiful chapels, Lincoln's Inn and the Temple, where the service is admirably performed ' in all the beauty of holiness,' as David would have said, and the preachers excellent—Lloyd, Heber, Maltby, and many other eminent men being the

preachers there, since I remember. I think there is much
in the association of the building, and our chapels at Oxford
are so beautiful.

April 21.—I cannot imagine anything more afflicting
than such a death as poor Mrs. W.'s,[4] and believe that in such
a case the only state of mind in which a person of real
feeling can find the least comfort is that of religious resignation. It matters comparatively little, as far as that con-
solation is concerned, how many erroneous dogmas and
absurd rites are connected with the religion, provided its
professors do not see through their fallacy. All I wonder at
is, that enlightened and inquiring minds can, even by aid of
early association, reconcile themselves sufficiently to some
parts of the Catholic creed and ceremonies. If it were not,
however, for the proneness of the people to superstition, and
their gullibility, I should like a good deal of ceremonial. I
like our English service so much better than the Scotch. I
know some religious Frenchmen, who say they would turn
Protestants if *we* did not believe so much the same as
Catholics, and that our creed is not much more philosophical
than theirs. I almost wonder how some of your German
friends can be such staunch Catholics and so tolerant. I
have sometimes thought that if all the world should become
Christian, and be as divided into sects as we now are, and as
tolerant as the Prussians, we should come round once more
to the opinion of the Romans, who most sincerely believed
' that all religions were true '—that the gods or saints of all
and every people were *their* gods, and that their forms of
worship were true, and worthy of veneration, though some
more and some less. That they were all, in short, modes of
adoring Him whom, as St. Paul says, ' we ignorantly worship.'
Talking of sects, a curious thing has happened here. The
Society of Useful Knowledge started a penny magazine a
month ago, admirably done and very entertaining, and with
a good moral tone. They sell 80,000 copies—an unexampled
success—well illustrated and remarkably cheap. The Dissenters have been so cut out in the sale of their innumerable
tracts, that they preached against the magazine, and thus
doubled its sale. The Church party could not help liking

---

[4] The wife of a Professor at Bonn.

what the Dissenters disliked, and so they rather patronise
the concern.

Wednesday.—A letter from Kinnordy from your father,
and was much delighted to find he was at last *domiciled*
there. He begs hard of me to sit for a portrait to Richmond.
When Mrs. Horner first asked it, I spoke to Wright, who
did those likenesses so well for Phillips, R.A., but I was so
pressed for time, I gave it up. If I do it, I shall go to
Wright, at your mother's old request. A north-country
attorney called this morning with a brief. He expressed great
admiration of my work on geology, and said it was thought
much of in the north. I declined the brief, though he tried
to persuade me that as it was merely an opinion wanted on
a simple road question, I might without scruple pocket the
fee. Of course I might have got a friend to do it, but I
have cut my cables, and now I will stand firm by one
business, and one only. If nothing untoward happens, you
will see I shall turn it yet to profit as well as fame.

April 24.- A letter from Elizabeth. Very sorry Mr.
Horner was gone, very dull without him. He was a great
favourite.

May 1.—I am just returned from Hampstead. Have
walked home with Herman Merivale, whom I like as a well-
informed man, and particularly in a *tête-à-tête*. He is, I
think, very remarkable for his age, in the accuracy and
extent of his acquirements. I like him, too, especially as a
brother Oxonian, for the warmth of his advocacy of the
more general cultivation of all branches of science, especially
against the Oxford exclusive system.

Wednesday.—Grand disputes at the Geological Society
about the propriety of admitting ladies to my lectures. Babbage most anxious to bring his mother and daughter and
Lady Guildford; Harris to bring Lady Mary Kerr ; and so on.
I begged them all not to do so, and they promised ; but at last
Murchison said, ' My wife, however, must come. I promised
to bring her, and she would be much disappointed. I will
not bring her till the doors are closed.' Then they all
declared they would too, and so bring the affair to a crisis
one way or other.

Friday.—Prepared yesterday a new lecture. The ladies,

Dr. Fitton, and others have been assailing Dr. D'Oyly and others of the King's College Council to admit ladies, and they seemed disposed. At all events I am likely to have a large audience to-day. Only nineteen subscribers up to yesterday, but I met with Spedding, who said, ' I am going to subscribe. Ignorant as I am, I followed you so well that you have really *forced* me, at great inconvenience to my law studies, to attend.' His and many other names are not down yet.

Tuesday.—I have most completely succeeded in my second lecture, which was liked much more than the first. I tell you these things freely, because you know it is not from vanity, and only between ourselves. I worked hard upon the subject of the connection of geology and natural theology, and pointed out that the system which does not find traces of a beginning, like the physical astronomer, whose finest telescope only discovers myriads of other worlds, is the most sublime ; and as there is no termination to the view as regards space of the acts of creation, and manifestations of divine power, so also in regard to time, &c.; concluding with a truly noble and eloquent passage from the Bishop of London's inaugural discourse at King's College, in which he says that truth must always add to our admiration of the works of the Creator, that one need never fear the result of free inquiry, &c. I did not the least, in the opinion of Fitton and Babbage, compromise the utmost freedom of the philosopher ; while Lonsdale, the Archbishop's secretary, Archdeacon Cambridge, Otter, and others were most exceedingly pleased. Mrs. Somerville and Mrs. Murchison were there. The professors mustered very strong in honour of me to-day, and the number of gowns, to an *old Oxonian,* looked most respectable and dignified. You don't know how, from old association, we attach respect, order, academical rank, &c., to these insignia. For my part, though I don't care about them at King's College, I cannot believe that I should look back to Oxford with as much regard but for these dresses. They become so interwoven with all our ideas of college, as much as a military costume to an officer. I slept ill last night, but no coming lecture will ever discompose me again. Jelf the clergyman, and his brother the lawyer and others, pressed me to publish the lecture. D'Oyly thanked me for the benefit I had conferred on King's College, &c.

May 7, 1832.—Went up to Hampstead, did not find Mrs. Mallet, but wrote her a note, offering her free entrance to my lecture-room, with two other ladies. Dined at Chelsea, and asked Mrs. Somerville to come to my lectures, and her two daughters. She said she was charmed with the effect my first lecture produced on Martha and Mary. They never would read before, but when they came home they set to work in earnest at my two volumes, and examined each other on the points of the lecture. The Somervilles are to come *regularly.* Then to Murchison's *soirée*—very full. Many had objected, or rather *some*, to the last part of my lecture, as too free ; but Babbage and others had converted almost all to think with them of the good I had done, &c. Walked home with Stokes, rather late.

I begin to flatter myself that I am, as Lubbock said last night, when he came down from the R.S. chair, doing good to science by these lectures, which at all events are talked of over London, but politics are of course the chief and absorbing subject. I am going to Uncle Lloyd's to dine, so if I do anything to-day to the book, it must be between this and dinner ; so let me not play the dog in the fable, and lose the substance by catching at the shadow—the solid reputation in many countries which my work may earn, for the exciting ready-money profit and applause of a lecture-room.

A letter just arrived from J. Murray, jumping at my proposed reprint, and only doubting between 750 and 1,000 copies for next fourteen months. Bravo ! Lord Tyrconnel has been begging me to give a *glossary* of such words as *tertiary*, and says it will sell hundreds more. I suspect it will, but fear delay. I shall set Martha and Mary [5] to make a list of hard words. I am in for thirteen lectures in consequence of repeating the first : if I can cram in three in one week I will do so.

Last night, in the House of Commons, Sir Robert Inglis declared that if the Duke of Wellington, now Premier, and trying to form an administration, takes up the Reform Bill as reported, all character and morality in public men will

[5] The Miss Somervilles.

for ever be destroyed. Baring said no administration could be formed but by reform. In short it seems impossible for the Duke to make an effective cabinet, and not impossible that Lord Grey may be recalled. Of course with the Tories so divided, and a majority of eighty in the House of Commons for reform, great concessions must be made. Sedgwick being at my lecture, I took a shot at him, which amused Fitton, and made him come up laughing after lecture, and shake me by the hand. Murchison said to-day, ' I don't know when they will leave off saying the last is the best lecture of all.' Every one can hear every word, even most distant benches. Sotheby came to-day, Malthus Junior, the Dean of Carlisle, only Dr. and Misses Somerville: she came not, being rather unwell. Such inquiries for her by strangers, as since her work was reviewed in the Edinburgh (for Herschel's in Q. R. is not yet out) she has been the great lion in town. How strange that people never knew before that she could have done all this ten years sooner! They knew it in France. I have told Murchison I cannot be President. He remonstrated, and was sorry, but I was firm, and I suspect he thinks me right. It is just one of those temptations the resisting of which decides whether a man shall really rise high or not in science. For two more years I am free from *les affaires administratives*, which, said old Brochart in his late letter to me, have prevented *me* from studying geology *d'une manière suivie*, whereby *you* have carried it already so far.' The Duke of Wellington has been unable to make a ministry: so much the better, so Lord Grey and the rest are back again.

Your father will tell you what an actor-like sort of celebrity my lectures obtain me, but I will look to something more solid. I am not improving my work and my own mind by them so much as by tranquil uninterrupted study. As I have renounced being President of the Geological Society, so will I write to Otter that I do not pledge myself to give any lectures next year, waiting till the second edition of my *whole* work is ready for press, and I will give him all my reasons—being all connected with my pursuing my original researches without disturbance, until this work is accomplished, and got out in its permanent form. Murray says it must be printed in a cheap form, which I am glad to hear.

Fitton said, 'I'm so sorry I missed you last time, when the
world says you floored Buckland.' As Babbage expressed it in
a note, 'tore Buckland's theory to tatters before his face.'
But I must say B. showed his good sense, for he has been
more good-humoured since. But he and Sedgwick & Co.
blaze away at me so, I can but retaliate. I pronounced my
éloge on Cuvier to-day, and a digression on the comparative
state of advancement of natural history in France and
England, and quoted Herschel and Babbage on the decline of
science, &c. The episode was, I fear, rather too long in proportion to the geological part of the lecture, but I am rather
glad to have unburdened myself.

May 27.—You cannot think how Marianne and Caroline
were charmed with Mrs. Mallet for her frank manners, and
the soul with which she entered into everything. Babbage
wrote me a note about my last lecture which I was glad of.
I doubted whether my *éloge* of Cuvier, and my comparison
of French and English systems of cultivating science, had
been popular. It seems they were, especially the former,
and the latter cried up by Babbage & Co. with enthusiasm.

One can only please one party in some things. He wants
me to print it, but I shall not be drawn into controversies
of the kind. Babbage in his note, enumerating distinguished German *savans* to whom we cannot show any equals
in their respective branches, reckons up many names of
astronomers and others, whom I never heard of, and at last
one, who he says has just married and become rich, and all
agree is doing and likely to do *nothing*. We have at least
no danger on one score, that of being *rich*, which I am sure,
much as money is wanted in science, does stop men's careers
more than anything, and gives them innumerable duties, by
which they become stewards of their property, rather than
men who have time to devote to philosophical pursuits. By
withholding the temptation of the President's chair and the
lectures at K. C. next year, I shall, I think, have done two
virtuous acts of self-denial, but they will require to be
backed by still greater; for precisely in proportion to my
not having any of those public and ostensible engagements
to plead, will be the difficulty of fighting off the importu-
nities to idleness with which I shall be assailed. Did I tell

you to what a fit of desperation the interruptions of genial
Oxford have at last driven Buckland ?   Literally obliged to
hire another house out of town, five miles, and to leave his
library and other conveniences !   Had he not got the 1,000*l.*
we should never have had another volume from him ; but,
luckily for his fame, it became at last his duty, and he was
driven to the plunge.   The loss of time in travelling to his
library, and going for books of reference, will be immense.
I think I should have given out that I was dying, and fee'd
a physician to have given bulletins.   But then one's relations
would not have kept the secret.   I reckon that the loss of
time, of reference even now and then to one book, as far off
as G.S. from me, is so great, that it is cheaper in general to
buy.   Only think of his going five miles from his books !

Otter said to me, ' You will find that when you are
married you will do a great deal more in geology than as
a single man, particularly as you will be able to follow it up
most successfully as a profession.'

CHAPTER XVI.

JUNE 1832–APRIL 1834.

HIS MARRIAGE—TOUR TO SWITZERLAND AND THE ITALIAN LAKES—RETURN
TO LONDON—LECTURES AT THE ROYAL INSTITUTION—HEIDELBERG—VON
LEONHARD'S LECTURE—DR. SCHMERLING'S WORK IN THE CAVES NEAR
LIEGE—DR. BOWSTEAD—SYDNEY SMITH—THE DRUMMOND LIGHT.

CORRESPONDENCE.

To CHARLES BABBAGE, ESQ.

2 Raymond Buildings, Gray's Inn: June 3, 1832.

MY dear Babbage,—I proposed to Fitton to take up the
lectures on geology, as my friend and representative at King's
College next year, while I am completing my education and
work ; and in a kind spirit he has at first sight of it agreed.
He could begin with Maestricht beds just where I end, and I
really know no one whom I would allow of all the London
geologists to teach in my place so soon as Fitton, as he
enters so fully into the spirit of the ' modern cause ' system.

Otter is delighted with the idea, having paid me the
compliment of being quite down in the mouth at my not lecturing next year. Fitton will consult you, so take care you
are not the cause inadvertently of his gibbing. You know
how many years we have been in getting him to let out his
knowledge, and this would be a most effective vent, and we
should all profit. To me, of course, it would be falling on
my legs most agreeably.

You will be glad to hear there are symptoms of repentance in regard to the hasty resolution of the council at K. C.
against the admission of ' women.' I have said nothing in
remonstrance, but they have made the discovery.

I have worked in the Temple of Serapis, and will soon send back the books, &c. The *Balanus* is *B. sulcatus*; the *Arca* I have not yet got at.

Ever truly yours,
CHARLES LYELL.

To G. MANTELL, ESQ.

June 14, 1832.

My dear Mantell,—My lectures were splendidly attended. I have shown the public, and made the discovery myself, that I *can* do the thing, and it may yet provide travelling money. But I shall not give much time to it, at least for several years. When the course was over, I received from some of the class whom I never saw, most enthusiastic letters, expressive of their admiration and gratitude. This is worth much—especially as they had the good taste not to applaud in the theatre.

Buckland was really powerful last night on the Megatherium—a lecture of an hour before a crowded audience: only standing room for a third. Lots of anatomists there; paper by Clift; the gigantic bones exhibited, and still to be seen there, but likely to be removed by-and-by. Buckland made out that the beast lived on the ground by scratching for *yams* and *potatoes*, and was covered like an armadillo by a great coat of mail, to keep the dirt from getting into his skin, as he threw it up. As he was as big as an elephant, the notion of some that he burrowed under ground must be abandoned. 'We may absolve him from the imputation of having been a borough-monger; indeed, from what I before said, you will have concluded that he was rather a *radical.*' He concluded with pointing out that as the structure of the sloth was beautifully fitted for the purpose for which he was intended, so was the megatherium for his habits. 'Buffon therefore, and Cuvier even, in describing the sloth as awkward, and Cuvier the megatherium, erred. They are as admirably formed as the gazelle,' &c. It was the best thing I ever heard Bucklan do.

Most truly yours,
CHARLES LYELL.

To HIS SISTER.

My dear Maria,—The grand day [1] is at last fixed for the
12th inst. Professor Sack, the Protestant clergyman (or
Lutheran, as they say here), agreed to put off his lecture in
order to oblige us. He appears to be a very superior man ;
indeed, I should think, from what I have seen, that the different departments were generally very well represented in
the university here, and the Government seems very liberal.
They tell me the king gives 14,000*l.* sterling a-year for the
Botanic Garden. The Professors have to lecture nine months
in the year—too much, I should think, for allowing time for
due advancement of the teacher. On Monday we had a
grand expedition to Godesberg. All we *ten,* Miss Parker
included, Mrs. Barton (widow of a General Barton), her
daughter, Mr. Frederick Windischmann, a gentlemanlike
young son of a Bonn Geheimrath, and Count Schlabrendorf,
eldest son of a rich Silesian noble, a young student. Godesberg is looking most delightful now, and the scenery, both
the home views and the distant, are scarcely equalled by
those of any single spot I know.

I have exchanged calls with Schlegel, and as I hear he
is reading Rossetti in earnest, and much interested in it, I
hope he will give my father the benefit of some of his own
comments upon it.

Your affectionate brother,

CHARLES LYELL.

To HIS SISTER.

My dear Eleanor,—I hope to send this letter from Strasburg to-morrow, and shall first take up the diary from Heidelberg, which we stayed at on the 18th. Leonhard was
very attentive, showed me part of his collection, and begged
his wife to take *mine* to a fine view from a neighbouring
hill; then went with us to the castle, showing me by the way
some geological sections, which, added to my short excursion

_____
[1] His marriage with Miss Horner.

to the Felsenmeer, have enabled me to obtain something like a fair notion of the Odenwald, both its scenery and geology. I then introduced Mary to Bronn, Professor of Natural History, and learnt some geology from him of the country in a different department from Leonhard's. Next day, the 19th, to Carlsruhe, making a delightful *détour* on the road, up a small valley leading from the plain up into the Odenwald hills, where I went to see a singular deposit, called 'loess' provincially, filled with recent species of land shells, and which is peculiar to the Rhine valley, found at Bonn, Strasburg, and hundreds of intermediate places. The Bergstrasse, I should tell you, is a line of road running nearly north and south along the foot, first of the Odenwald, as from Darmstadt to past Heidelberg, then of the Schwartzwald, another chain of mountains, which we have seen all to-day in coming from Carlsruhe to this. The plain of the Rhine is nearly as flat as that of the Po, and like it composed of horizontal beds of gravel and sand, hundreds of feet deep at least, but depth unknown. The mountains are of inclined strata of older rocks. If you keep in the plain it is like sailing on the sea and having a distant view of an adjoining isle, or mainland; and Mary was not sorry that my geological tastes have led me several times to *land*, and make short trips into the interior. This morning we saw the palace at Carlsruhe, and then returned to the Bergstrasse, making an excursion up the valley of the Murg, and from that to Baden-Baden, the pleasantest German watering-place I have seen.

With love to all, your affectionate brother,

CHARLES LYELL.

To His Mother.

Strasburg: July 22, 1832.

My dear Mother,—Voltz, a mining engineer of this place, and a good geologist, gave us a famous view of the Museum here on our arrival yesterday; after which I went with Mary up the tower of the Cathedral, and saw a fine panorama of the great plain of the Rhine, with the Schwarzwald on one side and the Vosges on the other, rising rapidly from the

plain. On the north the plain is uninterrupted, but looking
south you see a small group of rocks rising like an island
from a sea midway between the Vosges and Black Forest.
This group is the Kaiserstuhl, which I am told is an ancient
volcano—a submarine one though, and ancient enough. I
was thinking of seeing it, but as it would cost a day or two
more, I will not go. Voltz has recommended me to go by
all means to Berne, as Studer is now there, who is by far the
best geologist of Switzerland. As the excursions into the
Odenwald and Black Forest have repaid us so well (me in
geology, and Mary in seeing the country), I was glad to see
a little of the opposite coasts, or the Vosges. Voltz gave
me a plan of a trip for to-day to Soulz-aux-bains, or Soulz-
bad, which is at the foot of the Vosges, where I saw many
quarries of rocks, which compose a great part of the Vosges.
It was a delightful drive, first over the plain, then a hilly
country, behind which the mountains rose boldly. The costume of the peasants amused Mary exceedingly, and I was
delighted by a dance which they had at an inn. The music
was excellent, and the waltzing and gallopading by no means
to be despised. They seemed so happy and so superior to
our peasants in the refinement of these holiday pleasures,
that I could not but dwell on the unfavourable contrast. I
always fancied that Strasburg was on the Rhine, and mar-
velled to find that we left the river and found that the city
only communicated with it by means of a small canal. The
number of the National Guard (6,000), together with several
regiments of the line, gave a very gay appearance to the
town, and so many soldiers mixed with the costume of
the peasants made the roads appear as if there was a national
fête.

July 24.—From Strasburg, along the Bergstrasse, past
the Kaiserstuhl (which became a considerable group as we
approached) to Freyburg in Brisgau, a town larger than
Bonn, with a University and a most beautiful Cathedral,
both in the exterior, which resembles Strasburg a good deal,
and the interior, where there is good Gothic architecture and
painted glass. From Freyburg we have passed through
what the Germans call the 'Hölle Thal,' or valley of hell, a
defile in granitic mountains, but which is too beautifully

covered with fir wood to deserve the name. To-morrow we
hope to arrive at Schaffhausen.

Believe me, my dearest mother, your affectionate son,

CHARLES LYELL.

To HIS SISTER.

Sumiswald : July 28, 1832.

My dear Fanny,—We are here, on our way to Thun and
Interlachen, at a small inn. This route, which we take as
the easiest, and not requiring pedestrian or equestrian exertions, like the Brunig, has shown us so much quiet liveable
home scenery, and cultivated gently undulating ground, as
I really hardly knew to exist in Switzerland, where one
usually keeps so close to the mountains. We see many new
houses of the best style built in every part of Switzerland,
which bespeaks the growing prosperity of the country, and
which is never seen in Rhenish Prussia. The part of the
Bernese territory which we have been in, looks particularly
flourishing.

Berne: July 30.—I was much disappointed at finding
Studer absent, and mean to go with Mary in search of him
to-morrow at Gurnigel, five leagues distant. At Thun we
saw the Jungfrau in great beauty, and though the view of
the Alps from the ramparts here was not so clear as that we
saw in 1818, it was nevertheless very fine, and only half
covered with white clouds mingling with the snows. They
have now four bears in the moat here, and a herd of deer.
The town and everything in the canton looks most flourishing.
As my rapid tour has served to excite, rather than satisfy
my curiosity about the geology of Switzerland, I want to
ask Studer a hundred queries. He is just about to publish
on the geology of the country, and ought to stand a good
examination.

Fribourg: August 1.—We are stopping here for a twelve
o'clock dinner, having breakfasted at six o'clock. The view
of the Alps as we left Berne was very fine. Yesterday was
a well-spent day. We started early for Gurnigel, the highest
watering-place in Switzerland, among the fir woods. Found
Studer, a good, honest, intelligent Swiss, not quite read up

to Saturday night, but not so far behind as to be ignorant of a certain work, which made him receive me well, and express himself grateful for my visit. He took me to two ravines, where I saw a considerable section of the flanks of the Alps, very like the Pyrenees. He also gave me useful geological hints for the Simplon, Lago Maggiore, &c. The valley which led to Gurnigel was splendid, and the view of the Alps from it.

Ever your affectionate brother,
CHARLES LYELL.

To HIS FATHER.

Simplon : August 5, 1832.

My dear Father,—We have just reached this village, after ascending from Brieg in a most beautifully clear day, and are sure I think of seeing the descent, which I thought so much more beautiful, in most favourable weather.ʹ They say they have not had for a long time so little travelling over the Simplon—absolutely no English. Wars, cholera, and other fevers seem to have put an end to travelling. In the fine parts of Switzerland many French families driven away from Paris by cholera have come and sojourned, which has been some compensation, but none to this noble pass. In spite of this, however, no expense has been spared in keeping up the road here, which is in beautiful order, at least the Swiss side.

Domo d'Ossola, evening.—We reached this by six o'clock. Mary has been amused at the costume of the peasants, at the stone floor of our bedroom, and other novelties to her ; among the rest, the announcement that no milk, not a thimbleful, can be purchased in the whole town. And such herds of cows and goats as we passed. So we solaced ourselves with a repast which costs no more, and is perhaps more suited to the climate, an ice and a roll of bread. In my way up the Vallais I just stopped at Bex to call on Charpentier, the geologist of the Pyrenees, now superintendent of the salt mines there. We took a car and drove to his villa. He told me something of the geology of the country, and said if I would return and examine it he should be glad

to show it to me. We find the people here occupied with a
fête ' per l'onore della Vergine:' soldiers firing, band playing,
and a great pole in the market-place, with prizes for the
boy who climbs to the top; but they have put up so tall a fir
from the mountains, that no one can get a fifth of the way
up. But the soap will be rubbed off in time, and by midnight
some one will mount. The balconies are full of ladies gaily
dressed, and the crowd laughs loudly as often as any one
slips down the pole. They call this *la cocagna.* At Vevey,
I wrote to my geological friend Necker, to announce our
intended visit to Geneva and Chamouni.

Milan: August 7.—We saw the Boromean Isles in great
beauty, and had the gardener, who was so fair a botanist, to
show us the trees and shrubs. I am grieved to hear that
Boromeo's eldest son (aged about forty), a geologist to whom
I had letters last time I was here, and who *in spite* of his
rank is really advanced in scientific knowledge, is gone to
Aix in Savoy. At Arona, yesterday, we went up to San
Carlo Boromeo's statue, really a fine thing, and interesting
also from the great deception, produced I suppose by its good
proportion; for when you are at the foot of the statue you
would not believe that your arm could go into the nose,
whereas half your body, with head and arms, may be packed
into it. You mount by a long ladder to half the height,
then by an iron bar within, not a thing for *ladies,* nor indeed
for gentlemen who have not good nerves. I went up, and
got into the *nose,* at least part of me. Mary was glad when
I was down again, though really there is no danger, and, on
the other hand, no recompense for the trouble and fine of
four francs. After we had boated it most agreeably to the
isles, and got off in the carriage, a storm, the first rain for
two entire months, came down and laid the dust. It has
much refreshed the country. The Lago Maggiore is nearly
of the same colour (cabbage-green if I may be so unpoetical)
as the lakes of Switzerland which send their waters to the
Rhine. Geneva Lake is a deep blue, and, if I remember, much
more like the Mediterranean, to which its waters flow. Now
that they have made out that the sea contains iodine, as do
many mineral springs, and that there are innumerable mineral

springs in the Alps, I suppose they will find a solution of
their colour, but I forget Davy's conjectures on the subject.
With love to all,

Believe me your affectionate son,

CHARLES LYELL.

To CHARLES BABBAGE, ESQ.

16 Hart Street, Bloomsbury : Saturday, January 5, 1833.

My dear Babbage.—This year I am determined to give
the Temple of Serapis in grand style, and am preparing
scenic illustrations.

Would you be so kind as to lend me again all the plates
on that region which you let me have last year, and which I
was unable then to make use of.

The Royal Institution, hearing that the King's College
will not admit *ladies,* have sent me an invitation of a pressing
kind to give them some six or eight lectures after Easter,
offering me what I believe they consider very favourable
terms as far as emolument. I am inclined to accept, but
must think of it. I shall have time certainly.

If not at home when the bearer leaves this, you can look
out the plates and leave them out for me, and I will send
again. I should like to show you Sir W. Hamilton's plate
of the place where I think you found the serpulæ on the
cliff south of Puzzuoli, to ask you if am right.

Stokes and Broderip are to dine here on Tuesday at six
o'clock, no one else. Could you not join the party, and
renew your acquaintance with my wife? I cannot tell you
how much pleasure it would give us, to see you here at our
first party. Ever yours most truly,

CHARLES LYELL.

To GIDEON MANTELL, ESQ.

London : April 30, 1833.

My dear Mantell,—Like all the world, I have had the
influenza, which went through my house, servants, visitor and
all, save Mrs. Lyell, who remained well to nurse us. This
embarras, and getting *out* vol. iii. and *up* two courses of
lectures, must be my excuse for not having written to you or

anyone. My introductory lecture at the Royal Institution last
Thursday was attended by 250 persons. This morning, the
first King's College lecture by 200 persons, or, as some compute,
250. Nevertheless, I doubt whether I shall have any class,
scarce any having as yet entered ; and Jones, the Professor
of Political Economy, a very eloquent man, having had an
audience of 200 the first day, had on the third not *one individual,* and now he goes on lecturing to five, which I would
not. I was very happy to have my 200 to preach the true
doctrine to, gratis.

I am glad you have got your book out so soon. I mean
this summer to geologise in the Tyrolean Alps, and make
up my mind on Von Buch's dolomisation theory.

I start June 9, with Mrs. Lyell.

I have been taking lessons with her on shells from G.
Sowerby, who is a bankrupt, and now wishes to be a teacher
in Natural History, for which he is better qualified than for
shop speculations.

Believe me, ever most sincerely yours,

CHARLES LYELL.

To DR. FLEMING.

16 Hart Street, Bloomsbury, London : May 1, 1833.

My dear Dr. Fleming,—I am at a loss to decide how to
send you a copy of my third volume, which has been out this
week. I think you told me you should soon establish some
regular mode of communicating between London and Edinburgh, and that you would let me know how to forward a
parcel. I long to hear whether you have made progress in
the Fife coalfield, especially in the fossil shells and plants,
and *vertebrated* animals of the period, when only 'the
simplest forms of organic life' were in existence. As Agassiz
the icthyologist was seized with a complaint in his eyes, and
his visit to England consequently deferred, I have made no
step towards settling the genus and species of my red sandstone fish, which I much regret.

Since I launched my work, I have been busy lecturing.
The King's College governors determined this year that
ladies should not enter my lecture-room, because it diverted

the attention of the young students, of whom I had *two* in number from the college last year, and *two* this. My class being thus cut down to fifteen, I gave them a short course, which is now going on, and I find them very attentive and in earnest, but I would not of course give the time to such a class again. The Royal Institution invited me very warmly to lecture in their theatre, and I consented to seven lectures, which have been attended by 250 persons (scarce ever less), about a hundred of them *men*. But I do not fancy this want of concentration of my lecturing labours, and shall either settle upon some one theatre, or stick to writing, which I believe is most improving and profitable in every way. I regret that the bishops cut short my career at King's College, as I should have had a splendid class this year, and thrice as profitable as at the Royal Institution, but there seems no way of having a large audience but by one of two methods—an academical class, or one open to women as well as men. Grant lectured *gratis* at the Zoological Society to overflowing rooms, but the moment he began at the London University, for a trifling fee, only about eight or ten students came. Yet people will still buy dearish books, and though willing to hear Turner lecture gratis on Chemistry or Geology, they will not *pay* at the rate of 1s. a lecture to hear him. My book has got on so rapidly, as to put me in great good humour with the public. Bakewell has just come out with a new edition, but I have not heard whether there is anything new in it. I get on in conchology somewhat, and Mrs. Lyell and my sisters, two of whom are now with me, make rapid progress. I am going this year to the Tyrol, to make up my mind about Von Buch's theory of dolomisation of the limestone of the Alps. Ever most truly yours,

CHARLES LYELL.

To HIS SISTER.

Paris : June 18, 1833.

My dear Marianne,—We are just returned from a visit to Montmartre, where I have shown Mary the underground streets or excavations made to extract gypsum—the same gypsum in which Cuvier's celebrated animals were found.

She had also an opportunity of collecting a fossil shell or two for the first time. The view of Paris from Montmartre is by far the finest there is; St. Denis seems very near you on one side. I was much amused at the Geological Society last night. My last volume has told here very much. Prévost assures me that the first two were far beyond them here. He says they are beginning to appreciate them, but that it will require several years before they see as clearly as he does the geological bearing of them. There have been several schemes to translate the work, but the German translation checks them; for they say that in France every one writes, and no one reads or buys, and that they depend regularly on Germany and England for the sale of two-thirds of a scientific work. Their sale is far more limited than in Germany. I am resolved to continue with zeal to collect fossils and study conchology. I am very glad to see Mrs. Somerville again, and hear her opinion on society here, which upon the whole is favourable, but not in all things. I know no one further from being prudish, and yet she tells me that she can scarce admit any modern French novel nor literary work into her house, so unfit are they for her daughters.

CHARLES LYELL.

To HIS MOTHER.

Heidelberg: July 30, 1833.

My dear Mother.—I am now for the third time upon the Bergstrasse, and admire it as much as when I first visited it thirteen years ago with my father. When we left Bonn, Dr. Thomson and his son were agreeable companions in the steamboat as far as Coblenz. We first stopped at Neuwied, leaving the boat and taking a carriage at that small village-metropolis, to cross the plain to the boundary hills. We went over an immense flat plain about five miles broad, which is composed of horizontal beds of volcanic matter, chiefly white lapilli of pumice, which we saw at the pits, and when we came to the small river Sayn. Then at the town of the same name we saw the iron-works, having an introduction to the Superintendent. The views of the plain from the hills and the old ruined castle at Sayn are beauti-

ful. We went on to Ehrenbreitstein, slept there, and next
morning rejoined the steamer, to be carried on to Bingen,
where we disembarked for good, and got a carriage for
Mayence and did much geology on the way, in what we call
the Mayence tertiary basin. On the 27th July we followed
the Rhine to Oppenheim, and then to Alzy, Eppelsheim, and
Worms, seeing many quarries, and at Worms much of historical interest, particularly as what I had read in Robertson
and Schiller has left a lasting impression. A few miles
from Worms is a large lime tree of great age, which stands
like a giant, all the other trees in the plain being younger
and of moderate height. This tree has been spared because
tradition has said that Luther, when he was on his way to
the Diet of Worms, sat under its shade to rest. It is called
' Luther's Linden.' Worms is a fine old place, standing in a
great plain. We looked down upon it, and in the distance
on the one side, the Hardt mountains and the Donnersberg,
which we had approached in our excursion the day before,
and on the other, the hills at the foot of which we now are.
Opposite the cathedral, on the very spot where the Rath-
haus stood, while the Diet was held, stands the Lutheran
church, in which the Lutherans and Calvinists now join
worship in the same service here (Darmstadt), as in Prussia.
The church is very simple and plain in its exterior, but
within is remarkably striking. It is not much more than a
century old. On the entrance side is a large picture of the
Diet of Worms, Charles V. in the chair, and on his right
cardinals and bishops ; on his left, nobles in their robes ;
before him, Luther standing and speaking against Eckius,
who is contending on the other side. There is a gallery on
one side only of the church, all along the front of which is a
double row of pictures, more than a hundred I should think,
painted I believe on wood, the upper row in illustration of
the Old, the lower of the New, Testament. Each tells its tale
admirably, and the effect of the whole as adorning the church,
the ceiling of which is coloured so as to harmonise, is quite
beautiful. The pictures are very amusing, and many of the
marvellous stories of the Jewish history are collected ; and
from the Revelations there are dragons with many heads
and other beasts, and Death on the pale horse, &c. I should

suppose that the children, and not a few grown-up folk, are
employed during service in divining what the pictures relate
to. Martin Luther himself would stand no chance with the
artist. For what he said on this spot I believe he was imprisoned : it is curious that his church should now stand
there, exactly opposite the old vast Catholic cathedral.

July 31.—After passing along the plain from Worms to
Manheim, we crossed the Rhine on entering the latter, and
climbed the tower of the observatory, where we had a splendid view. In the town, the juncture of the Neckar with the
Rhine ; in the distance, Spires on one side, Worms on the
other ; the Vosges and Black Forest mountains to the south,
and the Taunus to the north. Professors Von Leonhard and
Bronn have been very attentive here. Yesterday evening we
spent with the former, and this evening with the latter.
After the party at Leonhard's, Mary and I took a moonlight
walk over the ruins of the old castle, and stood on the splendid
terrace which overhangs the Neckar. I received a letter
from Count Münster of Bayreuth here, in answer to one
from me from Bonn, saying he will be glad to see me, and
will be at home, and pointing out things which he recommends me to see before reaching him. He has not only the
first collection in Germany of fossil remains, but is reputed
to know more about them than any one else. We go by
Heilbronn, Stuttgardt, Göppingen, Pappenheim, Solenhofen,
Nüremberg, Bayreuth (seeing the famous caves of Muggen-
dorf, &c., between the last places), Bamberg, Würtzburg,
Aschaffenberg, Frankfurt, Bonn.

With my love to all at Kinnordy, believe me, my dearest
mother, your affectionate son,

CHARLES LYELL.

To HIS FATHER.

Manheim, near Pappenheim, Bavaria : August 8, 1833.

My dear Father,—We have just reached this on our way
to the famous lithographic stone quarries at Solenhofen.
My last letter to my mother was just before quitting Heidelberg. I attended a lecture of Von Leonhard on mineralogy,
and was much interested. He has been the first to introduce

a modified Lancastrian system of instruction into his lectures. When he has given a general history of a particular
mineral, he describes a certain number of specimens to four
or five of the more advanced of his class, who have already
followed one or two courses. He then passes to one end of
the room, and a small group closes round him, each seeing
the specimen well. He goes on: ' *Epidote*, this pale green is
epidote, the white is quartz,' &c. The others go round the
room with their specimens at the same time, so the whole
class is for some time broken up into a number of private
lecture pupils, till the exhibition or ' demonstration ' is over,
when the whole are again intent on the general views of the
Professor, who holds forth in a pleasant, easy, extempore
strain. He only got through two minerals in the hour. At
Stuttgardt I got a hasty view of two collections and of Drs.
Hähl and Jäger, the latter an author on fossil remains. I
was lucky enough to meet Professor Voltz, of Strasbourg, at
Dr. Hähl's, who gave much information on the fossils.

We saw Danneker's studio, and the sculpture by him in
the palace. I do not admire D.'s Ariadne, which is so
much talked of. His Christ I thought very fine, and many
others beautiful. I went to Göppingen, at the suggestion of
Count Münster. Dr. Hartmann, physician at Göppingen,
has a magnificent collection of organic remains of secondary
rocks, and was very liberal in giving me specimens, and took
us to some localities where we collected plentifully.

Believe me, your affectionate son,

CHARLES LYELL.

To G. MANTELL, ESQ.

16 Hart Street, London : September 16, 1833.

My dear Mantell,—I have just arrived from Dover, having
passed through Belgium in my way. I saw there at Liege
the collection of Dr. Schmerling, who in *three years* has, by
his own exertions and the incessant labour of a clever
amateur servant, cleared out some twenty caves untouched
by any previous searcher, and has filled a truly splendid
museum. He numbers already thrice the number of fossil
cavern mammalia known when Buckland wrote his ' Idola
specûs ; ' and such is the prodigious number of the individuals

of some species, the bears, for example, of which he has five species, one large, one new, that several entire skeletons will be constructed. Oh, that the Lewes chalk had been cavernous! And he has these and a number of yet unexplored, and shortly to be investigated holes, all to himself. But envy him not—you can imagine what he feels at being far from a metropolis which can afford him sympathy; and having not one congenial soul at Liege, and none who take any interest in his discoveries save the priests—and what kind *they* take you may guess more especially as he has found human remains in breccia, embedded with the extinct species, under circumstances far more difficult to get over than any I have previously heard of. The *three* coats or layers of stalagmite cited by me at Choquier are true. Talking of the priests, they have obtained grants for new monkish establishments in Liege, while the University of Ghent falls to the ground, and the Protestant professors are cashiered, and King William's patronage of natural history excluded. The movement was mainly one of Catholic bigots against a king who wished to introduce schools, and who, whatever faults he committed, was of all European sovereigns the greatest promoter and most judicious patron of physical science. Leopold has nothing left for it but to lean on the priests, reinforced as they are by the Jesuits exiled from France. But as yet the Belgian press is free: with that there is always hope. I examined near Mons strata of the age of Maestricht, and collected shells of which 250 specimens have been found, and were shown me by Count Duchatel. I saw those newer beds reposing on white chalk with flints, and this on chloritic chalk and chalk marl or malm, very soft and clayey, full of upper G. Sand shells (pecten 5—costatus, &c.). The ancient shore of the chalk, then resting upon inclined mountain limestone, is still seen, as the quarries are worked out, strewed with *pebbles* of coralline limestone (mountain), of old red sandstone, coal shale, &c.; these pebbles covered with green sand species of *plagiostoma, patella, emarginula,* &c., the softer old-red pebbles bored by fistularia, of which the shells remain —the subjacent limestone seen *in situ*, perforated by large regular sarcophagous hollows. But better than all, rents and sinuous cavities are seen in the underlying limestone into which the sand of the 'malm' or 'firestone' beds, to speak

Sussex to you, fell, and into which innumerable shells of the cretaceous era were drifted. These shells escaped being stonified, and consequently 200 species the same as in the marl are in a state of perfection rarely equalled by *tertiary* shells. Terebratulæ loose, and with all the internal parts entire. It is a wonderful country, and well ransacked by Duchatel, who took me to the best localities.

Very truly yours,   CHARLES LYELL.

[In 1834, he made a tour of several months in Sweden, and wrote a journal to his wife, which was posted at intervals, from which large extracts are given in the following pages. In November of this year he read a paper to the Royal Society on the proofs of a gradual rising of the land in that country, which was published in the 'Philosophical Transactions' the following year, as the Bakerian Lecture.]

CORRESPONDENCE.

To C. BABBAGE, ESQ.

16 Hart Street: January 16, 1834.

Dear Babbage,—I am half afraid you are gone to Cambridge, but will take the chance of this note reaching you, as I am at a stand for want of an answer to what may be a simple question.

Daniell tells me that the melting point of iron, or rather the heat of *melted* iron, is about 4,000° Fahrenheit; and that so long as any fragment of solid iron remains in the melted mass it is impossible to raise the temperature higher—just as water will not be raised above 33° so long as any ice is floating in it. Now he (Daniell) asks whether, if this be the case, the interior of the planet can, as the central-heat men assume, be at temperatures far *above* 4,000° F.? Would this be possible so long as a solid crust above remains unmelted?

I objected that *pressure* might possibly allow of great increment of heat beyond 4,000° without liquefaction being produced. But Daniell doubts whether iron and other solids would be prevented from melting by pressure? At all events, would it not follow that the central parts of the earth, or all below the outer rind, are incandescent and solid rather than fluid? I see that Fourier and other mathematicians assume that it is consistent with the laws which

regulate the progress of heat through conducting bodies
that a cold, frozen, solid crust may cover a nucleus which
may be at *any imaginable* temperature.   Query.

Dr. Young, in his lectures on Natural Philosophy, says
that at the earth's centre steel would be compressed into
one-fourth its bulk, and stone into one-eighth.

Ever yours most truly,   CHARLES LYELL.

P.S.—If you were not a Professor I should have craved a
breakfast with you on this subject; but fly me a note, if
possible, by post.

To LEONARD HORNER, ESQ.

London: February 26, 1834.

My dear Horner,—Your paper is positively to come on
to-night—first my Pyrenees, then you, then Egerton.   I
stood out for this arrangement, as we had priority by right.
The answering was a heavy business.   Greenough not in
spirits, Buckland heavy, no Sedgwick.   I communicated
some information received the same day from Boué about
the Strasbourg meeting, and the Stuttgardt reception of the
savans, which the King of Wurtemburg is preparing.   I
also took occasion to recommend those who were going to
the Continent to visit Germany, and praised their geologists,
and stated how interesting the geology of Germany was in
relation to our own.   Lastly, I gave a rapid sketch of the
controversy about the level of the Baltic and rise of Sweden,
and announced that I was going there.

Fitton's short speech was the only one which made them
laugh.   Dublin and Dr. Fitton was the toast.   He descended
on Antrim, and said that Playfair told him that when he
(Playfair) visited the Giant's Causeway 'the sea was rough,
a swell, what in Ireland would be called a state of *agitation*.
The boatman proud, as all Irishmen are of their country,
and annoyed that the stranger gentleman should be disappointed, after two ineffectual attempts to get to the
Causeway, said, ' Your honour sees that it's not the sea that
is out of humour: the sea (say) would be quiet, your honour,
if the wind would but let her alone.'   The speakers were
Greenough, Buckland, Sir T. Ackland, Sir J. Johnston, Murchison, Fitton, De la Bêche, Lyell, J. Taylor, Dr. Turner.
Fitton's account of Greenough's speech was that ' more than

half of it was employed in pelting Lyell with nonsense, and
at the end summing up in his favour.'

Yours affectionately,   CHARLES LYELL.

To LEONARD HORNER, ESQ.

April 17, 1834.

My dear Horner,—I have been so often in danger of
being fairly run down by the printers the last fortnight, that I
have found no time to write and congratulate you on the favourable aspect of affairs as connected with the Factory Bill.
I begin to be sanguine that the Bill, after considerable modifications when new clauses come into play, will last our time
and do good, especially in respect to education, the chief
point in which I conceive governments to have a right to
interfere, or rather a duty.

I should hope that by returning again and again to the
same districts (when you have got the factories in train)
you will be able to effect much in geology. For the grand
secret is to revisit countries, and to compare them frequently,
after thinking over what you see in the interval. If you
write on volcanos you can bring in the ancient lava and
dikes near Edinburgh, and the volcanic rocks near Dundee.
I wrote to Bowstead,[2] to say that if he had any notice on the
löess of Andernach, I would read it when I read my paper.
After some time he replied that he can venture nothing, but
should have answered my letter sooner, had he not been employed this Easter vacation in *geologising* in Derbyshire, with
a Mr. Hopkins. Is it not delightful to have got such a man
fairly bitten? He concludes his letter with saying he shall
see us in June (when we shall be far away), and adds : ' I
was delighted to hear of Mr. Horner's appointment to the
important and responsible situation which he fills,[3] as I
shall ever preserve a grateful recollection of the kindness,
encouragement, and assistance which I derived from him in
my first feeble attempts at geological investigation. I
should long ago have written, had I known Mr. Horner's
address in this country, to congratulate him on his appointment.' Amongst numerous sallies of Sydney Smith, some of

---

[2] Afterwards Bishop of Lichfield.

[3] The Factory Act was passed in August 1853, and Mr. Horner was
appointed one of the four Inspectors of Factories that autumn.

which I presume have been told you, when I dined there, he wondered when 'Cockburn would come up to town, and be baptised,' &c. On my saying that he was already one of the most social of beings, a most excellent fellow, &c., Sydney said, 'I know him well, and his merits; but why should a man say I will be intimate and social only when I have known people fifteen years? One must be always beginning, and extending one's acquaintance, or be left alone. Life is not long enough for such a plan. A man must allow for something intermediate between cold reserve and the familiarity of schoolboys. It will not do, sir, depend upon it, for any man to say, " I will shut my mouth, and be reserved to all with whom I cannot play at *leap-frog.*" '

You will be glad to hear that Constant Prévost overcame the École des Mines and all the intrigues, and is now *our* French Pres. G. S. I have heard no particulars. Last night I took my brother Hal to see a comparison of the Drummond light and the most intense light of voltaic pile which could be made. The latter they say was most bright, but it was not sustained enough to allow of a fair comparison. Buckland was there, and very full of Hibbert's discovery of a Plesiosaur and *crocodile* in the coal of Edinburgh, or Burdie House quarries. Murchison said that he had not made out that it was below mountain limestone; but I said that, having visited Burdie House, I could not doubt that any saurian found there must have been contemporaneous with the coal plants.

I hope you will be obliged to go to the mills and manse of Clackmannan. I know of no one so able to help one in Scotch geology as Dr. Fleming, especially on the coal. If you are much in Edinburgh, I would determine in five or six years to have the finest collection of the carboniferous beds north of the Tweed. A few pounds annually expended would be justifiable as an *investment,* as Murchison said to me in 1827, after he and others had lost thousands in speculations, and when old Sowerby's family sold his collection well. ' What an ass I was, when I sold my land, not to invest in fossils ! ' But I know how dangerous it is to begin, unless one confines oneself to a very limited field, either to some one branch of paleontology, or a certain geographical district.   Believe me, affectionately yours,

CHARLES LYELL.

CHAPTER XVII.

MAY–JUNE 1834.

DEPARTS FOR DENMARK—LÜBECK—PICKS UP SHELLS ON THE BALTIC—
ARRIVES AT COPENHAGEN—EXPEDITION TO MÖEN WITH DR. FORCH-
HAMMER — THE CROWN PRINCE CHRISTIAN — VISIT TO THE COUNTRY
PALACE OF SORGEN-FREY—GEOLOGY—DEPARTS FOR SWEDEN—ENGAGES
AN INTERPRETER—LEGEND OF THE BÄCKE-HÄSTEN—CASTLE OF KALMAR.

JOURNAL.

To His Wife.

On board ship for Hamburgh : May 21, 1834.

My dearest Mary, —I have been ten hours without a word
with my love, but thinking of her more than half the time,
and comforting myself that she is less alone than I am. I
foresee much profit from a full conference with the Danes
and Swedes. I have learnt much to-day about the state of
Heligoland, and will try to get drawings of it. I have been
reading Forchhammer, and am quite sure that the geology of
the Danish isles, Moën and Seeland in particular, on my way
to Copenhagen, will be one of the most important things I
have to do. I remember that Deshayes told me that the
fossils of Scania were those of Maestricht and Ciply, and
from the Cranias and other shells mentioned by Forchhammer, I shall probably find Möen to be so too.

Hamburgh: Sunday, 25.—The scenery on nearing Altona is
very beautiful, and sections by the riverside which I sketched,
and had decided to see if possible. Landed, and went to Hôtel
de Russie. This is a curious old place, more like Nürnberg
than any other we saw together, with a likeness to a Dutch
town also, with its canals. Drove to Altona, to Prof. Schu-

macher's, who was not at home. Resolved to dash on in a
drosky to the section lower down the Elbe than Altona, at
about three miles from Hamburgh. Got to the very place I
sketched in the morning, at the Teufel's brücke—cliffs, 60 or
70 feet high. Filled three pages of note-book. Saw the
source of the great Holstein granite blocks—gathered shells
thrown ashore by the Elbe. Returned to Altona and saw Schumacher, Danish Astronomer Royal, a man who has done much
to advance practical astronomy as applied to navigation.
Went with him to a Mr. Parish, a rich Anglo-Hamburgh merchant, who took me where I wished in his carriage. I have
made progress already, I think, in making out the Holstein
formation, and by aid of a peasant got two shells which are
rare.

Copenhagen: May 28, 1834.—I have not had a moment of
leisure to write to you since the morning I left Hamburgh, in
a carriage which that kind man Schumacher sent me from
Altona, because he said my bones would be broken in a post-
carriage, as the road was so bad that twice my head went
against the roof or windows. There are two reasons for the
disgraceful state of a road which connects two rich and
active independent cities, which have their vote. The first
is a geological cause, loose sand, and huge boulders floating
in it. But why with their wealth do they not break up the
granite and macadamise? The reason is that it would half
ruin Denmark's king, through whose dominion the road runs.
His revenue is principally derived from the dues paid on
entering the Sound, and on these the last debt was guaran-
teed. If the commerce of Hamburgh was conveyed by a
good road to Lübeck, half these dues would be lost. So that
though Hamburgh and Lübeck offered to make the road, Denmark refused. The negotiation is still going on, and it is
hoped Denmark will be shamed into concession. After
getting through innumerable small jobs at Hamburgh, and
the horrors of packing with an increase of things, small
boxes of shells, &c., I at last got off. It blew a hurricane,
and the loose sand and the cold forced me to shut up at last,
by which I lost no geology. I worked at Rang,[1] and was
glad you packed him up, an excellent book. I had collected

[1] Work on Conchology.

shells on the banks of the Elbe, which excited my desire to
consult him. On to Oldesloe, about five German miles,
having baited one pair of steeds three or four times. Here
are salt works, which I saw: then to Segeberg. Here an extraordinary mountain of gypsum, &c., bursts up through the
loose sand. I delivered my letter from Schumacher to
Geheimrath Von Rosen, who was just returned, late in the
evening, with his wife and six young children, from a tour of
seven days. Simple mannered Germans. After tea he went
with me up the Hill of Gyps, and introduced me to a Norwegian Bergmann, who superintends the works. He was
quite my beau ideal of an intelligent, frank, independent
Norwegian. It was darkish, so it was arranged that between
six and seven next morning I and the Bergmann should examine the hill together, which we did, much to my edifica-
tion. He was delighted to find I had letters to Count
Wedel and other Norwegians. Five hours' ploughing of the
deep sands took us to Lübeck. On the way I gathered some
freshwater and land shells; read Rang, and Forchammer,
and De la Bêche, on chalk, &c.; wished you were in the carriage; and at last got sight of the towers of Lübeck, a very
curious old city. I had only three-quarters of an hour before
the steamboat started, which conveyed us sixteen miles by
water down the Trave, first half fresh, and last salt. By land
it was only eight miles, but the steam beats the carriage.
The river sometimes opened into a kind of estuary, then
contracted—very picturesque. A fleet of thirty fishing-boats
were met all together in full sail, going to the Baltic. In two
hours and a quarter we were at the small port of Lübeck,
Travemünde, where we found two fine steamers, one Russian,
just going to Petersburg, the largest; and the other Danish,
into which we and our luggage were all transferred.

When my luggage was safe, we had still three-quarters
of an hour, and I was ambitious of picking up a shell or two
on the borders of the Baltic. I landed, as we were alongside
the pier, in our new boat—it was low water. The shore was
covered with *Turbo littoreus, Mytilus edulis, Cardium edule,
Mactra, Tellina,* and quantities of a minute *paludina,* I suppose
P. ulva. This association of the *paludina* with the marine
shells delighted me, as analogous to the Mayence basin, which

was no doubt brackish too. After we had started, I stood washing my new treasures, the shells, in a basin, and packing them up, to the admiration of the crew, who some of them begged the loan of my magnifying-glass. A Russian gentleman, who afterwards proved to be Mr. Eperlim, Russian Consul at Helsingborg, offered, as I seemed to him a naturalist, his services if I wished any information in my travels in the North. This was *àpropos* enough, and I profited greatly on the voyage by his hints and encouragement as to travelling cheaply and easily in Sweden. As we sailed out of the Gulf of Lübeck, we saw the Mecklenburg coast on our right, evidently composed of the same loam, sand, &c., as Holstein, and producing a similar aspect of country, like parts of England : the cottages and hedges of Holstein are much like the south of England. I found my little German there most useful, and should have learnt little without it, either at Oldesloe saltworks, or from the good Bergmann of Segeberg. The steamer was a good one from Lübeck, and we began with ten miles an hour, which was much reduced by a north wind steadily against us all the rest of the voyage. I ordered them to wake me when in sight of Möen. Up at seven o'clock, sketching the white chalk cliffs of that isle, which I hoped to visit again. To bed again. Then up, and read for some hours. Left Möen at last, and in sight of the white cliffs of Seeland, which are much lower. Those of Möen are 400 feet high, and look like the white cliffs of Dover. The first point of Sweden which I saw was the end of the small promontory of Falsterbo—low, only with a line of trees growing as it were from the water. Then the long line of the Swedish coast, and the Danish isle of Amager. Then the celebrated city of the Baltic, which, seen from the sea, disappoints one ; but when we had entered it, the squares and streets were fine. Went to Forchhammer. It being a great object for me to get Forchhammer to go with me, I wrote him a letter explaining my views. . . .

May 29.—After six hours' work with Forchhammer in the museum, he agreed to go to Möen, Seeland, &c., with me. We start to-morrow at six o'clock. Dr. Beck, the Prince's naturalist, a man with whom I am delighted, accompanies me to Lund on Wednesday next, to see Nilsson. There is much

doing here, which is unknown in England and France. I am
more than ever struck with the extreme slowness with which
science travels, what with the multiplicity of languages,
douanes, &c.

Copenhagen: June 1, 1834.—I have just returned from an
expedition of three days with Dr. Forchhammer, in which we
have travelled by land, posting upwards of two hundred
English miles, first from Copenhagen to Stevns Klint on the
east coast of Seeland, and then to the small ferry which
separates the isles of Seeland and Möen, visiting the quarries
of Faxo on our way : then to the tall white chalk cliffs of
Möen. The first day was cold and rather rainy, but on the
whole good working weather—the results will make alone an
interesting memoir illustrated by several drawings. I have
seen the junction of the upper chalk and Ciply or Maestricht
beds, and satisfied myself that Dr. Forchhammer has mis-
interpreted the Möen cliffs and their geological relations.
But the case was a most difficult one, with which I need not
trouble you yet. The scenery was often beautiful, friths and
bays running into the green isles ; a low undulating country
running in Möen into hills five hundred feet high, where
the chalk is covered by splendid beech trees ; cliffs near four
hundred feet high, covered with wood at the summit, and
with a fine shingle beach of flint at their base. My companion well informed on many branches in which I wanted
instruction, and who has taught me much of the geology of
regions I have just been seeing, and am to see. It opens a
new world coming here, and the Danes, I find, can tell one
much of Iceland and Greenland, as well as of the Baltic.
The night before this expedition I went to Mr. Macdougal,
who was very civil, and on my saying I was going to Dr.
Beck, the Prince's naturalist, said he would go with me, as
he knew him. I found Beck very strong in conchology.
He has, in 'his Highness's' collection, one of the finest in
Europe, if not the best, even up to the last new shells of
Cuming and others in London. Beck offered to go to
Lund with me, which I most gladly accepted, the more so as
he is intimate with Nilsson. My scheme is to try to get
some naturalist or student of Lund to go with me to Stock-
holm, and thus I should escape a servant, whose expenses

would be as great, and I should get instead an intelligent companion, such as Nilsson would recommend. It was strange, on returning by a different road to Copenhagen this evening, to find this capital a fortified town. Forchhammer tells me that one of Charles XII. of Sweden's dashing enterprises was to cross the ice with his army and cannon, marching down the Sound to attack Denmark, since which they have fortified Copenhagen. Forchhammer speaks and understands English like an Englishman. On my return, I found a letter from Councillor of State Adler, saying that the Prince Christian had ordered Count Vargas and Dr. Beck to attend me at his Highness's collection at ten o'clock to-morrow morning, and that the Prince had charged him to invite me to dine at the Palace to-morrow. I wish it had been any other day, for I fear I shall get late, if at all, to Forchhammer's *soirée*, where the famous Oersted, and the travellers, Pingel and Olafsen, and two other good men, whom I want to see as much of as possible, are asked to meet me.

Malmö in Sweden: June 4.—I was at the museum in the Palace at the appointed hour, and tried in vain to get Count Vargas, who was to show me the minerals, not to wait on me, as I wished to set to at the shells and fossils. I first got Beck to name the shells I had picked up; among others, a true *Neritina,* which abounds in the brackish water of the Baltic. Beck has decided that it differs from *Fluviatilis* and others, and says he has found it in the Elbe, as well as the Baltic, and would give it any name I chose, so I christened it 'subsalsa.' I was soon informed by the Secretary Adler, a gentlemanlike man, that the Prince wished to have an audience, so I was taken away from business to wait with others in the ante-room, and introduced to Baron Rumohr, also in attendance, who has written, I was told, on Grecian works of art, of virtu, and on *gastronomy.* When I had waited some time, long enough to be impatient when I thought of what I had left, I was introduced. The Prince was standing, decorated with orders, and playing the kingly part very well; good-looking rather, and of good height. He was very gentlemanlike, and said, 'I shall come to the collection when this ceremony is over, and spend the morning with you, and learn what you have made of the contro-

verted points of Möen. Meanwhile, as you are to be here one
day more, I may say I only named to-day on that account;
but if you have any engagement, dine with me at my country
house to-morrow.' I told him of Forchhammer's party, and
it was deferred till next day. He immediately sent, as I
afterwards learnt, to Dr. Forchhammer and Oersted to meet
me. After I had worked some time with Beck, the Prince
joined us, and spent many hours while I went over the Faxo
shells, an ancient coral reef, where the genera, *Cyprœa, Oliva,
Mitra, Voluta, Conus,* &c., had already in the chalk era
begun to flourish as now in the tropics. The Prince has
studied the characters of all the genera of shells. I had
several presented to me by him, as we went over the drawers.
When Möen was discussed, my task was a delicate one. It
would have been easy to damage Forchhammer, for Vargas,
Beck, and others, who thought him wrong, were glad to hail
my decision, though in some respects I think them wrong
also. I enlarged on the difficulty of the case, and declared
that had I not studied both the Isle of Wight and Norfolk
coast, I should have perhaps fallen into what I called Forchhammer's mistake. That he was ten years stronger in
geology now than when he wrote on Möen. In short, I took
care to do Forchhammer justice, praising also the candour
with which he was now inclined to abandon opinions to
which he stood so pointedly committed. The Prince then
ordered Vargas and Beck to bring me out next day, seven
miles, in one of his carriages, and I resumed my inquiries of
Beck in the collection. Vargas is a good-humoured man, but
a complete courtier. Beck called to say he was in great
distress, that his wife was very ill, and he could not possibly
go either to the dinner, or next day to Malmö and Lund with
me. I set off with Vargas in one of the royal carriages.
He told me that Prince Christian, or 'the crown prince,'
has no more than 12,000*l.* a year, which there may equal
twice that in England; that Adler has no easy matter to
manage his finances, as the Prince could not resist when a
good shell or mineral was offered. The country place was a
small château, with beautiful grounds in the English style.
When we entered the party was assembled. I was introduced to the Princess, a very handsome and interesting

woman, daughter of the Duke or Prince of Augustenburg, whose family will I believe reign in Holstein, if this Danish stock dies out, as is expected, when the small kingdom will be split up. She has a stately manner, yet is agreeable, and talked of my tour in Möen, &c. A German prince of Bertheim Steinfurt, related to the Duchess of Cumberland, led her out; then the Crown Prince took out the Countess Saxogeborne, princess of some house on the Rhine; then Count Moltke, who was long Danish ambassador in London, took out Countess Yule, a *dame d'honneur*; then Vargas another of these ladies; after which came Count Blücher, who is related to the general of that name, who sat next me, talks English well, and has married for love an English lady without fortune; then came Oersted, with two orders of merit decorating him. There was much laughing and free conversation at dinner. After dinner, Vargas wanted to escape, which led to an amusing scene. We all walked about the shrubberies, and the Prince pointed out to me and others the points where the trees were finest, or views best; when Vargas told me that the drive to a hill half way to Elsinore, which was proposed, would make me far too late in my packing, &c., upon which I agreed to return with him, provided he could do it with propriety. Meanwhile Forchhammer took it into his head, after talking with Count Holk, the Hof-Marschall, that Vargas was jealous of the Prince's attention to me, and Forchhammer accordingly begged Holk to take care that Vargas did not carry me off. So Vargas came to me and said, 'Now the Prince is with the Secretary Adler, answering letters—they will be for hours at it; and then comes the drive. Now I have only to get Holk to promise to make our excuses. Saying this, he led me to the ante-room of the Prince, where Holk and another were in attendance. On Vargas communicating the request, the master of the ceremonies laughed and said, 'Vargas, it is you that want to get home; go in yourself to the Prince.' An altercation ensued, with many a smart thrust and parry, when at last Holk said, 'Come, I will arrange it. There are three carriages and four, and one with two horses. Mr. Lyell and yourself shall be in the last, and when you get half way, I will contrive your retreat, and you shall be in

town by nine o'clock. Soon after this the Prince came. All
were ready. Holk called up the first carriage and four, and
the Prince put the Princess and some others into it, and
when the three were filled, and Vargas, being summoned,
had taken his place with a gracious bow as commanded, the
Prince turned to me, who remained alone, and said, 'You are
to go with me.' Up drove the carriage and two, and away
we went, preceded by a jäger as *avant-courier*, and with a
man on the box with two epaulets, and the coachman. The
train followed us, and made a gay *cortège*.

Lund : June 5.—My dearest Mary, the best way I can
comfort myself for the disappointment of having outrun the
tardy post, and found no letter from you here, is to continue
at once my journal, which makes me constantly think of you.
I had about three hours alone with the Prince. He asked me
to explain Faraday's last discoveries in electricity, of which
he heard I was the first narrator to Oersted. He entered
into it with pleasure. Then he inquired if I could explain
why beeches grow best after fir, and fir after beech, in his
country. I gave him Decandolle's theory of cropping, and
of the matter rejected by one plant being good for others ;
when he said, 'I will suggest another theory. Some plants
may absorb those gases and earthy matters which might
poison others—time may be required to purify the soil, and
then a tree of a different family may flourish.' I was so
struck with his way of putting it, that I asked Forchhammer
afterwards if it was not new. 'He got it,' said Forchhammer,
'from the very memoir of mine which he cited to you, but he
must have thought much of it, and has put it better, and
believes it is his own.' We had a splendid view from a hill,
to which all walked up, of Copenhagen, the royal parks, the
sea, the castle of Kronsborg at Elsinore. Then we continued
in the same carriages, the Prince describing to me on the way
their peat bogs, and fossils in them, mode of preparing peat
fuel, after which he sent all the other carriages home, and
took me to see some newly-discovered heathen tombs, huge
stones from the great drift boulders of granite. He also
good-humouredly stopped at a quarry of boulders, while I
got out and geologised it ; but while I was there he got the
jäger to help him over the dike, and came too, offering his

speculations as to which blocks came from Norway, &c. When we got home, we found the rest thirsty and waiting for tea, which would not have been touched until the Prince came, though it had been midnight. After coffee, when Vargas got me off, the Prince said before them all, ' If I did not know that your curiosity was excited about this country, and that you must return to visit Bornholm Island another year, I should be unwilling to let you run away in such haste.' His lady added a civil speech, and away we went in our carriage. Vargas amused me, as his ideas are so drilled to a court. He confessed that his only reason for cutting geology was ' that it was a science for those who could afford to be *independent* in their opinions, which he could not, for the controversies led to quarrels, and these to loss of place ! ' But he was in manner quite the gentleman.

Lund, in Scania : [2] *June* 5, 1834.—This evening, as Nilsson is unable to see me, I shall indulge in reminiscences of my proceedings during the last two days, which I have not yet sent you. I asked Prince Christian how Sweden, having only a few inhabitants, and so little a literary country, had done so much for science. He said, ' Because the universities were endowed with much of the wealth of the church, and there is an independence for men who, if they love science, may devote themselves for life to it.' He asked me whether I lived much in London, but he added, 'I need not ask, for I suppose you still hold your appointment in the University there.' I explained to him why I gave it up. He said, ' You do well, if you can dispense with it, to do so ; for after all it is a small circle you address as a *teacher* there, whereas by travelling and writing you can teach all the world.' This was all talked in French during our drive. At dinner the conversation was French and German, now and then running into Danish and English in different quarters. The Crown Prince has the character of being very aristocratical, tempered with great intelligence and information ; and that he has good men always about him, Vargas being the only thorough courtier.

[2] Scania, the ancient name for the southern provinces of Sweden, Malmö, and Christianstad.

Whenever people wished to be very civil in Denmark, they
began 'Herr Professoring' me, which goes for more there
than even in Germany.   In Denmark a title is called a ' character,' and here in Sweden I have been already asked
'what was my character?'   'When they ask me a favour,'
said Macdougall,'they say, " Now, good *Mr. Librarian* Macdougall,"' &c.   Prince Christian was King of Norway for a
short time, and would have remained so, had not the Norwegians been constrained by the Allies to be bound, much
against their wishes, to Sweden.   'There are twelve courts
to keep up,' said Vargas; and then he enumerated the different brothers or near relations of the King and Crown
Prince who had establishments.   The young Prince who is
banished for a season to Iceland is a son, by a former
marriage, of Prince Christian, and the mother is separated,
and at Rome—*mad*, as her son is thought by many to be at
times.   So was the father of the present King an avowed
madman, although it is a fact that he was named guardian
of the Duke of Ploën when he went mad, together with two
other guardians, the Emperor Paul of Russia and George III.
of England!   The young Prince is twenty-three, and married
several years, without children.   After him Holstein would be
separate.   As the monarchy is the most despotic in Europe,
the coming to the throne of such a man would of course be
a calamity.   Some say it is in prospect of this that the
present King and his councillors have wisely given a sort of
form of a constitution, which was proclaimed the very day,
May 28, our steamboat arrived at Copenhagen, but without
exciting any feeling.   It is, however, a very important event,
as it appoints four places, Holstein, the Isles, Sleswick, &c.,
where deputies representing the people are to meet; and
though they are only to consider provincial matters, it is a
beginning, and they will be the organs by-and-by of a legal
reform if the country is pushed by a madman to a revolutionary movement.   Good men, therefore, hail the gift with
satisfaction.   Meanwhile the King has retained absolute
power.   He is much liked and respected.   'He *was* liberal,'
said Oersted to me, 'and we cannot accuse him of having
changed; but the times have changed, and have now gone
far beyond him.'

Our late return from the country palace called 'Sorgen-frey,' or 'free from sorrow,' made my packing rather hurried, and I should never have got through it before the packet sailed, if I had not got the assistance of a *laquais de place* who talked English. We sailed over the Sound to Malmö in about five hours very pleasantly, though the boat was so small, that had it rained or been rough, it would have been wretched. The low island of Salholl was on our right nearly all the way. A Danish naval officer who talked English helped me, and I got successfully to a humble inn at Malmö, and walked on foot along the shore to the lime-works at Linnhaven, and back in a peasant car, filled with dry seaweed. On the sea-side I remarked the extremely small size of the *Cardium edule* and *Mytilus edulis*, which I have since learnt is characteristic of all the truly oceanic species which contrive to live (both fish and mollusca) in the brackish waters of the Baltic. They miss their proper dose of brine, and the farther from the Sound the smaller do they become.

Hurfra : June 8.—I must return to Malmö, where I began to feel the want of the language (even at so flourishing a port where the English trade) a very serious inconvenience. As to transacting the bargain of a carriage, my courage misgave me, when in stepped a gentlemanlike sort of a young man, talking good English, and introducing himself as Mr. John Robert Johnson of Malmö, a Swede, whose mother was English, and offered his services as interpreter. I liked his manner much, and after purchasing with his assistance a covered carriage on springs, with harness included, for the moderate sum of about 9*l.*, taking it first upon trial to Lund, I asked him if he would go to Stockholm. He said he should be glad to try and arrange it, so he went with me in the light carriage to Lund, and on the way told me his history. His father died when he was a boy, and the widow kept the English Hotel at Stockholm. She was soon ruined by the knavery of the waiter who managed the inn, and was then invited by Mr. Dundas, a Scotch gentleman who bought an estate near Lund, to be his housekeeper, which she was for five years, when she died. Her only son, who was thirteen, upon this quitted Stockholm, was first sent to school, and

then to study at the university of Lund, being intended for
the Church; but becoming rather deaf, and expecting to get
worse, he gave up the idea, and became teacher of the English
language at Malmö, where the Swedish merchants have need
of a teacher, and keep him to it ten hours in the day. I
was referred to Dundas, who was hourly expected in Malmö,
but having consulted a Malmö merchant on whom I had a
(letter of) credit, I felt no fear. After talking over the
matter at Lund with Professor Hill, a chemist, I offered
Johnson a little more than I should for a servant, and his
expenses back to Malmö, which he agreed to joyfully, requiring nearly the whole in advance for outfit. He returned
to Malmö to pay for the carriage, being absent only two
days, which I spent most agreeably and profitably with
Nilsson. He showed me his fossils, and gave me a valuable
collection of duplicates. We made an excursion together
through a country of greywacke with orthoceratite limestone
and schist, containing a curious zoophyte called graptolite
in great abundance, and a few shells. These rocks are
seen by the banks of a small brook in a place called
Fogelsång (pronounced song), or 'song of birds,' where I
heard the nightingale and cuckoo. The vegetation was
superb. It was here that Linnæus first studied botany when
a student at Lund. I had often wondered that in his tour
in Lapland he contrasts so much the plants and insects of
the northernmost climates with the south, thinking that
Sweden was far north; but Scania is like England, splendid
woods of beech and oak, and flowers innumerable.

Körby.—We have just made another stage from Lund of
nine miles, and are waiting a few minutes while a blacksmith mends the tyre of the wheel of my *bargain*; but really
the machine promises well, and will I hope sell for near
what I gave for it, so as to cover any extra expense of
Johnson, who comes on very well.

Nöbblö.—Another stage, and in danger of waiting an
hour for horses. After Christianstadt I suppose I must
have a forebud.[3] The country to-day from Lund by
Hurfra, Körby, and Wren to this place is like England,

---

[3] Forebud—a Swedish term for a messenger or *avant-courier*, sent on to
order horses from stage to stage.

and scarcely what I had expected. The fir trees have been rarer I think than even in the south of England, and oaks and beeches and a variety of other hard wood of handsome size have been spread over a country much like the New Forest, with the same glades opening here and there, but greener and much less heathy in Scania, and with the difference that the country here, though hardly more hilly than Hants, is strewed over with innumerable blocks of granite and other crystalline rocks, so as sometimes to resemble the Forest of Fontainebleau. The weather delicious, roads good, though not quite equal to England. It is most creditable to the Swedes to have made such roads without turnpikes. Every here and there are gates opened by children, who are pleased with the sixth or even twelfth part of a farthing. The juniper has been nearly as plentiful to-day as heath in England or Scotland. All the floors of the rooms are strewed with it. Between Hurfra and Körby we were for hours in sight of a noble lake, by the side of which are several country seats of noblemen and gentlemen, which look much like England. I went to the shore of the lake, and found it covered with stones, on which were abundance of the common *Succinea amphibia.* A squirrel was sitting on a great stone close to the road when we passed: it played about when we were very near, and looked at us. I saw a great many bogs, as in the New Forest, and horses feeding in them up to their knees in mud.

Christianstadt.—This is a neat fortified town; horse artillery exercising; passport demanded for the first time. Stopped at the gate while a blacksmith repaired the iron of the wheel of the carriage. Symptoms also of one spring of the said vehicle letting one side of our seat down a little; but it runs well, and in this cheap country the two repairs, which at Strasburg might have produced a heavy bill, were only 2*s.* 2*d.*

Karlshamn: June 10,—Johnson comes on very well, and is an excellent whip. After Christianstadt I made an excursion quite out of the way to visit the lake of Ifösjon, in the middle of which is the isle of Ifö (pronounced Ivon), where beneath the sand and boulder formation, under which in general the country is buried, appears in steep cliffs a calcareous green sand of the chalk era, with numerous belemnites, some oysters, and a few cranias. Of the belemnite, a species— which I believe is only Swedish—was so plentiful that I filled some small boxes full. Getting boats and cars for this expedition, and waiting for each, cost much time. There is a village on the isle, which is rich and beautiful, and we found an intelligent clergyman, who insisted on our dining with him, and while his horses were sent for to take us to a cliff, his daughter, who had waited at dinner, played from a book some pieces on a piano, with a kind of flute accompaniment, for it was half an organ. I cannot say much for the melody. The priest, as Johnson called him, has the reputation of speaking many languages, but he knows nothing but Swedish. I tried him in vain in Latin. He gave me a fossil or two, and offered me some stone hatchets. Before Christianstadt we passed a huge block of granite, as large as a good-sized house, which even in this country of boulders is a wonder, and is called the phantom stone 'Trollesten,' the subject of many a legend of witches.

Karlshamn: June 10, 1834—I mentioned the Trollesten in my last. When talking of those legends we asked the innkeeper whether the peasants about him believed in the 'Brook-horse,' or ' Bäcke-hästen,' pronounced Bake-hasteten, the *e* in the first word as in German. I had seen in the morning many horses and pigs up to their middle in the Scanian bogs, feeding, and thought how elephants of old and other cattle got mired for the benefit of our modern anatomists. This evil spirit corresponds exactly, I conceive, to the water-kelpie. The host began by saying how rapidly all these superstitions were disappearing; and then confessed that very lately, when a cow of his was put out to grass, and came home lean, and with her back rubbed and raw, nothing would persuade his servants that anything could be done for the beast. It was the Bäcke-hästen which had been riding it : there was nothing else the matter. This demon tempts the cattle into mires, and the peasants after them, and carries them out to where they sink, and drives the cattle, &c. Among other stories I am told that, as finding of buried treasure is of no uncommon occurrence still, there are some persons gifted with a magic glass, which,

when they take it out with a lantern at night, shows them
where the hidden money is. If anyone gets unaccount-
ably rich in trade, &c., it is attributed to his having a glass.
But as dragons guard the treasure, it is dangerous to take
it. If you ask what sort of dragons, they say like the
great wooden one in the church at Stockholm, which St.
George is killing. About four years ago, a live alligator in
very good condition was found in the fields near Lund, and
was taken by the peasants for one of these dragons; so that
it was not till they had collected in considerable numbers
that they ventured to seize him and carry him to Lund,
where he now appears in the museum as a stuffed specimen.
Malmö, the nearest port, was twelve miles off, but he is
supposed to have swum ashore from a wrecked vessel at
Helsingborg, thirty miles off, his line of march being traced
by the slaughter of geese, which had been attributed to the
unusual rapacity of the foxes. As it was the heat of
summer, the creature had probably formed a tolerably good
opinion of Scania.

I give the story as told me, but unluckily did not hear it
before I left Lund, so could not get from Nilsson an authentic
account. We get on very slowly, though the roads are good,
but the delays are so numerous. They often send the merest
children to bring back the horses, who have hardly strength
to hold the reins, so it is fortunate I have a good driver as
a companion. For three years he was a probationary
preacher, having, as is usual, obtained leave from the
bishop to that effect. My collection of belemnites excited
much curiosity. In the country where they are common the
people call them candles, saying that the witches burnt
them. When transparent, they are not very unlike the
amber so common on the shores of the Baltic.

Höby : June 10.—It seems to me we shall never get to
Stockholm at the rate we move on. We have only yet met
two other carriages since Malmö—posting. I only wonder the
inns are so good. What a vast country this is ! Bläking
is very beautiful, granite covered with wood, and blocks of
granite among the trees. The rye bread is sometimes nearly
white, and sweet and palatable. Along the road are black
painted boards, with the initials of the peasant who is to keep

it in order. I should be quite lost without an interpreter. Johnson is a perfect gentleman in his manners and feeling, and, when we were with the priest's family in Ifö, played his part admirably.

June 11.—Yesterday we slept at Rönneby, a Swedish watering-place situated in a picturesque granite country. Nilsson had begged me to inquire about the discovery of shells near the river at Rönneby, which I did from a fine old clergyman, whom they call the Dean, and who is much respected for his learning. Dean Weltergrund visited the spot with me, and cross-examined the men who dug the trench. This morning we were off by half-past six to Karlscrona, where I failed in seeing the docks, which are not shown to strangers. They are said to prove that the sea has not fallen there. The northern part of Bläking, which, like Scania, once belonged to the Danes, has afforded the first fir woods I had yet seen. The juniper is very ambitious there, and aspires to be a tree, rising sometimes to near fourteen feet ; and putting aside its shrubby, bushy growth, it is then quite handsome, a sort of northern cypress. We have also passed some woods of birch, but never saw large trees.

Kalmar : *June* 12.—To-day I have been examining and taking a sketch of the ancient castle here, now a prison, and it affords proof, I think, of the very slight, if any, change of relative level of land and sea since 500 years. It is a fine old castle, and stood many a good fight with the Danes. We saw the hall in which the treaty of union of Sweden, Denmark, and Norway was signed. I have been about the latitude of Edinburgh and Fife, and the climate and natural productions seem very similar. We have had no great heat nor cold, but the best kind of weather both for travelling and working. I am just returned from an examination of the beach. It seems as if the Baltic, being neither good for fresh nor salt-water shells, had not yet received a peculiar brackish tribe of its own. I only found a few small cockles and mussels, one *Tellina Baltica* and *Lymnea ovata?* the latter alive and in abundance.

June 13.—We are penetrating into the interior from Kalmar. All the houses have been of wood, almost through

our whole route, even in many towns.   It is a great country
of peasants, all of whom have a little land.   They invariably
take off their hats as they pass us.   This morning some poor
children brought out each of them a pottle of fine ripe wood
strawberries, to sell for a halfpenny each—the baskets in-
geniously made of the bark of the birch sewn together.
With rich cream at the post-house, and some sugar, we had
a fine dish for two, for twopence halfpenny.   We are now
passing numerous lakes, in a flattish granite country, covered
with fir wood, and here and there a piece of better land
with rye upon it, fenced in with fir fences.   I have just got
a new light in regard to good roads and cheap posting.   The
peasants, it seems, hold their lands upon the tenure of keeping the roads in order, and providing horses.   This last,
though paid for, is generally looked upon as a tax, and only
on some roads, where they are very poor and the travelling
frequent, do they profit.   They take turns in turning out
steeds, which are sometimes taken from the plough.   This
is a sort of feudal service, and the traveller is admitted to
share the privilege of a 'lord superior,' and have the services
of his vassals.

Norköping.—Yesterday I pushed on about a hundred
miles, finding the horses always ready, but I was on the road
from five o'clock in the morning till half-past nine o'clock.
A country of rock, fir wood, and peasantry.   It seems to me
that I have never seen a shop since I entered Sweden.   Even
at Linköping and this place there only appeared things to
sell in one or two windows.   I have not been at all annoyed
by dirt, but then I am not always inquiring for it, *à l'Ang-
laise.*   Twice I have been slightly bitten by the insect tribe.
To-day I went a round by Berg and the west side of Lake
Roxen, to visit limestone and slate quarries.   The locks of
the great canal, which are the wonder of Sweden, appear
nothing to an Englishman.   The cost, however, was severe
on this poor country.   The numerous smaller lakes of Oster-
gothland reminded me of Windermere and the English lakes.
The rooms are strewed here with the leaves of the birch tree.
Every room I have seen in Sweden is strewed with this, or
juniper, partly to scent the rooms, and in part to conceal dirt
on the boarded floors.

CHAPTER XVIII.

JUNE–DECEMBER 1834.

STOCKHOLM—BERZELIUS—REVIEW OF THE FLEET BY PRINCE OSCAR—
UPSALA — HERR ADJUNCTER MARKLIN — LINNÆUS'S GARDEN — LIBER
ARGENTEUS—MEASURES THE SEA-LEVEL—RETURNS TO STOCKHOLM—
GOTHEBORG — SAILS TO HULL — EDINBURGH SCIENTIFIC MEETING —
AGASSIZ.

JOURNAL.

Norköping : June 15, 1834.—It is now twenty-five days that
we have been separated, and I have often thought of what
you said, that the active occupation in which I should
constantly be engaged would give me a great advantage
over you. I trust, however, that you also have been actively
employed. At leisure moments I have done some things
towards planning my next volume. It will be necessary for
us to have a work together at fossils, at Kinnordy first, and
then in town, and then in Paris. When at Kinnordy, if you
could get some disciples to teach them fossil conchology from
Deshayes' work, it would be a great step. I picked *Lymnea
auricularis* out of the Gotha Canal, where it flourishes, as at
Poppelsdorf. I find that the work at the loëss shells has
made me acquainted with nearly every freshwater and land
shell I have yet met with here, which is encouraging.
Nilsson gave me his little work on the shells of Sweden.
When I consider how thinly the population is scattered
over this vast country, and that rich natives and foreigners
are passing along the high roads, and that the woods contain no robbers, I am lost in wonder. But then every peasant has something; there are scarce any paupers. The
peasants drive along in their light cars, with loads of deal
planks, at the rate of six or seven miles an hour, and often

kept up with us. They are a well-grown race. We saw a great number of jays in the woods, and I observed some insects which are common in the south of England, such as *Papilio rhamni*, which do not go so far north as Scotland. It is enough that they have a little very hot weather, and they care not for cold, as they can be dormant. If I had not seen them, I should hardly have thought that it was hotter here than in Scotland in summer. Much rain in the last two days, but only in showers. Drivers spoke with glee of ten cubs of wolves just caught, but no old one taken.

Stockholm: June 19.—I have been two days in this capital, and like what I have seen of it much. But before I begin upon it, I must endeavour to fill up my journal. The Norköping steamer went first smoothly down the river, the same which drains off the surplus waters of Wettern, and several other lakes, banks covered with fir, and islands and a broad river. Scarce anything but wood and rock. We came in a few hours to the open sea, and a gust of wind caused a swell, which made the ladies ill. They were merchants' wives and daughters. After a few hours of this we entered a passage between an endless string of islets and the mainland. The water here smooth as a mill-pond. We passed swiftly on in deep water, close to the rocks, on the barest of which are a few firs in the clefts. These are evidently the summit of submarine mountains. We then in the night came to the Södertelje Canal, which joins the sea and Lake Mälar. We had to wait here several hours till the lock was open, and the time was spent in taking in wood.

No. 4 *Norre Smidje Gatan (North Smith Street), Stockholm.*—While we slept, a quantity of fuel was taken into the boat, not black, dirty coal, but sweet-smelling fir wood, which never makes the least smoke from the chimney that was visible, and leaves the sails quite white. Early in the morning of the 17th the gate of the lock opened, and we ascended into the lakes.

The banks of the lake were of granite, covered with fir and some oak and birch; beautiful, yet perhaps somewhat monotonous, after the similar scenery before, until the fine

city of Stockholm came in view, so superior in its site to
Copenhagen, from the uneven ground on which it stands,
and the numerous deep friths which intersect it, and
the strong, deep current which runs out of Lake Mälar
through it. The inns were full, but I soon got a lodging here in a good central place. Went to Berzelius,
and had from him a most kind reception, and to business at
once. ' You have time to do much,' said he, ' provided you
can escape *dinners*.' He went at once to the point, and
having told me much that he knew about changes of level,
offered to drive me next morning to some interesting spots
in his drosky. He also introduced me to the keeper of the
Museum of the Academy Royal of Sciences, of which he,
Berzelius, is the head, or secretary ; and they find him a
good mansion, and I suppose a salary besides. He lectures
on chemistry to the College of Physicians. He was the
son of a priest, of a good family, educated for medicine, but
never practised it, but early devoted himself to science. Was
ennobled by the present King, without any title, except some
orders, which however are only appended to his name, as he
is still ' Professor de (af) Berzelius.' He has officiated in
the Chamber of Peers at some Diets, but tells me he has
cut politics, and a Peer may, when the Diet opens, declare
whether he will or not act in it. The present Diet is an
extraordinary one, called by the King to convert the paper
circulation into coin. The bank-notes descend now to the
sum of three-pence English ! and are almost all in rags. On
the morning of the 18th I spent five whole hours with
Berzelius, *téte-à-téte*. After all kinds of hints, which I noted
down for my tour, he drove me to see some alluvial phenomena near the Observatory, and in the beautiful royal parks
a few miles north of this city, of Haga and Ulricsdal, in
both of which are fine lakes and woods and country palaces.
Then to Solna, where marine recent shells are found some
thirty feet above the sea. Berzelius entertained me with
results of his last analysis of meteoric stones, which he
believes come from the moon. While I made drawings, and
collected specimens, Berzelius gathered grass for his well-fed
and handsome dun pony, which he is great friends with,
and can leave him quite alone. ' As he only gets hay,' said

Berzelius, ' all the year in the stable, every mouthful of this is a sweetmeat to him.'

Stockholm : June 21.—After I came home from my ride with Berzelius, in the middle of the day, I dined at the Club, or Society, as it is called, a handsome building over-looking the ström, or grand current which issues from Lake Mälar. Here some of the gentlemen helped me to order dinner, which only cost about a shilling, beer included. Afterwards to Colonel Blom, who drove me in his carriage through the ' deer park,' to show me this royal park which is open to the public, in which he, Blom, has erected several of the ' transportable houses ' of which he was the inventor. They can be put up in a few weeks anywhere, well heated with stoves : are made of wood, painted with oil. The King had one in a great hurry once, to give a *fête* in the park, and was so pleased with it, that it is now cased in stone, and forms a country palace, of which Blom gave me a draw-ing. On the 19th I saw Hisinger for the first time, for he is much taken up with committees of the Diet. He is oldish, good-humoured, but has withal rather a *triste* manner. I expected more of a devotee, and suspect it is an unfavourable time when a man is worked with political affairs. I got an old plan of Stockholm from Blom, and we went through Professor Johnston's article in Jamieson [1] together, whose proofs of the fall of the water here are so bad and exagge-rated, and some of them absurd, that it is enough to provoke scepticism. I determined next day to examine the spots, which Professor Johnston cannot have done. I was intro-duced by Blom to Lieutenant-Colonel Lundstedt, who re-places Hällstrom, whose absence all are regretting, for my sake. Next day (yesterday) I went with my interpreter in a *calèche* to all places cited in the article in Jamieson's Journal, a beautiful drive, and satisfied myself that, though the water may be falling, most of the historical proofs adduced are ridiculously nugatory. One case for example. Charles XI. built a fishing lodge 150 years ago, which stands thirty feet above water. This *near* the lake, but now far from it, and high above it; but I found a great oak, which

[1] Jamieson's *Edinburgh Philosophical Journal*, 1833. On the Rise of Land in Sweden, by J. F. W. Johnston, F.R.S.E.

is upwards of two hundred years old, and which I measured,. and of which the base is eight feet only above the lake. This oak then loved the water 150 or 200 years ago more than our willows do now. The hut must have always been, within a foot or two, as far from the lake as now. I spent two hours with Hisinger, who gave me four volumes of his works, with some plates and maps, all in Swedish. Read Swedish for the first time one hour with Johnson, who teaches pretty well.

Stockholm : June 22, 1834.—Stron, the Keeper of Woods and Forests, has offered to measure the height of Charles XI.'s oak, and has sent me in his reasons for saying it is four hundred years old, and much more valuable information. I am now in hopes, by the altitude of this tree, to fix a limit to the possible rate of the rise of land here. I must stay here a day or two more. A new man, Col. Nordewall, comes into the field to-morrow, willing and able to serve me. Berzelius says : ' I see you will extract what I could not get him to print, because he will see that you must have it at once, and he is a procrastinator.' This is justly called the Venice of the North, and now the sun is powerful, though hitherto the weather in Sweden has been delightful. Prince Oscar, who is Lord High Admiral, is to review the fleet to-morrow, and it has come into the suburbs, and four or five ships of war, among them one good-sized frigate, have actually cast anchor opposite the windows of the palace, such is the depth of these blue friths, which are yet so narrow as by no means to detract from the height of the houses. The bronze equestrian statue of Gustavus II., and those of other kings in the squares, are in a superior style. I shall only hear the cannon at a distance to-morrow, as I am determined to go to Södertelje again, and next day to Upsala. I have fully determined to return to Stockholm after a tour to Gefle, and a little to the north of Gefle, then Fahlun, Sala, and here. I am making myself master of the kind of evidence relied on for the change of level of the Baltic, and it is necessary to cross-examine both nature and man. The testimony of the former is strong; of the latter, I must say, so weak and contradictory, that I require to know the men, and find how they got their views. I must

endeavour to see Hällstrom. Old Nordewall stands me out
that a bed of *Cardium edule*, 100 feet high, proves that the
fresh water of Lake Mälar was once that much higher.

Your letter came to-day, and was most welcome, and
better than the *fête*, for which I partly stayed, knowing that
Johnson thought it a sin to be out of town. In its way it
was fine. The troops, both Swedes and Norwegians, well-
grown men. The parade very fine. The ships of war
opposite the palace covered with gay flags. Afterwards the
yards manned. All over in a short time. I afterwards
drove in a gig to talk with Berzelius, then to Nordewall,
then to the suburbs.

Upsala: *June* 26.—The Södertelje expedition told more
than anyone I have made in Sweden. Shells of the Baltic
nearly one hundred feet high. All that I got at Lübeck save
Mya arenaria, and perhaps that is not a Bothnian Gulf
species. But what think you of ships in the same formation?
—nay, a *house*. It is as true as the Temple of Serapis. I do
not mean that I discovered all this, but I shall be the first
to give a geological account of it. I came home in high
spirits at this prize, and the chilling influence of old Hisinger
was more than thawed. I think I have hit on his character.
Sweden is his preserve, and he wants no young poachers.
If you ask for instructions to such a place, he will say, ' Oh,
that is all known already.' I told Berzelius of what I
thought of Södertelje, and he was not staggered at my
requiring an enormous subsidence and re-elevation. Had
Nordewall or Captain Cronstrand only known that neither a
Cardium nor a *Tellina* will live in a lake, it would have saved
them a world of trouble.

Upsala: *June* 26.—We were sent off from two inns, and
this, into which we have found our way, seems excellent and
full. I went to bed without candles last night, after eleven
o'clock, and happening to wake at three o'clock, found
it daylight. This I believe is about the latitude of the
Shetland Isles, yet a nice summery-looking land in this
season. In a stream to-day, slow flowing, I found numbers of *Planorbis corneus*; at Lund, *marginatus* was equally
abundant.

June 27.—Wahlenberg exceedingly polite and gentle-

manlike, but seems sickly, which some here attribute to his
being a martyr to the homœopathic system.   As he is now
full of botany, he handed me over to Herr Adjuncter
Marklin, an assistant professor, who was a peasant, and only
began to read and write at the age of twenty-one.   Very
simple.   He tells me he has only 17*l.* sterling a year, but never
sells fossils or other things : an enthusiast, full of knowledge
and originality.   I have learnt much from him already, and
on finding he was delighted with Rang, I gave it to him,
and he has given me duplicate copies which he had of Dal-
man and Wahlenberg, on trilobites, and has presented me
with shells collected on the coast of Norway, and with some
transition fossils of Sweden, of which his collection is beautiful.   We talk German together, and regularly have
recourse to Latin when either of us are at fault.   He is
about fifty, and when I mislaid my gloves, he remarked that
he was fortunate in *never having had any.*   I should not be
surprised if they who search for the shirt of the happy man,
should find no such incumbrance on him.   Certain it is that
when the City of Frankfort offered him four or five times
his present income to be keeper of their Museum, &c., he
declined, partly, he confessed to me, because the German
diet makes him ill.   He has a collection of insects, in which
this country seems very rich, having many species which we
consider more southern.   I have been with him through
Linnæus's Garden, where the hedges of fir, clipped, stand as
he planted them (*Pinus abies*).   Linnæus had five children :
the son died young, and all but the eldest of the four
daughters are married.   They are much delighted at
foreigners coming to see the house in which their father
lived, especially at Hammerby, in the country.   I may perhaps have scarcely understood part of the story, but Marklin
was speaking of some Englishman who obtained from Miss
Linnæus the cups out of which the old gentleman drank,
and sent in return from England a service of plates with the
Linnæa borealis painted on each article.   I saw his lecture-
room, and Wahlenberg showed me his statue in another
place.   There are nine hundred students here.

   June 28.—Besides zoologising yesterday at Marklin's, I
went to the hill, a hundred feet high, on which the tower

stands, to examine marine shells. All of Baltic species.
You remember that in the half-hour between the two steamboats at Lübeck, or rather Travemunde, I collected shells
by the quay. Not one fossil have I found newer than the
chalk in Sweden that was not in the number of those
found living, in that half-hour. To-day, after studying with
Marklin, I saw Col. Bruncrona, chief of the Swedish pilotage
for many years. I am to get some hints from him to-
morrow, I hope, about the marks at Gefle. Tornea (says
Wahlenberg) is not so well worth examining as Gefle for
this point. Afterwards I went, at Wahlenberg's recommendation, to Ulfva, three or four miles off. I went in style,
with my driver, my interpreter Johnson, and, thirdly, my
naturalist Marklin, and a good pair of steeds to my carriage,
which after, on the whole, about four guineas repair, seems
fit to go the world over.

June 29.—I have learnt much geology here. It is fine
agreeable summer weather. Linnæus's Garden looking very
like one in Hampshire, and no signs of a latitude north of
Cape Wrath.

June 30.—I am stopping at a house in a fir wood, between
Upsala and the great iron mines of Danemora, while horses
are sent for. Nothing could have *told* more than my stay of
three days at Upsala. I have got a date for some of the
largest erratic blocks, and shall quite overset the *débacle*
theory, and I expect bring in ice-carriage as the cause. I hope
to stay two hours only at Danemora. My point now is to
examine several places where marks were made in the rocks
about fourteen years ago by Bruncrona. One of these is
near the port of Oregrund. I had a famous elementary
lesson in trilobites from Marklin, and wish you had been
there. All the recent animals most analogous were put
before me, and the principles on which the fossils have been
classified explained. At Ulfva I found recent Baltic shells
in clay. I visited on the way home Gama Upsala (Old U.),
where there is a church, in which Marklin says the
Catholic worship was for some time performed in one part
of the ancient temple, while the heathens sacrificed human
victims to Thor and Odin in the other. Near the church
are three immense tumuli, in which those two idols, and

Frigga, Odin's wife, were buried. It is the custom to drink
a glass of mead (honey and wine, or spirit from barley, I
believe) on the top of Thor's tomb, and as the King had
lately done so, said the peasant who drove us, he supposed
I should, which accordingly we did. The beverage was not
bad. The tradition of the joint worship is believed. The
priests of Odin and Thor have taken advantage of a natural
ridge of sand and gravel, and by cutting part of it into
gigantic hummocks, have skilfully enough given the appearance of a stupendous work. This was a trick worthy of
some of the Catholics who supplanted them. The iron mine
at Danemora is a grand sight, but for want of a good guide,
whom I could understand, I did not gain much, and there
appears to be no interesting geological phenomena to be
seen there. I am beginning to tire of fir woods again,
though now and then varied by birch. It has been rather
monotonous, whereas between the capital and Upsala it was
much the reverse. It is a *détour* of a day nearly that I am
making to Oregrund, to see a mark in the rock. I had a
very pleasant walk with Wahlenberg to see a meadow near
Upsala, where several marine plants, as *Glaux maritima*,
which require salt, still flourish. The sea does not come
nearer now than Stockholm : there is salt in the soil.

Oregrund : July 1.—A nice little port, with ships lying at
anchor. Just returned with the Lieutenant of the Pilotage
from the first examination of marks in the rocks. Results,
as far as our observation goes, very decided and satisfactory.
I am exceedingly glad I came here before Gefle. I think of
making a report to the Royal Academy at Stockholm. Went
in a boat about three and a half hours' pull, with four rowers.
The place of the mark, and the mode of marking, well
chosen. I had with me the pilot who made the mark
fourteen years ago. It seems true, as Galileo said in a
different sense, 'that the earth moves.' I found a *Neritina*
on the rocks where the mark was, four specimens. We are
just going to set off on our way to Gefle, through more fir
woods. In these woods the quantity of ant-hills at the foot
of the fir trees is quite extraordinary, and of great height,
but scarcely, I think, larger on the whole than those in the
New Forest, only higher. I have been determining that in

October, while I attend Daniell's lectures, I shall go on with zoology steadily, and we will study together. In a day or two, my dearest love, I shall be again turning towards you, and with the satisfaction of having accomplished the object of my journey, and with having, I am sure, reaped the fruit of the sacrifice we have both made in the separation, so far as my scientific career is concerned.

Gefle : July 2.—I forget whether I told you that at Upsala I saw the famous Liber Argenteus,[2] of which I fancy you read more than I could tell you in Clarke and elsewhere. I have found a new locality for Baltic shells. I begin to have great hopes that I shall feel strong enough in original matter and in decisive results to read my paper to the Royal Society.

Fahlun : July 5.—I had the chief pilot at Gefle for a day and a half, and visited two of the principal marks made in 1820 at Löfgrund and Esjko Sound. A retired sea captain of Gefle volunteered going with us. He had traded much with Hull, and has the reputation of knowing English well, as he was so much in England. Partly not to expose his great ignorance of the language, and partly to please his own reserved humour, he hardly uttered a word the whole time ; but he was so good-humoured and willing to do just as we liked, and to help to steer, &c., that we were as well with him as not. The first day it was very cold, with the wind against us, and slow work. I learnt much about the action of ice in these seas from the sailors and pilot, and we landed at Löfgrund, and drank milk in some summer fishing huts, where we found a nice peasant's family. I measured the stone at Löfgrundet, where there was one mark made in 1731, several feet above the sea. One of the fisher huts had been broken to pieces by the fall of a fir tree upon it. We next landed on the island of Edskö klubb, but could not hear news there of a mark which the pilot thought there was. Then I measured the line cut on St. Olaf's stone in Edskö Sund, and, it being late, determined to sleep there. The accommodation was scanty enough, but we had taken provisions with us. I slept well for four hours, and then

---

[2] The *Codex Argenteus*, containing the four Gospels in silver letters, a MS. of great antiquity.

waking, and seeing the sea quite calm, I got up, woke the pilot, and told him that as he said the evening before that my observation was incomplete from the state of the wind and sea, perhaps we now had a good opportunity of rectifying it. He was a sprightly, good-humoured young man, and very intelligent. He immediately jumped up, and declared that the sea was in so favourable a state, just at its mean level and quite calm, that no time should be lost. With great despatch all the things and his men were ready, and the old captain, who was never so happy as when sitting in the boat, was down among the first. As for Johnson, who sleeps hard, as he says, he was bundled out of bed, wondering that 'having ordered coffee, I should go without it.' Away we rowed down the sound, and revisited the huge stone, and I found the sea, as the pilot had predicted, several inches lower. I returned then to Gefle, much pleased, having gone about forty miles by sea in our excellent boat, and which sailed very well against wind. Had I taken a common fisher boat, I might have been beating about three days or more. Got off the same day towards Fahlun, and penetrated farther than I had ever done before into the *interior* of Sweden. A land of goats, none of which I had before seen nearer the coast.

July 6.—Johnson, who gets on in many respects very well, is so slow at getting up in the morning, that I have taken up the thing in earnest, having lost on the whole journey several days by want of method in posting. I have ordered the carriage to be greased over night, and have got up some phrases in Swedish, to order the man myself. He is astonished to see already the difference; but I will show him yet, when I have a few more words, that we can do a stage at least more in the same day's work, for when the horses come he has a number of things to do, and the bill is never paid. We are now stopping at a stage between Fahlun and Sala, having crossed more than once the Dal Elfven, one of the largest rivers in Sweden, of which we have seen much. I suppose I mentioned that, at Skjerplinge, the falls of that great river over the granite are very grand. Niagara in miniature, I suppose; but the island covered with firs which divides the waters is beautiful. The people of Dalecarlia.

which we entered yesterday, are a fine-grown race, and the cottages of the peasants quite handsome. Wood painted red, with very large panes of glass in the windows, and good stoves in the German style in every room. The beds look poor and not perhaps (except the sheets) very clean ; but I sleep sound, am rarely bitten, and on the whole find the plain diet agree with me.

6, *Evening.*—I have just seen the great silver mine. The metal is in a beautiful rock of crystalline limestone.

Stockholm : 7.—I feel now what I was very sensible of when correcting my last edition, that I was not justified in writing any more until I had done all in my power to ascertain the truth in regard to 'the great northern phenomenon,' as the gradual rise of part of Sweden has been very naturally called. You will see by-and-by how important a point it was, and how materially it will modify my mode of treating the science, and how much it will advance the theory of the agency of existing causes, as a key to explain geological phenomena. I should much like next year to find Deshayes' manual just coming out, or just finished. My plan is to take up by degrees the zoology of geology in all its branches, but to make shells the great business.

To LEONARD HORNER, ESQ.

Stockholm : July 9, 1834.

My dear Horner,—I shall be at Hull, unless some unforeseen obstacle occurs, by the packet which should arrive there on the 25th of this month. I am longing for Mary's company again, and hope to join her just within the ten weeks, or seventy days, which, as Mr. Kennedy told her, was a generous concession in her to make, by way of leave of absence. The sacrifice any other way would have been great, for to do what I have, would have required nearly perpetual absences, in boating excursions and others, and much roughing on the road, which I have not felt, because not only excited, but in the field all day. But the country inns and poor peasants' houses would have been wretched abodes for a woman, unable to keep up with the geology. I have seen a glorious country, and have examined many of

the marks made in 1820, all now some inches *above* actual
sea-level. The observations may be slighted by some,
because this sea rises and falls two or three feet in the course
of the *year*, but when the season and other circumstances
are taken fairly into account, it has made on me a strong
impression. My interpreter has enabled me to cross-examine
pilots and merchants and fishermen. Another line of
research has been the huge drift blocks, or Baltic boulders
or ' erratic blocks,' which cover all Denmark and Sweden.
Their size is often enormous. Some I have ascertained have
been placed where they are in times exceedingly modern,
geologically speaking, certainly late in Newer Pliocene period.
I believe that ice has brought them. I have questioned the
pilots closely about the agency of ice, in which they believe.
I am persuaded that ice can do much for us. The examination of Baltic species of shells above the sea-level at various
heights, some at new localities which I have found, is a grand
fact. I have still to see Uddevalla. I was glad to hear that
you had been *fishing*, or lizard catching, with Lord Greenock,
who has that greatest of all merits, loving the subject. The
Swedes seem to me more like the English than either the
Germans or French. They have most of our faults and
merits. Not so many simple characters as in Germany.
The heat is now very great here, and Stockholm looks almost
like an Italian city, and feels like it. I have had nearly
constant dry weather, scarcely three days with any rain,
which, with only three hours of night, has enabled me to do
much in a short time. I think of writing a paper for the
Royal Society on the elevation facts, and for the Geological
Society on Danish chalk.

I have not said in my letters to Mary that I have looked
in vain for any fine natural woods : all the large old trees are
cut down, even where the country is fully wooded and a
continuous forest. I have visited the greatest copper mine
of Fahlun, of silver at Sala, and of iron at Danemora. I
cannot say that they have edified me much as to the causes
of the metallic phenomena, but I was ill prepared to profit by
them. The silver occurs with lead in an enormous mass of
pure crystalline saccharoid limestone, not got through, at the
depth of 180 fathoms—the calcareous mass in the midst of

gneiss. I am glad to hear Mary is looking better, and
hope it will be some years before I shall be called to settle
any other point in the northern regions, for settled it is as
far as my private opinion goes, and that I shall make future
tours with her.

Believe me, your affectionate son-in-law,

CHARLES LYELL.

JOURNAL.

Stockholm : July 10, 1834.—An expedition with Colonel
Hällstrom, to see one of the Baltic shell beds of Bränkyrka,
never before examined by any geologist, proved remarkably
how much it is necessary to know what to look for in order to
find. Hisinger said, ' As you have seen Södertelje, and other
places, what do you expect to find more ? ' I answered, ' Either
I shall find the freshwater species which now inhabit the Baltic
mixed with the marine, or I shall begin to think that that sea
of old was more salt than now.' Accordingly I found for the
first time at Bränkyrka all the marine shells before met with,
and in addition, the *Neritina* and *Lymnea* I had been *wanting.*
I so wished you had been there. Both the Colonel and Johnson
entered with spirit into the search (seventy feet above the
sea), and wondered, when the shells turned out, which I had
complained of missing elsewhere. 10 P.M.—I have been
with Colonel Blom, walking in the ' deer park,' and seeing
the site of an intended royal museum for works of art, to
be built after Blom's plans. Berzelius had been requested
by the King to see if the marble was good. I suppose he
had spoken of me, for when we talked, said Blom, of rapid
travelling, the King said, 'There is an English gentleman here
inquiring at what rate the sea is being turned into land ; ' and
when Blom said he knew me, the King said, ' You may learn
then from him how soon England will be joined to Sweden,
and when we may think of a railroad communication.'

Johnson has just told me a capital anecdote to illustrate
the superstition of the Scanian peasants. A lady he knew
asked one of them whether he would answer for a horse, he
wished to sell her, being not apt to take fright. ' That I
can,' said the peasant. ' He stood for three weeks in my

stable, night and day, *alone*, and I never could see that he
showed the slightest symptoms of fear.'

Marklin sent the fossils, such as he had time to pack up,
as he had promised, to Stockholm. He diverted me much
by his originality, and among other things, when lamenting
at Gamla Upsala that the wood had been so cut down,
young and old all cleared away, he remonstrated with a
peasant that they would soon feel it. ' Not in our time,' was
the reply, ' those woods are so near.' ' Well,' said Marklin,
' it may be you will not in your time *here*, but have you not
heard that there is another place where the wood will warm
you—when every tree wastefully cut down in this world
is to be burnt over again, for the punishment of those who
felled it ? '

July 12.—We came by Södertelje, Arboga, Orebro, and
are now within a stage of Marienstad, through a fine sunny
country, where they are carrying hay, which smells very
sweet, and where the rye and wheat have already a yellow,
autumnal colour. So much of Sweden is of granite and
gneiss blocks uncovered with any soil, that it must always
remain mere forest land, and the trees have no small
difficulty in growing, by finding a small rent in the rocks.
At one place I remonstrated against a horse who had a
wound in the leg. The ostler said a wolf had bit it, and had
not the other horses in the field come up to his rescue, he
would have been devoured by the wolves. When I asked the
other evening why they blow a horn in the woods at night,
the answer was, for the wolves. They will not come near the
cattle when they know that the herdsman is awake. This
horn, which I have heard often, sounds savage, and like
an alarm for an enemy.

Marienstad: 13.—This town is finely situated on that
inland freshwater sea Mälar, and good-sized vessels, such as
can pass the Trollhätten canal, are here. The dry weather
has made one of the wheels in a bad way, and we are con-
sulting with a blacksmith about a repair. It is a fortnight
or more since any repair has been called for. We met to-
day the diligence from Stockholm to Gotheborg, the first
public conveyance of any sort I have yet seen on any road
in this country.

Lidköping : 14.—I have had a grand day on Kinnekulli, a mountain which rises nobly on the shores of Lake Wener, to the height of 800 feet above the level of that lake, composed at the summit of volcanic rock or greenstone, and below of perfectly horizontal shale and limestone, in which are trilobites, orthoceri, &c. I worked hard, and got a pretty good harvest. After I wrote to you at Marienstad, we were overtaken by the first heavy rain we have had in Sweden, and arrived very wet, at least some of the things, at a wretched poor cottage, the post-house at Forhsäm, where nevertheless I slept sound and unbitten. But to-day we have had splendid weather, and a view I should have liked you to see. We could have got on farther to-day, but at this place were told we must wait three hours for horses, so I resolved to go to bed very early, and be off at four o'clock to-morrow morning. It is quite amusing to see how splendid the maypoles are here, with flower garlands, and egg-shells blown and gilded, round a tall fir-tree pole.

15.—Went to see some mountains near Wenersborg, called Hunneberg and Halleberg. Saw a place, a great precipice, where *dévots* of old threw themselves down, in order to go to heaven to Odin. There is a pond below where the corpses were washed, and urns are dug up, in which the ashes of the burnt bodies were buried. These places are called 'ättestupa.' The place is in a defile between the mountains before mentioned and very romantic scenery. I then saw the Falls of Trollhätten, with which I was altogether disappointed. Mere rapids, a succession of them to be sure. The canal and locks are well worth seeing. I was very glad to get safe over the ferry of the Gotha Elf, as one of the horses liked the look of the rift as little as I did.

Gotheborg : 20.—Arrived very late last night. This tour has answered far beyond my expectations. Uddevalla was a great treat, a perfect Dammerie [3] of the recent period. The boating to the isles of Orost and Gullholmen, and afterwards farther south to Marstrand, also enabled me to do much geology, and ascertain about the levels, which I think will be valued by the Royal Society. Shells I saw and collected, and I think understood, in the same places. Four of the

---

[3] A noted place for the tertiary shells of the Paris Basin.

most energetically spent days I ever remember, a trial for
the carriage, and for Johnson's power of exertion. Both
stood it very tolerably, better than could be expected. I am
surprisingly browned and disfigured with my week's work
and boating, &c., but am thankful that without an accident,
or day's ill health, I have accomplished my objects.

July 21.—Got your letter of the 1st of this month the
moment the post office opened this morning. Last night
found the shore of the river covered with *Strombus pes-peli-
cani,* a *Donax, Cardium edule, Mytilus* do., a *Tellina,* a *Cardita,*
&c., recent species. Tasted the water, quite fresh; but it
was salt, as the shells speak to that. Asked Vice-Consul
Harrison, and was told that the sea is always retiring.

Cornubio Steam Packet : July 22.—The timbers tremble, so
that I must write a shaking hand. Off the low sandy
coast of Jutland. The steamer too much loaded with
Swedish iron to sail well, but wind quite in our favour. To return to my tour, I found Uddevalla an agreeable, picturesque
place, and accommodation good, because people come there
to some hot springs in the neighbourhood. Hired a coachman and horses to drive us about to the places where the
shells occur, and got some of the barnacles attached to the
solid rock of gneiss, and some other shells of which you must
tell me the genus. The next morning travelled south to a
ferry, which took us to the island of Orost, which is made of
mica-schist, much barren rock, with heather here and there,
and great mosses, and the soil of the Balentore part of
Catlaw.[4] The islanders very active and obliging. The pilot
dead who had made the mark at Gullholmen : unlucky. His
widow petitioned me to get Col. Lunstedt to get her the pension she was promised. Visited rocks which had emerged in
her lifetime. Hired the services of a smith to make a mark
at the water's edge : $\dfrac{\text{C. 18. L.}}{18.\ 7.\ 34.}$

July 24.—We have been getting on, with a good wind,
eight miles an hour, and are within one hundred miles now
of Hull.

Edinburgh : 27.—I have got my place outside of the

---

[4] The Grampian hill near Kinnordy.

Defiance coach, so I hope it will not rain. I shall put this
in at once.   Send to Glammis.

Your affectionate husband,

CHARLES LYELL.

CORRESPONDENCE.

To G. MANTELL, ESQ.

16 Hart Street, London : October 1, 1834.

My dear Mantell,—I am here again, returned with my
wife a few days ago from the Edinburgh meeting, which
went off in my opinion very well. After my return from
Sweden I passed some weeks at my father's, in Forfarshire,
where among other visitors, during my father's absence in
Paris, I had to entertain Robert Brown, the botanist, for
several days, who was good company, and had lately returned
from a short excursion with Fitton in Portland, where he
had collected the ' cycas,' as he now admits we may call it,
for he has discovered new points of identity in its structure
with the modern cycas. In Sweden I satisfied myself that
both on the Baltic and Ocean side, part of that country is
really undergoing a gradual and insensibly slow rise. A sketch
of my observations which I gave in Edinburgh to the geological section is already printed by Jamieson, and the
detailed paper on the subject I mean to read to the Royal
Society. When in Edinburgh, I promised to accompany
Agassiz to Brighton to see your collection, and I expect him
here in a week ; but he will be here some ten days, I suppose,
before leaving for you. He has found about a hundred
new species of English ichthyolites, making seven hundred
fossils in all. He made out that *all* the supposed saurians of
the carboniferous period of Scotland were only sauroidal *fish*,
and that Hibbert was so far out, but they were very curious
and new forms of fish. His readiness and knowledge are
surprising, and you will find him very skilful in reptiles,
though he does not profess to know anything about them,
except that he maintains he can prove the pterodactyls to
have been swimming, and not flying, animals ! I have been
sketching and making some progress in a single volume

which two years ago I promised Murray, a purely elementary work for beginners on Geology, and which I find more agreeable work than I expected.

I have begun fossil ichthyology, and am attending Daniell's lectures on chemistry, so I have enough to do.

Believe me, ever faithfully yours,

CHARLES LYELL.

To G. MANTELL, ESQ.

16 Hart Street: December 10, 1834.

My dear Mantell,—I ought to thank you for the warmth of your congratulations on my receiving the R. S. medal, which was sent me through my sisters, and assure you that the pleasure they received from it was doubled by having you to talk it over with, and witnessing your *ecstasies* at the news. You will be glad to hear that Greenough, who presided at the Geological Committee of the G. S., obtained by letter the suffrages of many absent geologists, how many I know not; but a letter of Fitton's which was read by him to the Council had no small influence, and assisted in overruling G.'s own doubts, which were strong on theoretical grounds.

The medal was given me distinctly for ' the Principles,' not for my paper on Sweden, which is half read, the remainder to be heard on Thursday, 18th.

Sedgwick's prebendal stall is seven hundred a year, two months' residence, and he is now giving occasional sermons in the Cathedral at Norwich, and very popular with the Dean, whom I saw the other day.

Ever sincerely yours,

CHARLES LYELL.

CHAPTER XIX.

[He was elected President of the Geological Society in February
1835, and in the course of this year he made a tour by the Jura into
Switzerland, returning by Bonn on the Rhine, where the German
Naturalists' Association met.

In the summer of 1836 he went to Edinburgh and Kinnordy.]

CORRESPONDENCE.

To REV. DR. FLEMING, PROFESSOR OF NATURAL PHILO-
SOPHY, KING'S COLLEGE, ABERDEEN.

16 Hart Street, Bloomsbury, London : January 7, 1835.

My dear Dr. Fleming,—I was made truly happy by again
receiving a letter from you, for I had been disappointed at
not seeing you at the Edinburgh meeting, where your name
not only occurred in print, but some were ready to declare
they had seen you in town. Sedgwick, in one of the animated discussions in the geological section, happened
incidentally to come upon you, and in his rapid extemporising
way said, 'Dr. F., a tough antagonist to deal with, and with
whom I never yet had to do in this, or other matters of controversy, without getting the worst of it;' or some words to
that effect. I think we were upon the fish scales in Old Red.
There was something droll as well as candid in what fell
from him, which made us all laugh at the time, and yet,
when I wanted afterwards to recall it, I could neither get
from Buckland nor anyone else the precise point or expressions.

There was so much done in debate, that one may be excused
for being a bad reporter. The sections answered well. The
evenings badly—too much display to suit with my notions of
what philosophers should do. The platform always reminded
me of the hustings, on which a set of political speakers were
holding forth to ladies and gentlemen, and this I think was
a pretty general feeling. I am using my wife's hand in
penning this part of my epistle, for my eyes, which, you remember, were never of the strongest, although now behaving
better than formerly, have been rather fatigued this week,
by having candles in the daytime during the fog, and I have
to read as Secretary at the G. S. to-night, Hamilton having
gone to canvass the Tory voters of the borough of Newport.
As you said nothing in your letter of Mrs. Fleming, I must
beg you to write soon again, and give me some domestic
news. We are quite well, and thinking of going to Paris
and Switzerland next year. I am writing an elementary
book on Geology, and wish I may disappoint my friends, who
all agree that I shall make it too deep. It is to be the size
of one of my four volumes of the third edition. I have been
reading Dr. Boase's work on Primary Geology, an attack
upon the Plutonic theory by one who was never out of Cornwall, and who seems to make light of all but Cornish observers. It is good, however, to read what an antagonist has
to say, who has worked hard at Cornish granitic and
mineral veins. You should read it when you have time to
take in hand the granite veins near Aberdeen, which Dr.
Knight took me to see. Pray make drawings of them when
you can, and anything which makes for or against my
hypogene and metamorphic views, or heresies, as I suppose
Boase would have it. I also wish you much to send a paper
to be read at the G. S. here, on the *great Flisk dike.* I never
saw a more splendid case of the passage from the granitiform
to the trappean mixtures. Don't ask me to define what I
mean by these terms. De la Bêche's last book, called 'Theoretical Researches in Geology,' is by far the best thing he
has written, but in his statement of the question as to the
origin of the crystalline or primary schists, he suppresses or
blinks the numerous facts of altered fossiliferous strata in a
sub-crystalline, or, as Werner would call it, transition state.

The key to their origin ought, I think, to be sought in such passages. I mean to go to Glaris this year, the slate of which, looking like primary roofing slate, contains, according to Agassiz, chalk fish. It has been said sneeringly that fossils may prove that black is white, and white black, but I have no objection to these black slates having once been mud of the chalk epoch. Buckland's Bridgewater Treatise is only promised to us at Easter. We are curious to know what he will do about his diluvial theory, and if I hear true he has more than once changed his plan. He told me, however, that his mode of reconciling Geology and Genesis in his B. Treatise had been approved of by the Oxford Professors of Divinity and Hebrew! Mantell, whom I visited lately at Brighton, has made a bold professional stroke in removing there, which you will be glad to hear is likely to succeed, in spite of the misgivings of many of his friends, who had not the confidence which I always had in his genius. He is, in fact, a man of great medical skill, and tact so great, as to triumph over the drawback of his having so fine a museum and so much fame in certain branches of geology. You will rejoice to hear of an anatomical triumph which he lately obtained.

Amongst the very numerous species of large saurians found in the Wealden beds, there were many bones without owners, some of which he claimed for his Iguanodon, rejecting others, and on several points Buckland presumed to take the opposite side. A large block of limestone was lately found, in which a great number of bones were together, accompanied by the tooth of the Iguanodon. They consisted exclusively of those which he had referred to that animal, vertebræ, femur, &c., and not one of those which he had refused to admit as belonging to it. The position of these bones is geologically very singular. They occur above the freshwater beds of the Wealden in the inferior part of the Lower Green Sand, thus connecting the fossils of the Wealden and cretaceous groups.

I hope you will see Mr. Horner when he is next in Aberdeen. With my kindest remembrances to your family, believe me ever most truly yours,

CHARLES LYELL.

To G. MANTELL, ESQ.

February 1835.

My dear Mantell,—The dinner[1] went off famously, more than a hundred present. After the toasts had been given of the King, Royal Family, Geological Society, late President, and President, I gave you. I send you a copy of my speech almost word for word as delivered, and on looking over my notes, I found I had not omitted any of the material points which I had intended to speak of. I assure you I had the feeling of the meeting with me, and in some respects it produced a better effect than if you had been there. It was by far the longest toast given, but I am sure they were not tired. Lord Lansdowne, who was on my left hand, asked all about you. I got him to give Oxford and Buckland. Fitton gave Cambridge, answered by Sedgwick; Sedgwick the Royal Society, answered by Lubbock; Buckland the Linnæan; I, the Astronomical, answered by Baily; Greenough the Geographical, answered by Murchison. We then drank Burnes, the traveller, who made a good speech. Warburton also held forth as Vice-President, and I wound up by an eulogium on Lonsdale, which I did *con amore*, and it was received with enthusiasm. We adjourned late to hear Greenough's address, a matter-of-fact abstract of the proceedings of last year. On my right hand at the dinner I had the Belgian minister, Van de Weyer. Among others, there were Hallam, Stokes, Lord Cole, Sir Charles Lemon, Duncan of Oxford, Sir A. Crichton, Ingham, M.P., Drinkwater, &c. &c.

I received a paper in the morning from Greenough containing the award in your favour, as also a new 20*l.* note and two sovereigns. I cannot tell you how soon the medal, which is of the value of ten guineas, will be ready.

Believe me, ever most truly yours,

CHARLES LYELL.

To G. MANTELL, ESQ.

April 13, 1835.

My dear Mantell,—I have been getting Dinkel to figure for me some fossil eggs of a turtle, found in the island of

---

[1] Anniversary dinner at the Geological Society, when the Wollaston medal was awarded to Mr. Mantell.

Ascension, imbedded in a hard rock something like that of Guadaloupe which contains the human skeleton. It is clear that the eggs were nearly hatched at the time when they perished, for the bones of the young turtle are seen in the interior with their shape fully developed, the interstices between the bones being filled with grains of sand, which are cemented together, so that when the egg-shells are removed perfect casts of their forms remain in stone. On my showing the specimen containing seven eggs to Owen, of the College of Surgeons, he remarked to me that they were hollow, whereas the bones of reptiles want the medullary cavity. Struck with this remark, and with the extreme hollowness of the bones, only to be compared to that of some Tilgate specimens which you have often shown me, I got Owen this morning to dissect for me a young turtle, not a fœtus, but so young that the mark of the attachment of the yolk was still a large opening. He immediately showed me that the bones were not hollow, though we both remarked that the outside looked harder than the interior. After we had discussed the matter for about a quarter of an hour, we found, to our great surprise, that the matter in the interior of a small humerus, clavicle, scapular, and coracoid, which had been taken out, had dried up and become as hollow as the fossils, or as some of your supposed birds' bones, which may perhaps be the fœtal bones of gigantic turtles, for the dried bones which I have saved for you are more empty than those of birds; and when the ball-joint of the humerus shrunk away to nothing, on the evaporation of the spirits of wine, it was marvellous to see how a bone, previously so dissimilar in form, became the facsimile of one of my small fossil humeri. Owen has promised to get me a set of very young turtle's bones from the Zoological Gardens, and I am persuaded it will clear up a number of your difficulties.

Believe me, ever most truly yours,

CHARLES LYELL.

To SIR JOHN HERSCHEL.

16 Hart Street, Bloomsbury : July 6, 1835.

My dear Sir,—I heard some months ago from Whewell that you had, in one of your letters to him, expressed much pleasure at some parts of my book, which I think you read when on your way to the Cape. It has been so much altered, enlarged, illustrated, abridged, and I hope improved since the first edition, that I am anxious, if you ever refer to it again, that you should see it in its amended state. Some of my friends have read letters of yours which they had received, to me, and I rejoice to find that your grand scheme of visiting the Southern Hemisphere has answered so well.

When at Copenhagen last year Oersted, who was reading your paper on double stars, was talking of it continually, and trying to make me understand the *poetry* of some speculations, which only amused me, from seeing that it was deep mathematics with which he was delighted, as with a romance.

Murray has sold 1,750 copies of my book in the last ten months, so that I have the satisfaction of being much read. You will see by my paper in the Royal Society Transactions, and by a new chapter in Book IV. last edition, that I have recanted about Sweden.[2]

Hoping to have one day the pleasure of seeing you again in England, and that before two years have passed away,

Believe me, ever most truly yours,

CHARLES LYELL.

P.S.—Some chapters in your astronomy, on the theory of the earth's motion, for example, and on parallax, appear to me very successful, in the object of making such subjects intelligible to persons ignorant of mathematics.

---

[2] In early editions of the *Principles*, he expressed his doubts as to the validity of the proofs of a gradual rise of land in Sweden, while his visit to that country in the summer of 1834 convinced him that the evidence adduced in favour of the change of level was full and satisfactory.—See fourth edition of *Principles*.

To the REV. ADAM SEDGWICK.

Paris: July 23, 1835.

My dear Sedgwick,—I hope you will receive this when you are at the meeting at Dublin, and shall be obliged to you if you would give me some account of the manner in which it went off, and send your letter to Poste Restante, Bonn.

We shall get to Bonn by September 18. Many thanks· for the letter of friendly counsel which I received from you just before I left London, and by which I profited.

I forget whether you were with us in June last, when the question of the age of the crag was discussed, but you have probably heard that a great number of very perfect shells have been procured from a lower bed of the crag in Suffolk, and it was supposed that this bed which Mr. Charlesworth called the Coralline Crag, might belong to an older Tertiary period than the upper or red shelly crag. Phillips leaned to the opinion that the proportion of recent to fossil shells would not turn out to be so large as that which was deduced from the first collection of crag shells which I brought some years ago to Paris. Buckland also dwelt on the fact that all the fish of the crag which Agassiz had seen were extinct species. I wrote in vain to Mr. Wood of Suffolk, who it seems was travelling for his health, to try and get information from him, as he is said to have collected 400 crag shells; but I brought with me to Paris more than sixty species of shells, in excellent preservation, all collected by Mr. Charlesworth from the Coralline Crag, some abundant, others rare, in short a fair sample, and have had them carefully examined by Deshayes, and have compared them all myself, with shells in his collection. The result is that the proportion of recent shells (almost all inhabitants of the British or Norway seas) is actually greater than that obtained from the first set which I took to Paris, in short the proportion of these is actually more than 40 per cent. On the other hand, the extinct species are not Miocene, or those of the Faluns of Touraine or scarcely any of them. I think we may lay it down as a rule, that if in any given tertiary deposit, in which we have found a few species of shells only, of which half or a third, or even less are recent, and those recent ones inhabit the seas immediately adjoining, the formation will be Pliocene, whereas the recent shells of the Miocene strata have a more exotic and tropical form. I have been rather diffuse on this point, because I think some of our friends will be interested about it at Dublin. I should add that I have not yet had the corals examined, which at first sight look more like the Touraine fossils than the shells do.

I found here Von Buch, E. de Beaumont, Dufresnoy, Constant Prévost, Virlet, Boué, Alex. Brogniart, and have had much talk with all of them, and some warm discussions with Von Buch and de Beaumont. Of the first, I must say that I find much to like in him. As I had handled some of his opinions very roughly, and as he is too much accustomed perhaps to have unbounded deference paid to them by most of his own countrymen, and by no one more so than by E. de Beaumont, I had no right to expect a very cordial reception, but he met me with great frankness, and at once set me at ease by vehemently protesting against my numerous and crying heresies, none of which, not even the elevation crater theory, seems to have excited so much honest indignation as my recent attempt to convey some of the huge Scandinavian blocks to their present destination by means of ice. If you read my Swedish paper, which I sent you, you will remember that I proved that some of the huge Swedish blocks near Upsala must have travelled to their present destination, since the Baltic was a brackish water sea, so that those who maintain that there was one, and one only rush of water, which scattered all the blocks of Sweden, and the Alps, must make out this catastrophe to be as it were an affair of yesterday. As to the elevation crater business, Von Buch, de Beaumont, and Dufresnoy, are to write and prove that Somma and Etna are elevation craters, and Von Buch himself has just gone to Auvergne to prove that Mont Dore is one also.

Prévost's sketches of Vesuvius and the Lipari volcanoes are excellent, and some of his objections to Von Buch's theory which I had not made are unanswerable. I will give

you one.   In Mont Dore and Cantal there is a vast thickness of volcanic matter in the centre, which thins off on all
sides towards the base of the cone, so that the granitic
rocks crop out in the one case, the fresh water in the other.
So in Etna do the marine strata; and in Palma the volcanic
mass is thinnest round the outer margin or base of the
island.   But if these masses had been upheaved by elevation, it would be in the centre, not towards the circumference that the fundamental rocks would rather be seen, and
the volcanic mass would not be thinner towards the middle.
As to marine shells found on the slope of Vesuvius, it is no
argument, for there must be many on the slope of Stromboli.
Prévost made a drawing of the four dikes which I saw in
the modern cone of Vesuvius, dikes resembling those of
Somma, which are numerous in Stromboli, Etna, Mont
Dore, and other volcanoes, near the great craters of eruption, and are comparatively rare in volcanic masses far from
those centres.   I much regret that so undescriptive and as I
think theoretically false a name, as elevation crater, is to be
given by all Von Buch's school to every dome-shaped elevation.   In the latter, as in the Wealden and others, the
oldest rocks are seen in the middle, while it is just the reverse in Mont Dore, Cantal, Etna, and other volcanoes.

July 23.—To-day I have seen Van Breda of Leyden, and
after him no less a person than Berzelius, just come from
the North, and to the same hotel in which we are.   He gave
me a warm and hearty greeting, in the German more than
in the French style.   I have not had time to hear news from
him.   The map of France is now all but finished, and looks
very handsome.   To me who know more of French than of
English geology it has the effect of a beautiful picture, and
the great granitic boss of Auvergne, &c., forms a grand central feature, without which there would be a want of unity, as
compared to our map.   I am reading you and Murchison on
the Eastern Alps, as I am going so near your section.   Your
elaborate joint paper is now quite a treat.   Boué has given
me many Gosau fossils.   He is going to live four years in
Vienna, and next year to do the Balkan.   My wife says,
' Give my kind regards to Mr. Sedgwick, and tell him it is
dreadfully hot.'

Remember me to Murchison and all friends, and hoping to hear from you, believe me, ever most sincerely yours,

CHARLES LYELL.

To VISCOUNT COLE.

Meyringen : September 6, 1835.

My dear Lord Cole,[3]—I promised to write to you after I had been at Glaris, but as it will now be impossible for me to go there, I shall still let you hear of my proceedings and explain my change of plans, assuring you that I am determined at no distant period to profit by the instructions which you and Egerton gave me. As my eyes are not strong enough to allow me to write much by candle-light, I shall beg my wife to write a little, while I dictate.

I set off for Berne by Vésoul and Porrentrui, having a good work at the latter place with Thurmann, who has a collection which you should see, if ever you pass that way, as also that of M. Thirria at Vésoul. I made an excursion also to Neuchâtel, and got some fossils there of which I daresay Agassiz may have given you some duplicates. I made an excursion there into the Jura with M.M. de Montmollin and Coulon, both of whom have collected many fossils from the Neuchâtel chalk. After having examined the museum at Berne with Studer, I set off to the mouth of your old friend the Kander, determined that I would not be outfathomed by the lake, as you were. I took a line about five hundred feet long, and carried out my soundings in two directions from the edge of the delta. I have not had time yet to calculate the results of these observations, but I believe the slope of the new deposit under water is not so steep as we made out from the soundings which you and Egerton made. When I was thus far on my way, I had everything prepared for the Glaris campaign, having always meant to take the Jungfrau where the gneiss overlies the limestone with oolite fossils in my way, as also the Urbach Thal near this place, where similar things were to be seen. But I found when I attempted to understand the geology of the neighbourhood

[3] The present Earl of Enniskillen.

of the lake of Thun, even with Studer's newly published book and map and sections as he calls them in my hand, that I could not at all comprehend it, nor make out what he meant by his numerous formations. I therefore determined to make myself master if possible of the geology of this part of Switzerland, on which much more has been now written than on any other part of the Alps, before I made an attack upon less known districts. This I have in some measure accomplished; but in doing it and climbing the Jungfrau and the Urbach Sattel, I have spent the time which was to have been given to Glaris, which I hope, however, is only postponed. I must now proceed at once on my way to the Bonn meeting. I have just come down from a châlet at the southern foot of the Stellihorn in the Urbach Thal, where at the height of nine thousand feet I found ammonites, and got some more from a chamois hunter at whose châlet I slept. I daresay you know that the Swiss Alps have a bad name for fossils, so you must not expect to see me return with many. This morning I got what I suppose to be a Pentacrinite at Brienz, which I bought from a boatman, and I suppose they might get other things if they did not waste all their time in hunting for quartz crystals. I long to hear how the Dublin meeting went off, for I have not seen a word about it in any paper. Indeed I have only got sight of newspapers about once in a week or ten days, just enough to see that the everlasting Parliament was still sitting, till after the Association met at Dublin, which I am afraid must have interfered with the attendance of some good men. I cannot sufficiently congratulate myself that I am not in such a position as to make it a duty to take an active part in politics in such times, but that I can ride my hobby during the whole summer, without fearing either a call of the House or the hustings to stop me short. I took your hint, and instead of Interlachen went to Unterseen, where I stayed several days and found the charges very reasonable. I hope still I shall get some ' birds' tongues ' [4] at Utmarsingen. I saw some of them in the Berne museum, or at least from that neighbourhood. I expect to be back again in town before the end of September, and shall have a pretty good load of rocks to

---

[4] A local name for the palatal teeth of certain fossil fishes.

take back, though I shall not have much to boast of in the way of organic remains. By-the-bye, I got a new light this morning, touching these remains. The waiter here assured me that it was only in stones that had been for a long time in the foundations of houses that such figures of shells and other things were produced, so that unless one had the luck to be passing when an old house was being pulled down, there was no chance of getting anything.

I must now conclude, as to-morrow I must be off early with horses to cross the Brunig. Mrs. Lyell desires her remembrances, and

Believe me, ever most truly yours,

CHARLES LYELL.

To GIDEON MANTELL, ESQ.

London : October 14, 1835.

My dear Mantell,—I made out the greater part of the tour which I had planned most successfully, first visiting Paris, where I found Von Buch, whom I had never met before, and whom I liked well, notwithstanding my opposition to many of his opinions, and although there are few persons who are more noted than he is in geology, for conceiving a personal hostility towards those who do not embrace the theoretical doctrines which he has once published. I took a collection of crag fossils from those lowest beds, which have been lately called the Coralline Crag, and they were carefully named by Deshayes, and compared with his usual patience with the fossils in his rich collection. It was a delightful lesson in conchology for me, and the result was that half of them were recent species, contrary to the anticipation of some of the worthies who had collected them or talked about them at our G. S. I entered Switzerland by Porrentrui, and there I had Thurmann for my guide, who gave me in a short time a beautiful insight into the structure of the Jura, on which he has published, and I was glad to verify his observations in the field, and to see his beautiful collection of Jurassic shells, and his attempt to assimilate the oolitic series and their fossils of the Swiss Jura with our English oolitic groups. I afterwards had a work in another part of the Jura with some geologists at Neuchâtel, where the chalk, as it appears to be by its fossils, fills the bottom of the valleys of the Jura limestone. I next had a work with Studer at Berne, and then had a work of about six weeks in that part of the Swiss Alps which is called the Bernese Oberland. I there saw alternations of the gneiss with limestone of the lias or something newer in the highest regions of the Alps, of which I shall give the Society an account during the next season. I also endeavoured to make out the order and age of the strata between the central granitic axis of the Alps and the great tertiary valley of Switzerland, in other words, the country around the lake of Thun and between the town of Thun and the Jungfrau. We then came down the Rhine to Bonn, and spent a week there during the meeting of the German Association, which was about six hundred strong. They elected four presidents for the geological section, Von Buch, Buckland, de Beaumont, and myself, and we presided in turn. For two whole days Constant Prévost and I fought Von Buch and de Beaumont on the craters of elevation, the discussion being in French, and with a crowded audience. I am as convinced as ever that their views are quite erroneous, and if I had to write over again to-morrow my chapter on that subject, I should have nothing to retract and many new arguments to add. We had Greenough there, and Omalius d'Halloy, Goldfuss, Count Mandelslohe, and Professor Walckner, Von Meyer, Alexander Brongniart, Adolphe Brongniart, Andouin, Berzelius, Schmerling of Liege, Hoeminghaus, Von Oeynhausen, Noeggerath. I mention some of those who first come to my recollection as having attended our section. On my return here I found that in spite of Lonsdale's exertions and Fitton's pledge to the Council, that his paper should be concluded in July, there is still many a green grain of sand yet to fall through the hour-glass. People seem to be in high spirits here on the success of the Dublin meeting. Agassiz looks in good health, and is satisfied with the great progress he has made, and looks forward with great pleasure to a first visit to Brighton. When I call, I find him sitting in his room with his two horse power, for he has had the spirit to get over a second German artist. I have been writing this by candle-light, after the labours of the day, and as you know my eyes of old, you will excuse its not being in my own handwriting. After all our wanderings we hope to be stationary here for the next nine months, and shall be always delighted to see you here, even for one of your flying visits, and my wife unites with me in kind regards,

Believe me, my dear Mantell, yours most sincerely,

CHARLES LYELL.

To PROFESSOR SEDGWICK.

London: October 25, 1835.

My dear Sedgwick,—I was very glad to hear your account of the Dublin festivities, which, like those of Bonn, were I suspect rather overdone in the way of dinners. I am persuaded that such meetings and displays of strength, and numbers, and convivialities, do much good, but really they are no small sacrifice to the real good and true workers in science, out of their summer's vacation, and when they have thought that they escaped from a bout of spring dissipation in London till the year following. This I felt on the Rhine, and should have done so at Florence-Court, though much good was done to our friend Lord Cole, who tells me that his father has been in great good humour ever since he had the satisfaction 'of seeing Murchison and some other guest *glorious,* and Sedgwick *comfortable.*' Depend upon it the building of the museum, and subsidies for what the old Lord once condemned as 'damned nonsense,' will go on with good spirit, after his finding that the hammer-bearers are such a jolly set. I had heard nothing about Agassiz' catastrophe system at Dublin, but no sooner did I try to get him upon fish when I met him here, than he let off upon me, and I was obliged to compromise half our time day after day most unprofitably in arguing with him. This I did the more zealously from his telling me that he was going to give some lectures on geology at Neufchâtel. He does not flinch at trifles, for finding that the whole cretaceous system from the bottom up to the Maestricht beds has the same fish (so he declares), he deposits the whole during one commotion of the sea. I am in hopes, however, that I have made some impression on him. His knowledge of natural history surprises me the more I know of him, and he has that love of imparting it, and that power of doing it with clearness, which makes one feel one is getting on, and that one has caught his enthusiasm. I feel this also strongly when in company with Deshayes, who continues steadily to cultivate his own branch, and that under somewhat discouraging circumstances, both pecuniary, and from being in an atmosphere where there are not many kindred spirits, who would continue devoted to their science for its own sake (as I am sure he would), and not as a means of living. In this respect he would be better off in Germany or England, or even in some of the provinces of France. I assure you that his examination of the 80 species of Crag shells was most conscientious, and I went into it thoroughly, and we had no idea, till we summed up at the end, what result we were coming to. As to Phillips doubting the identifications, recollect that many of these Crag species are the same as some brought from the North Cape and other parts of the northern ocean. They have not yet been described nor named, and whether Mr. Bean of Scarborough happens to have found them on the British coast or not, or whether they are in York Museum, is immaterial.

It requires a vast series of varieties to make out these identifications, and mark, the same method leads to consistent results, for Deshayes would agree with Phillips & Co., that scarcely any of the Eocene shells are recent, that the living shells of the French and English coasts agree in great part, while those of India and the Mediterranean differ. As to the fish of the Crag being extinct, I suspect that Phillips would admit with Lonsdale that Strickland's 25 or 30 species of freshwater shells were recent British ones, except perhaps one or two; and yet they occurred with an extinct hippopotamus, rhinoceros, &c. The vertebrate must have fluctuated more rapidly than the testacea, and if this be true of the first step, or newer Pliocene, is it not more consistent that it should hold on? Before I have done with Deshayes, let me remark that the School of Mines did homage to his knowledge lately, by offering him a sum of money to name their shells. It cost them no small effort, but there was no man in office, no successor to Lamarck's or any other chair, who could do the work for them. Deshayes declined, for he was deeply engaged with the booksellers. I tried to persuade

him to come over to England, but it is impossible. Having
begun without a farthing, and having dared to cut medicine
entirely, he must work hard the whole year to preserve his
independence. With all his faults, I am persuaded he might
have got on, and might now if he would condescend to be
more of a courtier, for court is too much exacted in Paris by
those who have the good things to give away.

As it is my wish to propose the Wollaston medal this
year to Deshayes, I wish you would have a talk with Agassiz
when he is with you about the matter, for I believe there is
no one in London who has seen so much of Deshayes, and
knows so well his acquirements and the difficulties he has
overcome as Agassiz. I am sure that if we could draw Deshayes over here, he would make a grand reform in our
Museums, like Agassiz, and that he would discover rich mines
of hidden treasure.

Did you see that Captain Fitzroy had borne witness in a
court-martial that the late Chilian earthquake had altered
the whole coast? that a north-westerly had become a
southerly current? that the island of Mocha was *upheaved*
10 feet? and this upon *oath*! and the captain of the 'Challenger' acquitted on this evidence! Give me but a *few* thousand centuries, and I will get contorted and fractured beds
above water in Chili, horizontal ones in Sweden, &c. I have
got from Denmark a notice by Dr. Pingel, on the gradual *subsidence* of Greenland. As to erratic blocks, I have no positive
general theory, but after seeing Sweden, I cannot for a
moment believe they are due to diluvial action. I saw none
in Italy, Spain, or Sicily, a few in the Jura and Alps,
hundreds in Denmark, thousands everywhere in Sweden,
and of stupendous size. It seems to me *a northern* phenomenon when best developed. Ice does annually carry
large stones on Lake Wener, and in the Gulf of Bothnia, not
so big as cathedrals I grant, but tolerable sized pebbles, such
as many men could not move.

But I must return to this when we meet, only remember
that icebergs float now much farther south than the Jura.
I have no room left for my Alpine observations, but hope to
give you a paper on them at G. S. Write when you can, or
come here, and believe me ever yours,

CHARLES LYELL.

To PROFESSOR SEDGWICK.

London : December 6, 1835.

My dear Sedgwick.—I have been drawing up a summary
of your short but pithy paper on slaty cleavage, and trying
how far I could reconcile what I saw this year in some slate
rocks in the Alps with your general rules, which in some
respects I find difficult enough, in as much as I found the
cleavage or slaty structure of fine drawing slate in the great
quarry of the Niesen, on the east side of the lake of Thun,
quite coincided with the dip of the strata ascertained by alternate beds of grey wacké. Now Murchison tells me, to whom I
showed specimens of this slate, that you would call its
cleavage ' the fine flaggy.' But as it is the best description
of drawing slate, and as divisible almost as mica into thin
plates, I cannot make out how to distinguish such a structure
from any which can be called slaty, and such an attempt
would I fear involve the subject in great confusion.

At the last meeting, Cuming's paper was read, denying
that in 1822 there was an elevation of the land at Valparaiso.
Greenough had been very anxious that this should come out.
I had already given an abstract of Cuming's evidence in my
4th edition. It is clear and curious, and although it
probably proves, as Basil Hall observed to me last year,
Cuming's want of optics, it is still worth recording, as his
attention was called to the fact of elevation by Mr. Graham's
paper, and he swears that he looked for the phenomena, and
found none. I believe Mr. Graham, my friend Cruikshanks,
and others *contra,* but it is a blessed thing in this world of
scepticism, that thirteen years only should have passed
away, before another and greater upheaving took place,
authenticated by Fitzroy, whose papers, which have not yet
reached the Admiralty, I am anxiously expecting. I remarked
at the last meeting, that if the notion of the rise in 1822 had
been as purely imaginary as Cuming, Woodbine Parish, and
Greenough pretend, that the chances were almost infinite
against nature's having performed a similar operation soon
afterwards on the very same coast as if expressly to give
credibility to the previous notion. How I long for the

return of Darwin! I hope you do not mean to monopolise him at Cambridge.

My wife joins in kind remembrances, and is curious to know whether her cousin Elizabeth Winthrop, sister of Mrs. Edward Pellew of Yarmouth, was able to come to the deanery to hear your lectures. If you see her, you will certainly fall in love with her.

Believe me ever, most truly yours,

CHARLES LYELL.

CHAPTER XX.

FEBRUARY–DECEMBER 1836.

ON MONEY FOR MUSEUMS BEING FRITTERED AWAY IN BUILDINGS—LETTER
TO SIR JOHN HERSCHEL ON THE ORIGINATION OF NEW SPECIES —KIN-
NORDY—DR. FLEMING APPOINTED PROFESSOR IN ABERDEEN—DR. FITTON
DECLINES PRESIDENTSHIP OF THE GEOLOGICAL SOCIETY—LETTER TO
DARWIN AGAINST HIS UNDERTAKING OFFICIAL WORK.

CORRESPONDENCE.

To G. Mantell, Esq.

February 10, 1836.

My dear Mantell,—I shall of course mention your Tilgate
birds with due honour, but as I have written my speech exclusively on the proceedings of last year, I would rather
defer till my second address your Swanage Gavial, and hope
before that time that you will send me a brief notice, however short, for our proceedings on the same, though I hold
that no public christening of any new-born fossil is valid
unless a description accompanies the new names sufficient,
as Dr. Beck would say, 'for a diagnosis.' I had not time to
tell you when you were here, how much I apprehend that
the architect will run away with whatever money a zeal for
science or friendship and regard for you may raise by subscription at Brighton. I remember that when some 8,000*l.*
had by a great effort been got together at Bristol, for
lectures, and for a collection of books and other useful aliment for the mind, as the misguided projectors thought ; in
came the architect, gave them a handsome building,
pocketed the cash, and left them with a room for the newspaper readers, and scarce a farthing to pay their invaluable
curator Millar. So it was with the London Literary

Institute in Moorfields—about thirty or forty thousand
pounds sunk, and a hundred other cases. Yet unwarned by
experience, the planners of the London University followed
in the same track, and spent several hundred thousand
pounds in erecting a huge and never-to-be-finished edifice
with a splendid portico, costing alone 30,000*l.* or 40,000*l.*,
and leaving nothing for professors but debt. Then came
the King's College, and another splendid subscription, for
there is no end to the gullibility of John Bull. Might not
anyone have told the poor parsons, who with an honest zeal
for their cause, put down their 5*l.* or 50*l.*, that it was not
science, nor learning, nor religion, nor anything but *architecture* that is encouraged by such munificent donations in
England. Had they hired a set of the ugliest houses in the
Strand, and bribed, with their two or three hundred
thousand pounds, the first teachers in Great Britain, they
would have carried everything before them. But what did
they do? Reared a huge wing of a building which swallowed
up all the money, and is now unfinished inside, although
part of it, contrary to the original plan, is fitted up for
students' private chambers. There is no hope for natural
history or science, or literature, until they *precede*, instead of
following, the architect. I sat next a pupil of the late G.
Rennie at dinner the other day, and he attributed the success
and large fortune of that engineer, not so much to his civil
engineering, though of course he was eminent in that, as to
his knowledge of men and of human nature. He told his
pupil that to *work* the original plans or designs was a small
affair comparatively—the great business was ' to work the
committee.' So some skilful architect at Brighton will be
found, who will have tact enough ' to work ' your sub-
committee, with your friend Horace Smith at the head of it.
He will not frighten them with too large an estimate at first,
but let it be at least some hundreds under the subscription,
for he knows that so many gentlemen of taste and leisure
will see improvements that may be made as the work proceeds, and to these he, who has the *working* of them, will
after a decent show of resistance accede, and then he may
lay to the charge of their departure from the first plan all
excess above the estimate. Having spent all the money and

mortgaged the handsome edifice, there will be nothing left for the Mantellian collection, and still less for lectures. Try and preach against this, though I have but a faint hope. The 'Challenger' frigate, when wrecked in Chili, and in the midst of danger and distress, fired a gun to warn a Swedish vessel off the breakers, and she was saved: it was a generous act. I am reading their diary. I wish some of our bankrupt scientific institutions would fire guns of distress.

Yours ever most truly,

CHARLES LYELL.

To SIR J. W. HERSCHEL.

London: June 1, 1836.

My dear Sir John Herschel,—If anything could have made your letter of February 20 last more welcome and useful to me, it was the time when it reached me, just as I was unexpectedly called upon by my publisher to interrupt my other scientific occupations by preparing a new or fifth edition of the 'Principles of Geology,' the fourth edition of two thousand copies having been nearly all sold in one year. The continued sale of the work is a sufficient earnest of its having been written in a popular style, in attempting to accomplish which I devoted so much time, that I have sometimes thought it was scarcely justifiable, and might have been better employed, but the interest you have expressed, and the fact of your having read the book three times, and your approbation of one of the newest parts of the treatise, 'the climate theory,' is a much better test to me than the sale, being something solid and substantial in my labours. I may truly say that when the Royal Society voted me a medal for my book, I was not more gratified nor more encouraged than by your full and interesting comments which have given me a feeling of strength and confidence in myself, which will assist me in my future studies. First I shall take this opportunity of thanking you for the manner in which when writing long ago in the 'Transactions of the Geological Society,' on the effects on climate which variations in the earth's eccentricity might have produced, you alluded to my notice of the variable quantity of light and heat received by the two hemispheres during the precession of the equinoxes. You contrived so to set me right in a great blunder, as to make it appear as a compliment, and I was able to correct it in after editions as you see. I went over this subject and your paper with Mr. James Bowstead, of Corpus Christi College, Cambridge, who made me understand with tolerable clearness the mathematical part of the reasoning, which appeared to me most beautiful, and made me regret that I had not given some of the years which I devoted to Greek plays and Aristotle at Oxford, and afterwards to law and other desultory pursuits, to mathematics. The life of a geologist is too short to repair that deficiency, and in this and other matters we must borrow of others. With this I send a copy of my first President's speech, in which I have dwelt somewhat at length on the controverted point of the elevation of land by earthquakes. Perhaps in my next address I shall limit myself less to the papers read at our meetings in the year, but as a rule I rather grudge the devotion of time to such reviews, and to official duties, which some people like, but which are not to my taste, and fritter away one's time and thought; and certainly if working men of science must perform such duties, no Presidentship ought to be more than biennial. I sometimes make some of my friends here very angry, by congratulating them as friends to science that we failed in making you perpetual President of the Royal Society. Just now the society's chair may be said to be in abeyance, but it continues to be a good and rapid publishing machine, and at all events it has not to answer for the sin of annihilating part of your time in choosing officers, writing *éloges*, attending councils, &c. &c., which I think a man of good fortune and tolerable abilities, without genius, or great acquirements, may perform very well. What we most suffer by now, is the too great facility of admission, which nothing can effectually check but the President's tact and firmness.

I shall send you my fifth edition as soon as it is printed, and you will see that your information about the boiling of seeds has been useful, and that I have attended to other corrections. Your very interesting volcanic theory was too much for me to attempt to grapple with, as an intercalation

into my book, at least this time ; but I mean to have a work
at it this summer as soon as I have got through the reprint.
It struck me on first reading as singularly like a speculation
of Babbage's, which he appended to a paper of his on the
Temple of Serapis ; and of which an abstract appeared in
the ' Proceedings ' of the Geological Society which I suppose
you have, although in case you should not, I will send you a
copy if I can procure it.   I have just seen an ' Observer '
Sunday paper, in which you are made to ' set at rest ' some
curious points in the planet which received Astolfo's brains.
Mrs. Calcott (Maria Graham) promised me a letter which I
think would be useful at Rio de Janeiro, but I have not yet
received it, probably because she has been so ill.   I am glad
to hear that your thoughts are bent homeward.   What you
say of the basaltic and granitic dikes is very interesting.   I
am so busy with my printer, that I shall not digest all that
till I get into the country, our geological session being just
over.   I go to my father's in Scotland with my wife for two
months, and afterwards away to the west of Scotland, and I
hope Arran for a month, and return by the end of September.
We dine to-day (June 4) with the Somervilles, and in the
evening Babbage gives one of his evening meetings which
are very brilliantly attended by fashionable ladies, as well as
literary and scientific gents, and where one meets with
persons high in all professions, and with distinguished
foreigners.   You remember Fitton's ' Philosopher Turnstile,'
but we must take men as they are, and Babbage must be
admitted to have really succeeded in commanding the
respect and admiration of persons not only of the first
talent, but not a little even of fashion and beauty, and I
maintain that he has done good, and acquired influence for
science by his parties, and the manner in which he has firmly
and successfully asserted the rank in society due to science.
I believe he works hard now in spite of the time he must give
to the arrangement of his parties, and to society.   I think
Whewell's report on mineralogy very good in ' British
Association Reports,' which I have just read.   Murchison's
book will scarcely be out this year, and as for Buckland's
' Bridgewater,' we are tired of waiting for it, as it has been
reviewed in the ' Quarterly ' two months.   He says it will

be out in six weeks. Babbage has promised a ninth 'Bridgewater,' in which 'the Devil is to have his due,' and he has
actually written a part of it. Sedgwick has been giving
very useful and popular lectures in geology at Norwich,,
where he is now Prebendary. Fitton is at last, we believe, at
the point of being delivered of the Green Sand paper, on
which occasion we trust that the shock of an earthquake
may not throw down our museum. I always made up my
mind it would be a posthumous production. He is as amiable as ever, and if he was not such a procrastinator would
have done much for us—as being so good a writer.

When I was at Copenhagen the year before last, Oersted,
who had just received your nebula paper in the 'Philosophical
Transactions,' expressed the greatest admiration of it, and
endeavoured to make me comprehend what he thought 'the
poetry' of some of the speculations, but it was far beyond
my depth in mathematics. You have not in Europe a more
sincere admirer than he.

I shall now conclude, and add in another hand[1] some
other remarks, and perhaps send the whole before I get the
introduction to Rio, which shall follow. I hope some day to
have the pleasure of introducing my wife to Lady Herschel.
In the meantime believe me, ever faithfully yours,

CHARLES LYELL.

P.S. In regard to the origination of new species, I am
very glad to find that you think it probable that it may be
carried on through the intervention of intermediate causes.
I left this rather to be inferred, not thinking it worth while
to offend a certain class of persons by embodying in words
what would only be a speculation. But the German critics
have attacked me vigorously, saying that by the impugning
of the doctrine of spontaneous generation, and substituting
nothing in its place, I have left them nothing but the direct
and miraculous intervention of the First Cause, as often as a
new species is introduced, and hence I have overthrown my
own doctrine of revolutions, carried on by a regular system
of secondary causes. I have not wasted time in any contro-

[1] Mr. Lyell had frequently to employ an amanuensis to save his weak
eyes.

versies with them or others, except so far as modifying in
new editions some opinions or expressions, and fortifying
others, and by this means I have spared a great deal of ink-
shed, and have upon the whole been very fairly treated by
the critics. When I first came to the notion, which I never
saw expressed elsewhere, though I have no doubt it had all
been thought out before, of a succession of extinction of
species, and creation of new ones, going on perpetually now,
and through an indefinite period of the past, and to continue
for ages to come, all in accommodation to the changes which
must continue in the inanimate and habitable earth, the
idea struck me as the grandest which I had ever conceived,
so far as regards the attributes of the Presiding Mind. For
one can in imagination summon before us a small past at
least of the circumstances that must be contemplated and
foreknown, before it can be decided what powers and
qualities a new species must have in order to enable it to
endure for a given time, and to play its part in due relation
to all other beings destined to coexist with it, before it dies
out. It might be necessary, perhaps, to be able to know the
number by which each species would be represented in a
given region 10,000 years hence, as much as for Babbage to
find what would be the place of every wheel in his new
calculating machine at each movement.

It may be seen that unless some slight additional precaution be taken, the species about to be born would at a
certain era be reduced to too low a number. There may be
a thousand modes of ensuring its duration beyond that time ;
one, for example, may be the rendering it more prolific,
but this would perhaps make it press too hard upon other
species at other times. Now if it be an insect it may be
made in one of its transformations to resemble a dead stick,
or a leaf, or a lichen, or a stone, so as to be somewhat less
easily found by its enemies ; or if this would make it too
strong, an occasional variety of the species may have this
advantage conferred on it ; or if this would be still too much,
one sex of a certain variety. Probably there is scarcely a
dash of colour on the wing or body of which the choice would
be quite arbitrary, or which might not affect its duration for
thousands of years. I have been told that the leaf-like ex-

pansions of the abdomen and thighs of a certain Brazilian Mantis turn from green to yellow as autumn advances, together with the leaves of the plants among which it seeks for its prey. Now if species come in in succession, such contrivances must sometimes be made, and such relations predetermined between species, as the Mantis, for example, and plants not then existing, but which it was foreseen would exist together with some particular climate at a given time. But I cannot do justice to this train of speculation in a letter, and will only say that it seems to me to offer a more beautiful subject for reasoning and reflecting on, than the notion of great batches of new species all coming in, and afterwards going out at once. Your facts about the boiling of seeds are very curious, and I have availed myself of them most gladly in my fifth edition.

I wish much you could measure exactly the temperature of the great hot springs which you mention in the Brand valley, or get some one to take their temperature at different seasons. The observations will perhaps only become of importance like some of those which you have made in astronomy, many generations or centuries hence, when a comparison would show an increase or decrease of heat. It would be well to point out to the Literary and Philosophical Society which you have founded at the Cape (as I heard from Major Cloete), the remarks which Forbes has made on the light which such observations may throw on the heat of the interior of the earth, when comparisons are made after considerable intervals of time. (See Forbes, 'Proceedings of Royal Society,' No. 24, p. 382.)

I forgot to remark on what you say of the argument of the originally limited distribution of plants being capable of being inverted. You will at least find that I have precisely anticipated the necessary occurrence of such ambiguous phenomena at vol. iii. p. 165, in the paragraph beginning 'If the views.' But I anticipate that geology will hereafter enable us to determine in many cases which of the recent species, now having a confined range, are the last survivors of species once more flourishing, and which are the newly introduced and infant colony destined to become a great nation in future.   C. L.

To Sir John Herschel:

16 Hàrt Street, Bloomsbury : June 7, 1836.

My dear Sir John Herschel,—A few days ago I sent to
Captain Beaufort a long letter which I had written to you,
in which I hoped to enclose some letters of introduction to
persons at Rio, as you wished. I now enclose them, together with the abstract of Babbage's paper to which I
alluded. Pray if you have time to look out any marginal
criticisms on parts of my book, send them to me. Yesterday I sat next Babbage at Miss Rogers' at dinner, and asked
for the abstract which I now send. Mr. Rogers, the poet,
was talking of your astronomy which he had read, as well
as the introduction to ' Natural Philosophy,' and with both of
which he had been much delighted ; and among other things,
with the manner in which you had alluded to certain papers
of Dr. Young's on light and colour, which Brougham had so
presumptuously and unmercifully cut up in the ' Edinburgh
Review,' as Arago has remarked in his *éloge* of Young.
I think it was Sydney Smith who said of Brougham that
he had made two great discoveries in the ' Edinburgh Review '
—the first was that Byron was no poet, the second that
Young was no philosopher—for Jeffrey had nothing· to do
either with the attack on Young or with that on Byron.

Hoping you will receive this and my former letter at one
time,   Believe me, ever sincerely yours,

Charles Lyell.

To G. Mantell, Esq.

Kinnordy, Kirriemuir, N.B. ; July 6, 1836.

My dear Mantell,—Here am I rusticating in a very beautiful country, not too hot, but with weather much like a fine
English spring. I am now and then devoting some stray
hours to my ' Elements,' like Buckland's ' Bridgewater,' long
promised—but not yet reviewed, thank heaven. I have received a very pleasant letter from Alexander Burnes, who has
returned to Cutch and re-examined the delta of the Indus.
He reports that the submerged tract which sank in 1819 is
in statú quo. He has sent me off some Cutch secondary
fossils, ammonites, belemnites, &c. His letter came in nine

weeks per steamer from Cutch !   I am glad to escape awhile
from the excitement of London, and wish I could hear of
your doing so from Brighton, but fear from your last letter
that there is no immediate prospect of that.

A letter from Dr. Silliman informs me that my 'Principles' are being reprinted at Philadelphia, and nearly ready.
John Murray was in hopes he had reduced the price so as to
prevent this happening.

I shall be here for more than a month, then go with my
wife to geologise in Arran, then return here, and afterwards
in September, or perhaps October, to London.   If you send
a letter to 16 Hart Street, it will be forwarded to me, and
do give me a full account.

To G. MANTELL, ESQ.

Kinnordy, Kirriemiur, N.B. : September 19, 1836.

My dear Mantell,—I have written to invite Dr. Fleming
to pay us a visit here, and he has accepted for to-morrow,
so when we meet, or when I next write, I will tell you how
he looks.   I am sorry to say he has had much ill health.   I
suppose you know that he got his living of Flisk—the smallest
in emolument in the East of Scotland, and where he had
been for most of his life, 'like a pony tethered on a highland
moor'— exchanged for a much better living in Fife, where he
was quite adored by his parishioners, but so great was the
duty that it left not a moment for a conscientious man to
devote to science or literature, and was almost too fatiguing
for his health.   It was therefore most natural that he should
have sighed for a professorship in a Scotch University,
which gives a six months' vacation every year.   After several
applications in vain for chairs more consistent with his
zoological and botanical acquirements, he accepted one in
Aberdeen, of Natural Philosophy I think it was called,
which no doubt has given him much fag to get up arrears
of mathematical knowledge.   But unfortunately, something
worse than the lectures fell to his hard fate.   Several University bills and a Royal Visitation caused tremendous secre-
tarian or clerk's labour to fall on the Junior Professor, who
is obliged to serve as secretary to the University.

You will see by this that you, my good friend, are not

singular in finding it difficult to gratify your liberal thirst for science, without interfering with professional profits. Really, as Milman says, it would be well for the country if, instead of abolishing prebendal stalls, they were given to clerical and lay cultivators of literature and science, who had shown that they would devote energy and superior talents to those departments. When Babbage was taunted one day by a Conservative with ' What do you mean to be when the revolution comes?' he said, ' Lay Archbishop of Winchester.' A few comfortable scientific sinecures would be good things, but I fear they might become useless like University fellowships.

Did you read those lines in the ' Morning Chronicle ' about the ' Aristocratodon,' and the ' Episcopus Vorax ' ? I wonder who wrote them! There is an attack in the ' St. James's Chronicle ' on Buckland, for having said at Bristol that the world was millions of years old. Mr. Horner tells me that Lonsdale is looking better, but thinking himself still ill—he has worked like yourself too hard ; half what either of you have done would have finished me.

I hear that Fitton's ' Greensand ' is printed, but still he is *polishing* it. I trust that for his and our sakes I shall find it done when I get to town.

Ever most truly yours,

CHARLES LYELL.

To HIS FATHER.

16 Hart Street, London: October 4, 1836.

My dear Father,—I hope my mother received a letter which I put into the General Post Office on Friday, on our arrival here. We called, after church on Sunday, upon Dr. and Mrs. Fitton in Portland Place, and saw them both ; he very well, and she wonderfully recovered from a severe illness. I had much talk with him about the Presidentship, which he would on no account accept, ' having the weight of eight children upon him.' He was glad to hear that I had offered the chair to Lord Northampton, which he said would not have occurred to him, and which was a proper compliment, and one of which Lord N. would feel the value, knowing that our Presidentship had not been conferred on *rank*.

I asked if he had thought of Broderip. He said yes, of course, and had felt, as I probably had, that he ought not to accept it, as knowing nothing of geology; I agreed to this, and wondered that it should be so. Fitton said, ' It is truly marvellous, with his love of natural history and deep knowledge, that he should not even have made any progress in fossils, that he should have served three entire years as secretary of the Geological Society, and so well,—should have edited a volume of the " Transactions," gone through your second volume of " Principles," and Buckland's " Bridgewater," and remain so ignorant of the " Elements "—indeed, he rather prides himself, I think, on the inaptitude of his mind for geological reasoning, and most dexterously has he cleared all the hoops of geology.' On my laughing at his simile, he said, ' You know that Pope has it so in the " Dunciad,"

' Never by tumbler through the hoops was shown
Such skill in passing all, and touching none.'

Upon this we told Fitton that we had guessed him to be the author of the verses in the ' Morning Chronicle,' on ' Aristocratodon,' &c. ' Oh no, you give me a great deal too much credit for powers of versification; they were by Tom Moore.' ' But how came he by so much geology ? ' ' Why, he got it up partly at Bristol, and partly from Babbage, who was much with him there.' He then told us, what Gardner the mapseller has since confirmed, that Buckland's edition of 5,000 of the ' Bridgewater ' is all sold, and 5,000 more printing, each of which editions, Fitton says, will produce the professor 2,000l.—a piece of news I am truly glad to hear, for from what I have read of the book, I think it will do much good in spreading correct notions of the science, and probably popularise it much. Murchison calls it ' Bridge-over-the-water; ' and really that part which is to carry us over the abyss of cosmogony is better constructed than I expected, though I should have been sorry to have had to trust myself upon it. Fitton thinks the moral of it to be, that words may mean anything we like, or that science may require a dangerous rule in its general application. However, the splendid sale shows that you were right in thinking that the newspaper attacks of the Nolans would prove mere fleabites. I have seen Stokes, who was delighted with the first batch of ferns, and has the *Botrychium lunaria* still alive, and thriving under a glass case, together with exotic ferns and New Holland orchidæ in full flower.

September 5.—Murchison drank tea here yesterday evening, and gave us an account of the Bristol doings, and his own tour with Sedgwick, and other geological news. I told him I thought of proffering the President's chair to Whewell, which he thought a capital hit, and I have accordingly sent a strong letter to Whewell to-day.

Every one is quite struck with the improvement in Mary's health and appearance, especially Mr. Murchison.

Mr. Stokes dines with us to-day, quite alone, and is to look over the ferns. We hear that Babbage is in town, among others, in this emptiest of seasons. Murray is looking well and prosperous, and asked politely after you. The Honourable Mrs. Norton came out of Murray's as I went in, and had been showing him the MS. of a publication of her own. She told him she was living with her mother —he had been much struck with the quiet and modest appearance of the authoress.

Mary desires her kindest love, and believe me, ever your affectionate son,

CHARLES LYELL.

To CHARLES DARWIN, ESQ.

16 Hart Street, Bloomsbury: December 26, 1836.

My dear Sir,—I have read your paper with the greatest pleasure, and should like to point out several passages which require explanation, and must have a word or two altered, but it would be impossible in a letter. I have made notes on them, and hope you will call here before you read the paper. Will you come up on Monday, January 2, and come and dine with us at half-past five o'clock, or come at five, and I will go over the paper before dinner? No one dines with us but Mr. and Mrs. Horner and one daughter, and Mr. Horner will be glad to renew his acquaintance with you.

We dine early, because we have one of our small early tea parties, and one or two are to be here, to whom I should like to introduce you, besides a few whom you know already.

If you cannot get here to dinner, you must if possible join the evening party.

The idea of the Pampas going up, at the rate of an inch in a century, while the Western Coast and Andes rise many feet and unequally, has long been a dream of mine. What a splendid field you have to write upon!

I have spent the last week entirely in comparing recent shells with fossil Eocene species, identified by Deshayes. When some great principle is at stake, all the dryness of minute specific comparisons vanishes, but I heartily long for some one here with a collection of shells, and leisure to talk on these matters with. Lonsdale is overpowered with work. Don't accept any official scientific place, if you can avoid it, and tell no one that I gave you this advice, as they would all cry out against me as the preacher of anti-patriotic principles. I fought against the calamity of being President as long as I could. All has gone on smoothly, and it has not cost me more time than I anticipated; but my question is, whether the time annihilated by learned bodies (' par les affaires administratives ') is balanced by any good they do. Fancy exchanging Herschel at the Cape, for Herschel as President of the Royal Society, which he so narrowly escaped being, and I voting for him too! I hope to be forgiven for that. At least, work as I did, exclusively for yourself and for science for many years, and do not prematurely incur the honour or penalty of official dignities. There are people who may be profitably employed in such duties, because they would not work if not so engaged.

Whenever you come up, you must be here on Wednesday, and if you like to dine at the club do so. There is no vacancy, but you stand the first of those who are knocking at the door for admission.

Yours very truly,
CHARLES LYELL.

END OF THE FIRST VOLUME.

Spottiswoode & Co. Printers, New-street Square, London.

Printed in the United States
By Bookmasters